Lecture Notes in Artificial Intelligence 8064

Subseries of Lecture Notes in Computer Science

LNAI Series Editors

Randy Goebel
University of Alberta, Edmonton, Canada
Yuzuru Tanaka
Hokkaido University, Sapporo, Japan
Wolfgang Wahlster
DFKI and Saarland University, Saarbrücken, Germany

LNAI Founding Series Editor

Joerg Siekmann
DFKI and Saarland University, Saarbrücken, Germany

T0238909

Nathan F. Lepora   Anna Mura
Holger G. Krapp   Paul F. M. J. Verschure
Tony J. Prescott (Eds.)

Biomimetic and Biohybrid Systems

Second International Conference, Living Machines 2013
London, UK, July 29 – August 2, 2013
Proceedings

 Springer

Volume Editors

Nathan F. Lepora
Tony J. Prescott
University of Sheffield, UK
E-mail: {n.lepora, t.j.prescott}@sheffield.ac.uk

Anna Mura
University of Pompeau Fabra, Barcelona, Spain
E-mail: anna.mura@upf.edu

Holger G. Krapp
Imperial College, London, UK
E-mail: h.g.krapp@imperial.ac.uk

Paul F. M. J. Verschure
University of Pompeau Fabra
and Catalan Institution for Research and Advanced Studies
Barcelona, Spain
E-mail: paul.verschure@upf.edu

ISSN 0302-9743   e-ISSN 1611-3349
ISBN 978-3-642-39801-8   e-ISBN 978-3-642-39802-5
DOI 10.1007/978-3-642-39802-5
Springer Heidelberg Dordrecht London New York

Library of Congress Control Number: Applied for

CR Subject Classification (1998): I.2.11, I.2, I.4-6, F.1.1-2, H.5, K.4, J.3-4

LNCS Sublibrary: SL 7 – Artificial Intelligence

Typesetting: Camera-ready by author, data conversion by Scientific Publishing Services, Chennai, India

Printed on acid-free paper

Springer is part of Springer Science+Business Media (www.springer.com)

Preface

These proceedings contain the papers presented at *Living Machines: The Second International Conference on Biomimetic and Biohybrid Systems*, held in London, UK, July 29 to August 2, 2013. This international conference is targeted at the intersection of research on novel life-like technologies inspired by the scientific investigation of biological systems, *biomimetics*, and research that seeks to interface biological and artificial systems to create *biohybrid* systems. The conference aim is to highlight the most exciting international research in both of these fields united by theme of "living machines."

The development of future real-world technologies will depend strongly on our understanding and harnessing of the principles underlying living systems and the flow of communication signals between living and artificial systems. The development of either biomimetic or biohybrid systems requires a deep understanding of the operation of living systems, and the two fields are united under the theme of "living machines" — the idea that we can construct artefacts, such as robots, that not only mimic life but share the same fundamental principles, or build technologies that can be combined with a living body to restore or extend its functional capabilities.

Biomimetics can, in principle, extend to all fields of biological research from physiology and molecular biology to ecology, and from zoology to botany. Promising research areas presented at the conference included system design and structure, self-organization and co-operativity, new biologically active materials, self-assembly, learning, memory, control architectures and self-regulation, movement and locomotion, sensory systems, perception, and communication. Biomimetic research was also being seen to drive important advances in component miniaturization, self-configuration, and energy-efficiency. A key focus of the conference was on complete behaving systems in the form of biomimetic robots that can operate on different substrates on sea, on land, or in the air. A further central theme was the physiological basis for intelligent behavior as explored through neuromimetics — the modelling of neural systems. Exciting emerging topics within this field include the embodiment of neuromimetic controllers in hardware, termed neuromorphics, and within the control architectures of robots, sometimes termed neurorobotics.

Biohybrid systems usually involve structures from the nano-scale (molecular) through to the macro-scale (entire organs or body parts). Important examples presented at the conference included: bio-machine hybrids where, for instance, biological muscle was used to actuate a synthetic device; brain–machine interfaces where neurons and their molecular machineries are connected to microscopic sensors and actuators by means of electrical or chemical communication, either in vitro or in the living organism; intelligent prostheses such as artificial limbs, wearable exoskeletons, or sensory organ-chip hybrids (e.g., cochlear implants and

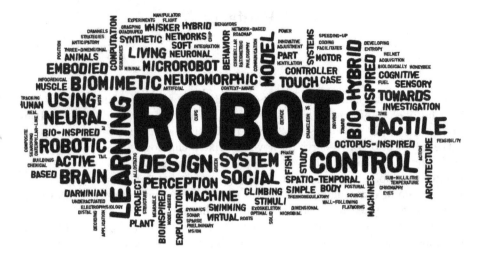

Fig. 1. Living Machines 2013 themes. The most popular 150 terms taken from the titles of papers are displayed in a word could. Evidently, terms such as robot, learning, control, biohybrid, neural, brain, design and neuromorphic feature prominently

artificial retina devices). Biohybrid systems were also considered at the organism level, including robot-animal and robot-human communities.

Leading conference themes are displayed with a word cloud in Fig. 1.

Five hundred years ago, Leonardo da Vinci designed a series of flying machines based on the wings of birds. These drawings are famous for their beautiful, lifelike designs, created centuries before the Wright brothers made their first flight. This inspiration from nature that Leonardo pioneered remains as crucial for technology today as it was many centuries ago.

Leonardo's inspiration was to imitate a successful biological design to solve a scientific problem. Today, this subject area is known as biomimetics. The American inventor Otto Schmitt first coined this term in the 1950s while trying to copy how nerve cells function in an artificial device. He put together the Greek words bios (life) and mimetic (copy) and the name caught on.

Why is nature so good at finding solutions to technological problems? The answer lies in Charles Darwin's theory of evolution. Life, by the process of natural selection, is a self-improving phenomenon that continually reinvents itself to solve problems in the natural world. These improvements have accumulated over hundreds of millions of years in plants and animals. As a result, there are a myriad natural design solutions around us, from the wings of insects and birds to the brains controlling our bodies.

Biomimetics and bio-inspiration has always been present in human technology, from making knives akin to the claws of animals. Curiously though, there has been a dramatic expansion of the biomimetic sciences in the new millennium. The same coordination initiative, the *Convergent Science Network (CSN) of biomimetic and biohybrid systems*, that organized this and last year's

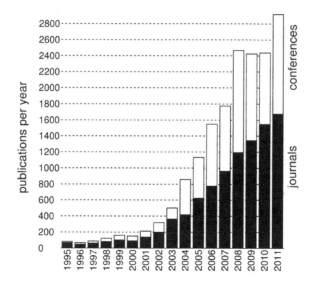

Fig. 2. Growth of biomimetic research. The bar chart plots the number of papers published each year in biomimetics starting from 1995. The *black bars* indicate the proportion of journal papers and the *white bars* the proportion in books and conference proceedings. Growth has been rapid since 2001 with no sign of it saturating. (*Reprinted with permission from Lepora, Verschure and Prescott, 2013: 'The state of the art in Living Machines', Bioinspiration & Biomimetics 8: 013001*)

conference on Living Machines, has also completed a survey on *The State of the Art in Biomimetics* (Lepora, Verschure and Prescott, 2013). As part of the survey, we counted how much work on biomimetics is published each year. This revealed a surprising answer: from only tens of articles before the millennium, it has exploded since then to more than a thousand papers each year (Fig. 2).

This huge investment in research inspired from nature is producing a wide variety of innovative technologies. Examples include artificial spider silk that is stronger than steel, super-tough synthetic materials based on the shells of molluscs, and adhesive patches mimicking the padded feet of geckos. Medical biomimetics is also leading to important benefits for maintaining health. These include bionic cochlear implants for hearing, fully functional artificial hearts, and modern prosthetic hands and limbs aimed at repairing the human body.

Looking to the future, however, the most revolutionary applications of biomimetics will likely be based on nature's most sophisticated creation: our brains. From our survey of biomimetic articles, we found that a main research theme is to take inspiration from how our brains control our bodies to design better ways of controlling robots. This is for a good reason. Engineers can build amazing robots that have seemingly human-like abilities. But really, no existing robot comes close to copying the dexterity and adaptability of animal movements. The missing link is the controlling brain.

It is often said that future scientific discoveries are hard to predict. This is not the case in biomimetics. There are plenty of examples surrounding us in the natural world. The future will produce artificial devices with these abilities, from mass-produced flying micro devices based on insects, to robotic manipulators based on the human hand, to swimming robots based on fish. Less certain is what they will do to our society, economy, and way of life.

The main conference, July 30 to August 1, took the form of a three-day single-track oral and poster presentation programme that included five plenary lectures from leading international researchers in biomimetic and biohybrid systems: Mark Cutkosky (Stanford University) on biomimetics and dextrous manipulation; Terence Deacon (University of California, Berkeley) on natural and artificial selves; Ferdinando Rodriguez y Baena (Imperial College, London) on biomimetics for medical devices; Robert Full (University of California, Berkeley) on locomotion; and Andrew Pickering (University of Exeter) on the history of Living Machines. There were also 20 regular talks and a 3-hour poster session (afternoon of August 1) featuring approximately 50 posters. Session themes included: biomimetic robotics; biohybrid systems including biological-machine interfaces; neuromimetic systems; soft robot systems; active sensing in vision and touch; social robotics and the biomimetics of plants.

The conference was complemented with two days of workshops and symposia, on July 29 and August 2, covering a range of topics related to biomimetic and biohybrid systems: the self and cognitive systems (Peter Ford Dominey and Paul Verschure); learning from the plant kingdom to invent smart solutions (Barbara Mazzolai and Lucia Beccai); neuromorphic models, circuits and emerging nano-technologies for real-time neural processing systems (Giacoma Indiveri and Themistoklis Prodromakis); emergent social behaviors in biohybrid systems (Jose Halloy, Thomas Schmickl and Stuart Wilson); and societal impacts of living machines (Tony Prescott and Michael Szollosy).

The main meeting was hosted London's Natural History Museum, a world-famous center for the study of the natural world homing many scientifically and historically important biological collections. Satellite events were held nearby at Imperial College, London, and an exhibition of biomimetic robots, technology, and art together with a poster session and banquet were hosted in the adjacent Science Museum. These museum venues with their outstanding collections of the natural world and technology were an ideal setting to host the second Living Machines Conference.

We wish to thank the many people who were involved in making LM2013 possible. On the organizational side this included Gill Ryder. Artwork, including the excellent conference poster (pictured), was provided by Martin Bedford. Sytse Wierenga assisted with the production of the website. Organization for the workshops was headed by Holger Krapp and for the exhibitions by Jen Lewis and Stuart Wilson. We would also like to thank the authors and speakers who contributed their work, and the members of the International Programme Committee for their detailed and considered reviews. We are grateful to the five keynote speakers who shared with us their vision of the future.

Finally, we wish to thank the sponsors of LM2013: The *Convergence Science Network for Biomimetic and Biohybrid Systems* (CSN) (ICT-248986), which is funded by the European Union's Framework 7 (FP7) programme in the area of *Future Emerging Technologies* (FET), the University of Sheffield, the University of Pompeu Fabra in Barcelona, and the Institució Catalana de Recerca i Estudis Avançats (ICREA). Additional support was provide by the ICT Challenge 2 project EFAA (ICT-270490). Living Machines 2013 was also supported by the *IOP Physics journal Bioinspiration & Biomimetics*, who this year will publish a special issue of articles based on last year's LM2012 best papers.

July 2013   Nathan F. Lepora
                                                              Anna Mura
                                                       Holger G. Krapp
                                                 Paul F.M.J. Verschure
                                                      Tony J. Prescott

PROGRAMME
The 2nd edition of Living Machines will highlight cutting edge research in biomometics and biohybrid systems and will have :
1 - A single track oral programme (30 July - 1 August)   2 - Six plenary talks   3 - An exhibition & poster programme

SUBMISSIONS
The LM2013 proceedings will include full papers and extended abstracts to be published in the Springer-Verlag LNAI Series.
Expected deadline 15th March 2013

SATELLITE EVENTS
Workshops and tutorials can be scheduled for either the 29th July or the 2nd of August.
If you are interested in organising a satellite event or to find out about our attrative sponsorship packages write to : info.csnetwork@upf.edu

Living Machines 2013 is supported by - The Convergent Science Network project & Bio-inspiration & Biomimetics IOPscience

Organization

Committees

Conference Chairs

Tony Prescott
Paul Verschure

Program Chair

Nathan Lepora

Local Organizer

Holger Krapp

Website

Anna Mura
Sytse Wierenga

Art and Design

Martin Bedford

Conference Administration

Gill Ryder

Exhibition Committee

Jen Lewis
Nathan Lepora

Tony Prescott
Stuart Wilson

Workshop Organizers

Lucia Beccai
Peter Dominey
José Halloy

Giacomo Indiveri
Barbara Mazzolai
Tony Prescott

Themistoklis Prodromakis
Thomas Schmickl
Michael Szollosy

Paul Verschure
Stuart Wilson

Program Committee

Robert Allen
Sean Anderson
Joseph Ayers
Yosepth Bar-Cohen
Lucia Beccai
Frederic Boyer
Darwin Caldwell
Hillel Chiel
Eris Chinellato
Anders Christensen
Frederik Claeyssens
Noah Cowan
Holk Cruse
Mark Cutkosky
Danilo De Rossi
Yiannis Demiris
Peter Dominey
Stephane Doncieux
Marco Dorigo
Volker Durr
Wolfgang Eberle
Mat Evans
Chrisantha Fernando
Charles Fox
Simon Garnier
Benoit Girard
Michele Giugliano
Paul Graham
Christophe Grand
Frank Grasso
Roderich Gross
John Hallam
José Halloy

Huoshong Hu
Auke Ijspeert
Giacomo Indiveri
Serge Kernbach
Mehdi Khamassi
Holger Krapp
Jeffrey Krichmar
Maarja Krussma
Cecilia Laschi
Nathan Lepora
Arianna Menciassi
Ben Mitchinson
Keisuke Morishima
Emre Neftci
Jiro Okada
Enrico Pagello
Martin Pearson
Andrew Philippides
Tony Pipe
Tony Prescott
Roger Quinn
Ferdinando Rodriguez y Baena
Jonathan Rossiter
Raul Rojas
Thomas Schmickl
Mototaka Suzuki
Reiko Tanaka
Jonathan Timmis
Barry Trimmer
Pablo Varona
Paul Verschure
Julian Vincent
Stuart Wilson

Table of Contents

Long Term and Room Temperature Operable Muscle-Powered
Microrobot by Insect Muscle 1
   Yoshitake Akiyama, Kikuo Iwabuchi, and Keisuke Morishima

Speeding-Up the Learning of Saccade Control 12
   *Marco Antonelli, Angel J. Duran, Eris Chinellato, and
Angel P. Del Pobil*

Sensory Augmentation with Distal Touch: The Tactile Helmet
Project 24
   *Craig Bertram, Mathew H. Evans, Mahmood Javaid,
Tom Stafford, and Tony J. Prescott*

Benefits of Dolphin Inspired Sonar for Underwater Object
Identification 36
   Yan Paihas, Chris Capus, Keith Brown, and David Lane

Time to Change: Deciding When to Switch Action Plans during a
Social Interaction 47
   *Eris Chinellato, Dimitri Ognibene, Luisa Sartori, and
Yiannis Demiris*

Stable Heteroclinic Channels for Slip Control of a Peristaltic Crawling
Robot 59
   *Kathryn A. Daltorio, Andrew D. Horchler, Kendrick M. Shaw,
Hillel J. Chiel, and Roger D. Quinn*

Design for a Darwinian Brain: Part 1. Philosophy and Neuroscience 71
   Chrisantha Fernando

Design for a Darwinian Brain: Part 2. Cognitive Architecture 83
   Chrisantha Fernando, Vera Vasas, and Alexander W. Churchill

Virtual Modelling of a Real Exoskeleton Constrained to a Human
Musculoskeletal Model 96
   Francesco Ferrati, Roberto Bortoletto, and Enrico Pagello

Where Wall-Following Works: Case Study of Simple Heuristics vs.
Optimal Exploratory Behaviour 108
   Charles Fox

Miniaturized Electrophysiology Platform for Fly-Robot Interface to
Study Multisensory Integration 119
  Jiaqi V. Huang and Holger G. Krapp

Property Investigation of Chemical Plume Tracing Algorithm in an
Insect Using Bio-machine Hybrid System 131
  Daisuke Kurabayashi, Yosuke Takahashi, Ryo Minegishi,
  Elisa Tosello, Enrico Pagello, and Ryohei Kanzaki

NeuroCopter: Neuromorphic Computation of 6D Ego-Motion
of a Quadcopter 143
  Tim Landgraf, Benjamin Wild, Tobias Ludwig, Philipp Nowak,
  Lovisa Helgadottir, Benjamin Daumenlang, Philipp Breinlinger,
  Martin Nawrot, and Raúl Rojas

A SOLID Case for Active Bayesian Perception in Robot Touch 154
  Nathan F. Lepora, Uriel Martinez-Hernandez, and Tony J. Prescott

Modification in Command Neural Signals of an Insect's Odor Source
Searching Behavior on the Brain-Machine Hybrid System 167
  Ryo Minegishi, Yosuke Takahashi, Atsushi Takashima,
  Daisuke Kurabayashi, and Ryohei Kanzaki

Perception of Simple Stimuli Using Sparse Data from a Tactile Whisker
Array 179
  Ben Mitchinson, J. Charles Sullivan, Martin J. Pearson,
  Anthony G. Pipe, and Tony J. Prescott

Learning Epistemic Actions in Model-Free Memory-Free Reinforcement
Learning: Experiments with a Neuro-robotic Model 191
  Dimitri Ognibene, Nicola Catenacci Volpi, Giovanni Pezzulo, and
  Gianluca Baldassare

Robust Ratiometric Infochemical Communication in a Neuromorphic
"Synthetic Moth" 204
  Timothy C. Pearce, Salah Karout, Alberto Capurro, Zoltán Rácz,
  Marina Cole, and Julian W. Gardner

Bacteria-Inspired Magnetic Polymer Composite Microrobots 216
  Kathrin E. Peyer, Erdem C. Siringil, Li Zhang, Marcel Suter, and
  Bradley J. Nelson

Generic Bio-inspired Chip Model-Based on Spatio-temporal Histogram
Computation: Application to Car Driving by Gaze-Like Control 228
  Patrick Pirim

Embodied Simulation Based on Autobiographical Memory 240
  Gregoire Pointeau, Maxime Petit, and Peter Ford Dominey

Three-Dimensional Tubular Self-assembling Structure for Bio-hybrid
Actuation   251
    Leonardo Ricotti, Lorenzo Vannozzi, Paolo Dario, and
    Arianna Menciassi

Spatio-temporal Spike Pattern Classification in Neuromorphic
Systems   262
    Sadique Sheik, Michael Pfeiffer, Fabio Stefanini, and
    Giacomo Indiveri

Encoding of Stimuli in Embodied Neuronal Networks   274
    Jacopo Tessadori, Daniele Venuta, Valentina Pasquale,
    Sreedhar S. Kumar, and Michela Chiappalone

Modulating Behaviors Using Allostatic Control   287
    Vasiliki Vouloutsi, Stéphane Lallée, and Paul F.M.J. Verschure

A Biomimetic Neuronal Network-Based Controller for Guided
Helicopter Flight   299
    Anthony Westphal, Daniel Blustein, and Joseph Ayers

Bioinspired Adaptive Control for Artificial Muscles   311
    Emma D. Wilson, Tareq Assaf, Martin J. Pearson,
    Jonathan M. Rossiter, Sean R. Anderson, and John Porrill

TACTIP - Tactile Fingertip Device, Texture Analysis through Optical
Tracking of Skin Features   323
    Benjamin Winstone, Gareth Griffiths, Tony Pipe,
    Chris Melhuish, and Jonathon Rossiter

Sensory Feedback of a Fish Robot with Tunable Elastic Tail Fin   335
    Marc Ziegler and Rolf Pfeifer

Leech Heartbeat Neural Network on FPGA   347
    Matthieu Ambroise, Timothée Levi, and Sylvain Saïghi

Artificial Muscle Actuators for a Robotic Fish   350
    Iain A. Anderson, Milan Kelch, Shumeng Sun, Casey Jowers,
    Daniel Xu, and Mark M. Murray

Soft, Stretchable and Conductive Biointerfaces for Bio-hybrid Tactile
Sensing Investigation   353
    Irene Bernardeschi, Francesco Greco, Gianni Ciofani,
    Virgilio Mattoli, Barbara Mazzolai, and Lucia Beccai

Learning of Motor Sequences Based on a Computational Model
of the Cerebellum   356
    Santiago Brandi, Ivan Herreros, Martí Sánchez-Fibla, and
    Paul F.M.J. Verschure

Bio-inspired Caterpillar-Like Climbing Robot 359
  Jian Chen, Eugen Richter, and Jianwei Zhang

The Green Brain Project – Developing a Neuromimetic Robotic
Honeybee 362
  Alex Cope, Chelsea Sabo, Esin Yavuz, Kevin Gurney,
  James Marshall, Thomas Nowotny, and Eleni Vasilaki

Efficient Coding in the Whisker System: Biomimetic Pre-processing
for Robots? 364
  Mathew H. Evans

Octopus-Inspired Innovative Suction Cups 368
  Maurizio Follador, Francesca Tramacere, Lucie Viry,
  Matteo Cianchetti, Lucia Beccai, Cecila Laschi, and
  Barbara Mazzolai

A Cognitive Neural Architecture as a Robot Controller 371
  Zafeirios Fountas and Murray Shanahan

A Small-Sized Underactuated Biologically Inspired Aquatic Robot 374
  Max Fremerey, Steven Weyrich, Danja Voges, and Hartmut Witte

Neural Networks Learning the Inverse Kinetics of an Octopus-Inspired
Manipulator in Three-Dimensional Space 378
  Michele Giorelli, Federico Renda, Gabriele Ferri, and Cecilia Laschi

A Minimal Model of the Phase Transition into Thermoregulatory
Huddling 381
  Jonathan Glancy, Roderich Groß, and Stuart P. Wilson

Towards Bio-hybrid Systems Made of Social Animals and Robots 384
  José Halloy, Francesco Mondada, Serge Kernbach, and
  Thomas Schmickl

Biomimetic Spatial and Temporal (4D) Design and Fabrication 387
  Veronika Kapsali, Anne Toomey, Raymond Oliver, and Lynn Tandler

A Swimming Machine Driven by the Deformation of a Sheet-Like Body
Inspired by Polyclad Flatworms.................................... 390
  Toshiya Kazama, Koki Kuroiwa, Takuya Umedachi,
  Yuichi Komatsu, and Ryo Kobayashi

Towards a Believable Social Robot 393
  Nicole Lazzeri, Daniele Mazzei, Abolfazl Zaraki, and Danilo De Rossi

Towards a Roadmap for Living Machines 396
  Nathan F. Lepora, Paul F.M.J. Verschure, and Tony J. Prescott

Acquisition of Anticipatory Postural Adjustment through Cerebellar
Learning in a Mobile Robot 399
  Giovanni Maffei, Ivan Herreros, Martí Sánchez-Fibla, and
  Paul F.M.J. Verschure

Using a Biological Material to Improve Locomotion of Hexapod
Robots 402
  Poramate Manoonpong, Dennis Goldschmidt, Florentin Wörgötter,
  Alexander Kovalev, Lars Heepe, and Stanislav Gorb

Angle and Position Perception for Exploration with Active Touch 405
  Uriel Martinez-Hernandez, Tony J. Dodd, Tony J. Prescott, and
  Nathan F. Lepora

Toward Living Tactile Sensors 409
  Kosuke Minzan, Masahiro Shimizu, Kota Miyasaka,
  Toshihiko Ogura, Junichi Nakai, Masamichi Ohkura, and
  Koh Hosoda

Virtual Chameleon: Wearable Machine to Provide Independent Views
to Both Eyes 412
  Fumio Mizuno, Tomoaki Hayasaka, and Takami Yamaguchi

Bioinspired Design and Energetic Feasibility of an Autonomous
Swimming Microrobot 415
  Stefano Palagi, Francesco Greco, Barbara Mazzolai, and Lucia Beccai

Climbing Plants, a New Concept for Robotic Grasping................ 418
  Camilla Pandolfi, Tanja Mimmo, and Renato Vidoni

Biomimetic Lessons for Natural Ventilation of Buildings:
A Collection of Biomimicry Templates Including Their Simulation and
Application 421
  David R.G. Parr

Sub-millilitre Microbial Fuel Cell Power for Soft Robots.............. 424
  Hemma Philamore, Jonathan Rossiter, and Ioannis Ieropoulos

How Active Vision Facilitates Familiarity-Based Homing 427
  Andrew Philippides, Alex Dewar, Antoine Wystrach,
  Michael Mangan, and Paul Graham

Embodied Behavior of Plant Roots in Obstacle Avoidance 431
  Liyana Popova, Alice Tonazzini, Andrea Russino, Alì Sadeghi, and
  Barbara Mazzolai

Motor Control Adaptation to Changes in Robot Body Dynamics
for a Complaint Quadruped Robot 434
  Soha Pouya, Peter Eckert, Alexander Sproewitz, Rico Moeckel, and
  Auke Ijspeert

The AI Singularity and Runaway Human Intelligence.................   438
    Tony J. Prescott

ASSISI: Mixing Animals with Robots in a Hybrid Society.............   441
    Thomas Schmickl, Stjepan Bogdan, Luís Correia,
    Serge Kernbach, Francesco Mondada, Michael Bodi,
    Alexey Gribovskiy, Sibylle Hahshold, Damjan Miklic, Martina Szopek,
    Ronald Thenius, and José Halloy

Chroma+Phy – A Living Wearable Connecting Humans and Their
Environment   444
    Theresa Schubert

Plant Root Strategies for Robotic Soil Penetration   447
    Alice Tonazzini, Ali Sadeghi, Liyana Popova, and Barbara Mazzolai

The Synthetic Littermate   450
    Stuart P. Wilson

Evo-devo Design for Living Machines   454
    Stuart P. Wilson and Tony J. Prescott

Preliminary Implementation of Context-Aware Attention System for
Humanoid Robots..   457
    Abolfazl Zaraki, Daniele Mazzei, Nicole Lazzeri,
    Michael Pieroni, and Danilo De Rossi

Author Index..   461

Long Term and Room Temperature Operable Muscle-Powered Microrobot by Insect Muscle

Yoshitake Akiyama[1,2,3], Kikuo Iwabuchi[4], and Keisuke Morishima[1,3,*]

[1] Department of Mechanical Engineering, Osaka University,
2-1 Yamadaoka, Suita, Osaka 565-0871, Japan
[2] Frontier Research Base for Global Young Researchers, Osaka University,
2-1 Yamadaoka, Suita, Osaka 565-0871, Japan
[3] Graduate School of Bio-Application and System Engineering (BASE),
Tokyo University of Agriculture and Technology,
2-24-16 Naka-cho, Koganei, Tokyo 184-8588, Japan
[4] Department of Applied Molecular Biology and Biochemistry,
Tokyo University of Agriculture and Technology,
3-5-8 Saiwai-cho, Fuchu, Tokyo 183-8509, Japan
morishima@mech.eng.osaka-u.ac.jp
http://www-live.mech.eng.osaka-u.ac.jp/

Abstract. This paper describes an insect muscle-powered autonomous microrobot (iPAM) which can work long-term at room temperature without any maintenance. The iPAM consisting of a DV tissue and a frame was designed on the basis of a finite element method simulation and fabricated. The iPAM moved autonomously using spontaneous contractions of a whole insect dorsal vessel (DV) and the moving velocity was accelerated temporally by adding insect hormone. These results suggest that the insect DV has a higher potential for being a biological microactuator than other biological cell-based materials. Insect dorsal vessel (DV) tissue seems well suited for chemically regulatable microactuators due to its environmental robustness and low maintenance.

Keywords: Microrobot, Bioactuator, Chemical stimulation, Insect, Dorsal vessel, Neuroactive chemical.

1 Introduction

Recently, mammalian muscle cells have received considerable attention as a novel actuator for microdevices [1]-[7], and reported bio-hybrid microdevices using mammalian heart or skeletal muscle cells include a pillar actuator [1][2] and a micro heart pump [3][4]. Muscle cells are well suited to work in a microspace due to their size and their high energy-conversion efficiency. For instance, muscle tissues and cells are soft and small, and they can contract using only the chemical energy in adenosine triphosphate (ATP). However, these devices require precise environmental

---

* Corresponding author.

N.F. Lepora et al. (Eds.): Living Machines 2013, LNAI 8064, pp. 1–11, 2013.

control to keep the contractile ability of the muscle cells. The medium must be replaced every few days and pH and temperature must be kept around 7.4 and 37 °C, respectively.

Tissues and cells of insects are generally robust over a much wider range of living conditions as compared to those of mammals. As an example, the characteristics of insect dorsal vessel (DV) tissue and rat cardiomyocyte (CM) are summarized in Table 1 based on literature data; values for the DV tissue were obtained from [8], [9], and [10] and the values for the rat CM tissue were obtained from [1] and [7]. We previously proposed to utilize insect DV tissue and cells as an actuator and we demonstrated a micropillar actuator which worked at room temperature for more than 90 days without medium replacement [8]. Surprisingly, the micropillar actuator could work at temperatures from 5 to 40 °C though the contracting velocity and frequency of the micropillar decreased with lowering of temperature and the actuator was irreversibly damaged at 40 °C [9].

There are several advantages when utilizing biological tissue and cells as an actuator. One of them is chemical controllability of the contractions. It has been reported that epinephrine, acetylcholine, and caffeine, which are physiologically active chemicals in vertebrates, have an effect on heart beat of Periplaneta americana [11]. On the other hand, crustacean cardioactive peptide (CCAP) has been found in the moth Manduca sexta [12]. We have already confirmed that CCAP has an ability to accelerate the heart beat of an inchworm in vitro [13].

In this paper, we demonstrate a pantograph-shaped microrobot (PSMR) as an example of an insect muscle-powered autonomous microrobot (iPAM). The PSMR using a whole DV will work autonomously at room temperature without any maintenance for a long time. At first, the PSMR is designed and its deformation by the contraction of the DV tissue is simulated using finite element analysis simulation software. Then, the frame of the PSMR is fabricated by molding. After that, the PSMR is fabricated by assembling the whole DV onto the frame. The PSMR is evaluated by measuring the moving distance from the side. Finally, we attempt to accelerate the PSMR movement by adding the neuroactive chemical, CCAP.

Table 1. Comparison of insect DV tissue and rat CM

	DV tissue	Rat CM
Lifetime	90 days	14 days
Contractile Frequency	0.2 Hz	1 Hz
Contractile Force	96 μN	3.5 μN
Medium Replacement	Not Needed	2 to 3 days
Viable Temperature	5 to 35 °C	37 °C

2   Design of PSMR

We designed the PSMR as shown in Figure 1. The contractile force of a whole DV has been reported to be 96 μN [10]. Based on this value, we designed a frame consisting of a pantograph-shaped body and four legs. The frame was made of polydimethylsiloxane (PDMS) and its width and height were each 200 μm. Two slits were made at each end of the body so as to hold the DV tissue in place when it was wrapped around the body. The diameter of each leg was 200 μm. The front legs were 1100 μm long and the rear legs were 600 μm long.

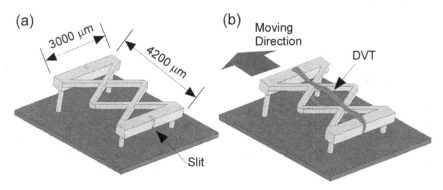

Fig. 1. Respective illustrations of PSMR (a) before and (b) after DVT assembly. The arrow in (b) shows the moving direction of the PSMR.

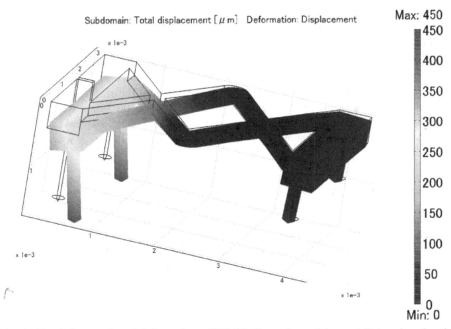

Fig. 2. Simulation results of deformation of PSMR. Young's modulus and Poisson's ratio of PDMS were set to 1.8 MPa [14] and 0.48, respectively.

Next, we calculated the deformation of the PSMR by the DV tissue contraction using the simulation software COMSOL Multiphysics. In this simulation, a linear elastic model was used. The contractile force was loaded onto the edges of the slits. The analysis results are shown in Figure 2. The maximum displacement of the front leg was about 420 µm. The distance between the front leg and the rear leg was reduced by 420 µm, which was more than the diameter of the leg. This result strongly suggests that the PSMR will move by spontaneous contractions of the DV tissue.

3   Experimental

3.1   Insect and Its DV Excision

The final stage larvae of the inchworm, Ctenoplusia agnata, were used in this study. The inchworms were raised continuously at 25 °C with only an artificial diet. Their DVs were excised under a stereomicroscope after surface sterilization in 70% ethanol solution. The excised DVs were cultured in the culture medium, TC-100 medium supplemented with 10% fetal bovine serum and 1% penicillin-streptomycin solution, at 25 °C.

3.2   Fabrication of the Frame

The frame for the PSMR was fabricated by molding PDMS (Figure 3). The mold was fabricated by machining a 1 mm thick poly(tetrafluoroethylene) (PTFE) sheet with a machining center (ROBODRILL, FANUC, Yamanashi, Japan). Then, uncured PDMS (Sylpod184, Dow Corning Toray, Tokyo, Japan) was poured onto the mold. Next, a slide glass was placed on the mold and they were pressed in a vise in order to remove excess uncured PDMS. After baking at 80 °C for 60 min, the frame with the slide glass was detached from the mold. Finally, to avoid tearing, the frame was carefully peeled off. In the case of the micropillar array, the PTFE mold was baked directly with excess uncured PDMS at 80 °C. After curing, the micropillar array was obtained by carefully peeling the PDMS film off.

3.3   Assembly of DV Tissue onto the Frame

The frame was hydrophilized using an oxygen plasma asher (PIB-10, Vacuum Device, Ibaragi, Japan) and was coated with Cell Tak (BD Biosciences, Franklin Lakes, NJ, USA). The excised DV was wrapped using tweezers onto the frame in the culture medium while being viewed under the stereomicroscope. The PSMR was then incubated at 25 °C without medium replacement.

3.4   Image Analysis for Evaluation of PSMR

The PSMR observations were made at 25 °C in all the experiments. Deformation distance of the frame and moving distance of the PSMR were observed with a digital

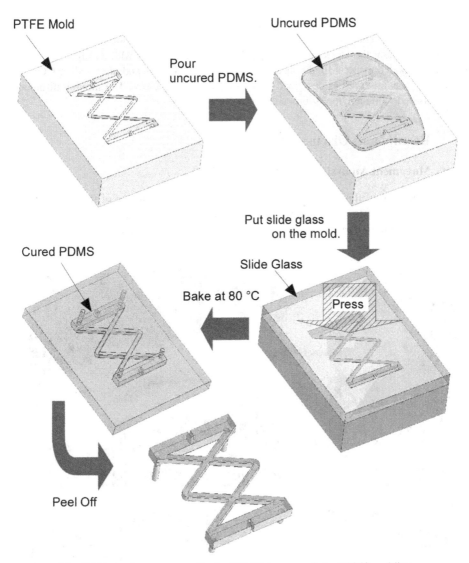

PTFE Mold

Uncured PDMS

Pour uncured PDMS.

Put slide glass on the mold.

Cured PDMS

Slide Glass

Bake at 80 °C

Press

Peel Off

Fig. 3. Fabrication possesses for the PSMR frame made by PDMS molding

zoom microscope (KH-7700, Hirox, Tokyo, Japan) and a zoom microscope (AZ-100, Nikon, Tokyo, Japan) equipped with a CCD camera, respectively. The obtained microscopy movies were analyzed with analysis software (DippMotion, Ditect, Tokyo, Japan).

3.5 Evaluation of Insect Hormone

We tried to regulate the PSMR by adding insect hormone. We used the PSMR within a few days after assembling. CCAP purchased from LKT Laboratories (St. Paul, MN, USA) was made up as stock solutions of 10^{-3} M using ultrapure water. The stock solution was stored at -20 °C. The concentration of CCAP in the culture medium was gradually increased by adding the stock solution or a diluted stock solution with TC-100 medium. CCAP was added to get a final concentration of 10^{-6} M. Before and after adding CCAP, the PSMR was observed as described in the image analysis above. The moving distance was measured with the DippMotion analysis software.

4   Results and Discussion

4.1   Movement Analysis of the PSMR

The deformation of the front and rear leg tips and moving distance of the PSMR was measured from a side view. Views of the relaxing and contracting PSMR are shown in Figure 4. Only part of the DV under the pantograph-shaped body contracted

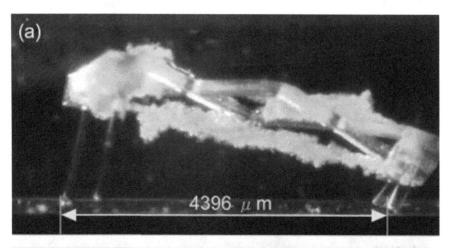

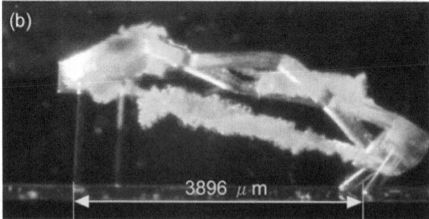

Fig. 4. Microscopic side views of the PSMR when relaxing and contracting

spontaneously, that action bowed the pantograph-shaped body. As a result, the distance between the front leg and the rear leg was reduced by 500 μm from 4396 μm to 3896 μm. The measured value was almost the same as the predicted value, 420 μm. We attribute the difference to the stiffness of the PDMS which depends on baking time and temperature and variability of the contractile force of DVs among individuals.

The movement of the PSMR was analyzed by image analysis (Figure 5). During 30 s, the PSMR moved 793 μm while the DV tissue contracted 12 times. Based on these results, the average stroke and velocity were calculated as 66.1 μm and 26.4 μm/s. The average stroke was much smaller than the reduced distance of 500 μm between the front and the rear legs. This shows the contractile force of the DV was utilized poorly. The efficiency could be improved by optimizing the shape of the leg tips. For instance, it is desirable that the shapes for front and rear legs allow the front legs to stick and the rear legs to slip when the DV contracts and conversely, the front legs to slip and the rear legs to stick when the DV relaxes. If the contractile force of the DV is evoked efficiently by improving the shape of the tip, theoretically, the velocity of the PSMR will increase to 200 μm/s.

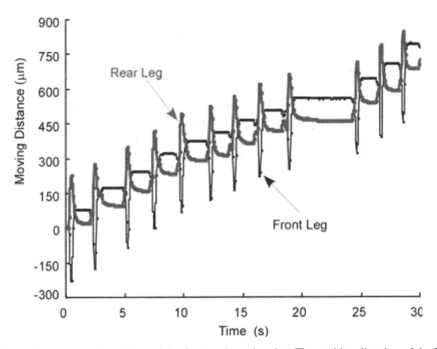

Fig. 5. Time course of positions of the front and rear leg tips. The positive direction of the Y axis was set to the designed moving direction from the starting position.

In general, the contractile force of muscle tissues depends on their own length and the maximum force is produced when they contract from the resting length. Therefore, it is important to wrap the DV onto the pantograph-shaped body with a small tension because the DV shrinks less than its resting length as soon as it is excised from the insect. On the contrary, the contractile force of the DV decreases when the DV is extended excessively.

The PSMR was observed with a scanning electron microscope after fixation with paraformaldehyde. As shown in Figure 5, the DV under the pantograph-shaped body was wrapped tightly and the DV over the pantograph-shaped body was wrapped loosely. In these experiments the DVs were wrapped manually, but it is difficult to produce a large number of PSMRs with exactly the same wrapping conditions. Further research is needed to identify the relationship between length and contractile force of the DV and to develop a way to assemble the DV onto the frame with the desirable tension.

Fig. 6. Scanning electron microscope image of the PSMR

4.2   Acceleration by Adding CCAP

The trajectories of the PSMR before and after adding CCAP were analyzed and compared and the results for the first 30 s are show in Figure 7. The contractile frequency of the DV and the moving velocity of the PSMR were clearly increased by adding CCAP. The moving distances for 30 s before and after CCAP addition were 114 μm and 723 μm, respectively. The result indicates that the velocity of the PSMR increased 6.3-fold by adding CCAP at the final concentration of 10^{-6} M. We also calculated that the moving velocities before and after addition were 3.8μm/s and

24.1μm/s. The moving velocity before addition is much lower than that of the PSMR in the previous section. This is because the contractile frequency of the DV used in this experiment was lower than that of the DV used in the previous experiment and the friction force between the leg tips and the bottom of the culture dish differed between them.

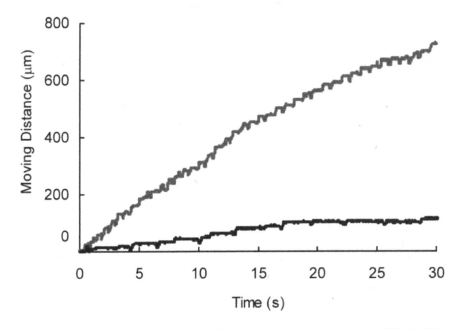

Fig. 7. Trajectories of the PSRM for 60 s before (blue line) and after (red line) CCAP addition. The frequencies before and after CCAP addition for 30 s were 0.43 Hz and 1.33 Hz, respectively.

5   Conclusion

We succeeded in fabricating the PSMR which autonomously moved at 25 °C. The PSMR was fabricated by assembling a whole DV onto the frame made of PDMS. The distance between the front and the rear legs was reduced by 500 μm when the DV contacted, which was almost equal to the value obtained by simulation. However, the moving distance in one contraction was 66.1 μm, which was much smaller than the reduced distance between the front and rear legs. The velocity obtained in the experiment was 26.4 μm/s. The velocity can be increased by improving the shape of the leg tips so as to make a difference in friction forces between the leg tips and the surface over which the body is traveling. We also confirmed that the PSMR could be accelerated temporally by adding CCAP, a kind of insect neural peptide. The velocity of the PSMR was increased 6.3-fold by adding CCAP. These results indicate that the DV is strong enough to be utilized as a microactuator which can be regulated by adding a chemical agent. Our group has also successfully light-regulated DV tissue contraction of a fly by expressing channelrhodopsin-2, a directly light-gated cation-selective membrane channel found only in muscle tissues, and then irradiating the excised DV tissue with blue light [15]. In conclusion, the results in these experiments suggest that the insect DV has a higher potential for being a biological microactuator than other biological cell-based materials. Insect dorsal vessel (DV) tissue seems well suited for chemically regulatable microactuators due to its environmental robustness and low maintenance.

Acknowledgements. The present work was supported by Grants-in-Aid for Scientific Research from the Ministry of Education, Culture, Sports, Science and Technology in Japan Nos. 21676002, 22860020 and 23700557, the Industrial Technology Research Grant Program (2006) from the New Energy and Industrial Technology Development Organization (NEDO) of Japan, and CASIO Science Promotion Foundation.

References

[1] Tanaka, Y., Morishima, Y.K., Shimizu, T., Kikuchi, A., Yamato, M., Okano, T., Kitamori, T.: Demonstration of a PDMS-based bio-microactuator using cultured cardiomyocytes to drive polymer micropillars. Lab Chip 6, 230–235 (2006)

[2] Morishima, K., Tanaka, Y., Ebara, M., Shimizu, T., Kikuchi, A., Yamato, M., Okano, T., Kitamori, T.: Demonstration of a bio-microactuator powered by cultured cardiomyocytes coupled to hydrogel micropillars. Sens. Act. B 119, 345–350 (2006)

[3] Tanaka, Y., Morishima, K., Shimizu, T., Kikuchi, A., Yamato, M., Okano, T., Kitamori, T.: An actuated pump on-chip powered by cultured cardiomyocytes. Lab Chip 6, 362–368 (2006)

[4] Park, J., Kim, I.C., Baek, J., Cha, M., Kim, J., Park, S., Lee, J., Kim, B.: Micro pumping with cardiomyocyte–polymer hybrid. Lab Chip 7, 1367–1370 (2007)

[5] Xi, J., Schmidt, J.J., Montemagno, C.D.: Self-assembled microdevices driven by muscle. Nat. Mater. 4, 180–184 (2005)

[6] Feinberg, A., Feigel, A., Shevkoplyas, S., Sheehy, S., Whitesides, G., Parker, K.: Muscular thin films for building actuators and powering devices. Science 317, 1366–1370 (2007)

[7] Kim, J., Park, J., Yang, S., Baek, J., Kim, B., Lee, S.H., Yoon, E.S., Chun, K., Park, S.: Establishment of a fabrication method for a long-term actuated hybrid cell robot. Lab Chip 7, 1504–1508 (2007)

[8] Akiyama, Y., Iwabuchi, K., Furukawa, Y., Morishima, K.: Long-term and room temperature operable bioactuator powered by insect dorsal vessel tissue. Lab Chip 9, 140–144 (2009)

[9] Akiyama, Y., Iwabuchi, K., Furukawa, Y., Morishima, K.: Fabrication and evaluation of temperature-tolerant bioactuator driven by insect heart cells. In: Int. Conf. Proc. Miniaturized Systems in Chemistry and Life Science, pp. 1669–1671 (2008)

[10] Shimizu, K., Hoshino, T., Akiyama, Y., Iwabuchi, K., Akiyama, Y., Yamato, M., Okano, T., Morishima, K.: Multi-Scale Reconstruction and Performance of Insect Muscle Powered Bioactuator from Tissue to Cell Sheet. In: Proc. of IEEE RAS & EMBS Biomedical Robotics and Biomechatronics, pp. 425–430 (2010)

[11] Krijgsman, B., Krijgsman-Berger, N.: Physiological investigations into the heart function of arthropods. The heart of Periplaneta americana. Bull. Ent. Res. 42, 143–155 (1951)

[12] Lehman, H., Murgiuc, C., Miller, T., Lee, T., Hildebrand, J.: Crustacean cardioactive peptide in the sphinx moth, Manduca sexta. Peptides 14, 735–741 (1993)

[13] Akiyama, Y., Iwabuchi, K., Furukawa, Y., Morishima, K.: Biological contractile regulation of micropillar actuator driven by insect dorsal vessel tissue. In: Proc. of IEEE RAS & EMBS Biomedical Robotics and Biomechatronics, pp. 501–505 (2008)

[14] Choi, K., Rogers, J.: A photocurable poly (dimethylsiloxane) chemistry designed for soft lithographic molding and printing in the nanometer regime. J. Am. Chem. Soc. 125, 4060–4061 (2003)

[15] Suzumura, K., Funakoshi, K., Hoshino, T., Tsujimura, H., Iwabuchi, K., Akiyama, Y., Morishima, K.: A light regulated bio-micro-actuator powered by transgenic Drosophila melanogaster muscle tissue. In: Proc. of IEEE MEMS Micro Electro Mechanical Systems (2011)

Speeding-Up the Learning of Saccade Control

Marco Antonelli[1], Angel J. Duran[1], Eris Chinellato[2], and Angel P. Del Pobil[1,*]

[1] Robotic Intelligence Lab, Universitat Jaume I, Spain
{antonell,abosch,pobil}@uji.es
[2] Imperial College London
e.chinellato@imperial.ac.uk

Abstract. A saccade is a ballistic eye movement that allows the visual system to bring the target in the center of the visual field. For artificial vision systems, as in humanoid robotics, performing such a movement requires to know the intrinsic parameters of the camera. Parameters can be encoded in a bio-inspired fashion by a non-parametric model, that is trained during the movement of the camera. In this work, we propose a novel algorithm to speed-up the learning of saccade control in a goal-directed manner. During training, the algorithm computes the covariance matrix of the transformation and uses it to choose the most informative visual feature to gaze next. Results on a simulated model and on a real setup show that the proposed technique allows for a very efficient learning of goal-oriented saccade control.

1 Introduction

Saccades are ballistic, fast movements that are used to gaze a visual stimulus. The movement can be as fast as 300 degrees per second and its execution is not modified by the visual perception, since the saccade is blind [3]. Considering the open-loop nature of this movement, it is important for the brain to have a good knowledge of the oculomotor plant. Indeed, several adaptive mechanisms were discovered to maintain the calibration of the saccadic generator system in humans [10].

In humanoid robots, learning the saccade control consists in converting the visual position of a stimulus into a motor position that allows the visual system to see the stimulus in the center of the image. The transformation between visual information and motor command, namely visuo-oculomotor transformation, requires the knowledge of the intrinsic parameters of the camera. This transformation is useful when the cameras have high distortion (e.g. log-polar sensors) and to create an implicit representation of the peripersonal space [6].

These parameters can be obtained by calibration procedures that are performed off-line using some known visual patterns. An interesting alternative for an active system is to describe the transformation with a non-parametric model

---

* This work was supported in part by Ministerio de Ciencia y Innovación (FPI grant BES-2009-027151, DPI2011-27846), by Generalitat Valenciana (PROMETEO/2009/052) and by Fundació Caixa-Castello-Bancaixa (P1-1B2011-54).

N.F. Lepora et al. (Eds.): Living Machines 2013, LNAI 8064, pp. 12–23, 2013.
© Springer-Verlag Berlin Heidelberg 2013

that is adjusted on-line after a movement of the camera. These techniques do not require a mathematical model of the camera and adapt to changes of the visual system.

The most common used non-parametric model for the saccade control are look-up tables [17,4] and neural networks [25,23,12,2]. Among the biological inspired models, an interesting solution are the radial basis function networks (RBFs). Indeed, they are often employed to simulate the neural activation of the cells of the parietal cortex of the human brain [20,6]. Moreover, a recent work shows that RBFs can also simulate the learning profile and the transfer of the adaptation typical of the human saccades [7].

The non-parametric models are usually calibrated on-line by means of ballistic movements of the camera. The learning process requires at least a visual stimulus as reference. After the movement of the camera, the visual position of the stimuli before and after the movement is used to update the model. In this way we can directly learn only the forward model, which describes how the stimulus moves due to the movement of the camera [12]. However, the saccade control requires the motion of the camera that brings the target in the center of the image, that is, the inverse model. One solution is learning the forward model and then inverting it [15,26,12]. However, the most common solution in the learning of the saccade control is to learn directly the inverse model [17,25,23,2].

Learning the inverse model suffers the problem of the lacking of a teacher. That is, the visual system does not know the movement that should be performed until when it is performed. This problem can be solved through random exploration [4,14] or by using some heuristics to drive a goal-direct movement [25,2].

The random exploration is easy to implement but requires that a random movement brings the visual feature near the center of the image. Given that it can requires a lot of time, some techniques have been proposed in literature to speed up the learning process. For example, we can track the history of the eye movements and when the eye land on the visual feature we can train the system for the all the "visited" eye position [4]. Another approach consists in diving the learning process into sequential stages. During earlier stages, the network is trained with suboptimal movements and the quality of the accepted training points is increased in later stages[14].

The goal-directed learning has more biological plausibility because it trains the system each time a new movement is performed, as primates do [18,5]. To do that, the visual error is converted into a motor error using some heuristics. A popular approach is the feedback error learning, which multiplies the visual error by a gain that is provided by a proportional controller [16,25]. A more sophisticated approach consists in locally inverting the Jacobian of the forward model of the transformation and using it as proportional gain.

Independently on the learning strategy, every methods assume that the visual system detects a target stimulus that can be used to train the inverse model. However, features detector algorithm and saliency maps usually detect more than one stimulus. Even if all these features can be used simultaneously to train

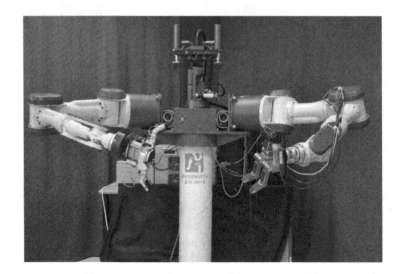

Fig. 1. Tombatossals humanoid torso

the forward model [12], only one can be used to train the inverse model because the robot can performs just a goal direct movement at time.

In this study we provide a feature selection procedure that allows the robot to speed up the training process. In our framework the visuo-oculomotor transformation is encoded by a radial basis function network that is trained using a Kalman filter [13]. At each eye movement, the Kalman filter updates the weights of the networks (state) and their covariance matrix.

Herein, we propose to exploit the covariance matrix to guide the visual exploration. For each stimulus, we compute the variance of the predicted saccade and then we choose to gaze the stimulus with the highest variance.

Preliminary results on a simulated camera show that the covariance matrix can be successfully employed to improve the exploration of the visual space. So that, our proposed approach can be used to speed up goal-directed learning strategies. Further, experimental results on the UJI humanoid torso (see Fig. 1) validate the proposed methodology.

The reminder of this paper is organized as follow. Section 2 describes the procedure that is used to learn the visuo-oculomotor transformation. Section 3 describes the RBFs and how they can be trained with a Kalman filter. Section 4 describes the experiments and the achieved results on a simulated model while the results with the robot are described in Section 5.

2   Learning of the Saccade Control

In this work we train the neural network to encode the visuo-oculomotor transformation required to perform a correct saccade. This transformation converts the visual location of the stimulus into the movement that is required to gaze to the stimulus. The transformation is encoded by a radial basis function network (see detail in the next section). RBFs were chosen because they can model the gain field effect observed in some areas of the parietal cortex that are related with gazing movement [8].

In order to strengthen the biological plausibility, the saccade control is learned through the interaction with the surrounding space [6]. We suppose that the visual system observes some salient features that can be used as visual reference to train the visuo-oculomotor transformation. The main idea is to move the camera to bring one of this feature in the center of the image. If the performed movement is not precise, the target feature will be visible in a position of the image that is not the center.

The displacement of the stimulus from the center can be considered as the error of the network. However, this error is observed in terms of visual position but the training of the network requires an error in terms of motor position. The motor error can be provided by an external teacher or can approximated by using a feedback gain [25] or a linear approximation [2]. A previous work [7] shows that the on-line learning of RBF can reproduce the saccadic adaptation effect that is observed in cognitive science experiments [24,9].

Independently on the strategy used to train the network, the system need to choose a target. Usually, the choice of the target depends on the some higher level task, such as object recognition or movement detection. However, if no other task are provided, the system can autonomously explore the environment.

The simplest exploration behavior consists in choosing randomly one target among the visual stimuli. However, better performance can be achieved if the robot tries to gaze and correct the visual region with bigger uncertainty. The latter solution is employed in this study.

So that, the exploration behavior is the following. Each time the agent observes some stimuli, it employs the visuo-oculomotor transformation to compute the movement (mean and variance) that is required to gaze every features. Then, it chooses to perform a movement toward the stimulus that has the bigger variance. After the movement the agent obtains the new visual position of the target and uses it to train the network.

3   Radial Basis Function Networks

Learning the visuo-oculomotor transformation can be treated as a function approximation problem. Radial basis function networks can potentially approximate any function with the desired precision [19], so they are especially suitable for encoding sensorimotor transformations [20,21]. The input of the transformation is the visual location of the stimulus while the desired output of the network is the ocular movement that is required to gaze to the stimulus itself.

Basis function networks are three-layer feed forward neural networks whose hidden units (h_i) perform a non-linear transformation of the input data, whereas the output (y) is computed as a linear combination (w_i) of the hidden units:

$$y = \sum_{i=1}^{n} w_i \cdot h_i(\mathbf{x}) = \mathbf{w}^T \cdot \mathbf{h}(\mathbf{x}). \tag{1}$$

Learning in the context of the radial basis function networks can be divided into two phases. An unsupervised phase sets the parameters of the network, such as the number of hidden units or the position of their centers, whereas a supervised phase adjusts the weights. In the proposed framework, in order to improve the biological plausibility of the model, we employ fixed centers, whose receptive fields do not move according to the input data.

We model the activation of the hidden neurons by using Gaussian functions. Each unit is characterized by its center of activation $(\mathbf{c}_i)$, whereas the spread of the activation (Σ) is equal for every unit:

$$h_i(\mathbf{x}) = h(||\mathbf{x} - \mathbf{c}_i||) = e^{-(\mathbf{x} - \mathbf{c}_i)^T \Sigma^{-1} (\mathbf{x} - \mathbf{c}_i)} \tag{2}$$

Using this setup, the learning process consists in finding the weights that better approximate the sensorimotor transformation.

We trained the visuo-oculomotor transformation using a Kalman filter approach [13]. The Kalman filter updates incrementally the state of the networks each time a new training point is available. Moreover, it keeps trace of the covariance matrix of the network that can be used to optimally integrate different sensory-motor modalities [1] or, as we show later, to drive the exploration of the space.

In the common notation of the Kalman filters, the state of the system is given by the matrix of the weights. Given that changes in the visuo-oculomotor transformation are not predictable, we model the state transition with an identity matrix and we add a forgetting factor (λ) to the covariance matrix to model such uncertainty. The matrix of the measurement is given by the activation of the hidden layer, while the observation is the desired eye movement.

At each step, the algorithm updated the weights $\mathbf{w}$ and the covariance matrix $\mathbf{P}$ using the following equations:

$$
\begin{aligned}
\mathbf{P} &= (1 + \lambda) \cdot P \\
e &= y - \mathbf{w}^T \cdot \mathbf{h} \\
S &= \mathbf{h}^T \cdot \mathbf{P} \cdot \mathbf{h} \\
\mathbf{K} &= \mathbf{P} \cdot \mathbf{h} \cdot (S + \sigma_m^2)^{-1} \\
\mathbf{w} &= \mathbf{w} + \mathbf{K} \cdot e \\
\mathbf{P} &= \mathbf{P} - \mathbf{K} \cdot \mathbf{h}^T \cdot \mathbf{P}
\end{aligned}
\tag{3}
$$

where e is the error of the network, $\mathbf{K}$ is the Kalman gain, σ_m^2 is variance of the measurement noise and S is the variance of the output of the network.

In general, σ_m^2 depends on the noisy observation of the motor position and on the visual error which is propagated through the hidden layer of the network. On the other hand, S depends on the visual position of the stimulus (more or less explored region) and on the quality of the measurement in its neighborhood. In this work we use S to drive the learning process through visual points of greater uncertainty.

At the beginning, the weights of the network are set to zero while the covariance matrix is set to a diagonal matrix. The value of the diagonal was chosen in order to ensure that, at the beginning of the algorithm, the standard deviation of the output is one third the range of movement of the camera.

4 Simulation Results

4.1 Setup

Preliminary results are obtained on a simulated model of a pan-tilt camera. Even if the model is simple, it allows us to validate the proposed approach.

We simulated a camera with a resolution of 1024×768 pixels and the focal length of 1075 times the pixel size. Using this model a visual feature in the periphery (512 pixels) is center with a rotation of approximately 25.2 degrees.

The network is composed by 11×11 centers uniformly distributed in the input space (image surface). Some neurons were placed out the visible space, in order to reduce border effects in the results. We also introduce a model of the noise obtained by the motor position.

4.2 Exploration Behavior

The aim of first experiment is to test how the proposed exploration behavior speeds up the learning process. To do that we compare the convergence speed of the system subjected to a different number of visual stimuli. The number of the stimuli is chosen between 1 and 15. With only one stimulus, the performance of the system are the same of the random choice.

At each iteration, the features appear at a random position of the image and the agent selects the feature with higher variance as saccadic target. Given that this experiment is focused on the performance of the exploration behavior and not on the training strategy, after the movement, the network is trained with the ground truth signal.

To compare the performance, we measured the number of iterations that the system takes to converge. The convergence is reached when the root mean squared error (RMS) of the network is lower than 0.5 pixels. The RMS is computed on a dataset composed of 1000 input-output points. Figure 2 shows the convergence time (number of iterations) of the networks as function of the number of visual stimuli. The convergence time is averaged on 500 trials and the dashed line represents the standard deviation.

Figure 3 shows the evolution of the learning performance in a scenario with 10 stimuli. We compare the behavior of three decision rules that are: choosing

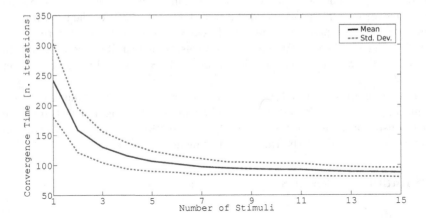

Fig. 2. Convergence time of the networks as function of the number of visual stimuli. The converge time is averaged on 500 trials. Dashed line represents the standard deviation.

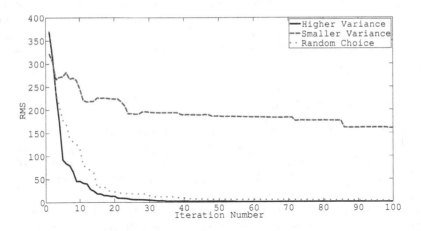

Fig. 3. Evolution of the learning process with different decision making strategies

the target randomly, the target with the higher variance and the one with the lower variance. Choosing the feature with a bigger variance allows the system to explore the region with higher uncertainty and, as expected, it converges faster than the other methods. Choosing the feature with the lower variance forces the system to explore already known regions, so that it does not converge and it represents the worst case. The performance of the random choice is in between the other two.

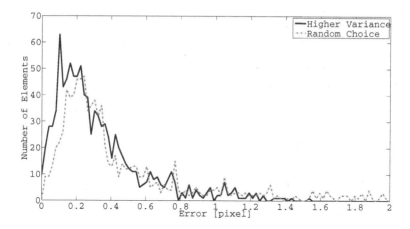

Fig. 4. Histogram of the error after a saccade. The error is computed as the distance of the stimulus from the center of the image.

In order to better compare the performance of the network, we plot the histogram of the error after the first hundred training iterations. The error is computed as the distance of the stimulus from the center of the image. From the histogram we can note the error has a more compact distribution near zero when the algorithm chooses the stimulus with the higher variance. Indeed, by choosing the higher variance, the 99.5% of the tested points have an error smaller than 1 pixel and the maximum error is 9 pixels. On the other hand, by choosing a feature randomly, just the 78.9% of the tested points have an error smaller than 1 pixel and the maximum error is 34 pixels.

4.3   Comparing the Learning Strategies

The aim of this experiment is to test different learning strategies. We compare the performance of the training algorithm by using as teaching signal the ground truth, the linear approximation and the feed-back learning.

We trained the network with the three strategies to reach an RMS error smaller the 0.5 pixel. At each iteration, some visual features are visible in a random position of the image and the robot gazes to it. After the movement the networks are updated.

The achieved results are reported in figure 5. As in the previous experiment, the results are obtained as average of 500 repetitions of the learning algorithm and the performances are computed on a dataset composed of 1000 points.

In the case of the ground truth (GT) we compare the movement performed by the network with the desired movement. As expected, the ground truth approach provides the best performance but it is not really applicable on a real scenario because of the lacking of a teacher.

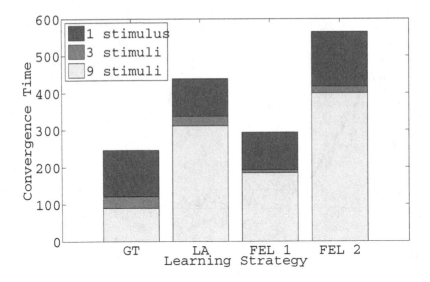

Fig. 5. Convergence time of the networks depending on the learning strategy: ground truth(GT), linear approximation (LA) and feedback error learning (FEL) with two different gains

In the case of the feedback learning, we transform the visual error in motor command by multiplying it by a constant gain. Best results are obtained with the gain set to the inverse of the focal length (FEL 1). It is the expected result because, for small magnitude of the visual error, it is the linear approximation of the input-output transformation. We also tried the feedback error learning with a gain that was the 20% bigger than the focal length (FEL 2) and the consequence is that the performance decreases notably.

Finally, the linear approximation (LA) considers the performed movement as the desired output for the virtual input provided by the visual displacement of the stimulus [2]. Linear approximation performs better than the feedback learning only when it has a non-optimal gain. However, it does not requires any parameters and it just exploits the monotonic proprieties of the visuo-oculomotor transformation.

The experiment was conducted with 1, 3 and 9 stimuli (see Fig. 5). As in the previous experiments, a higher number of visual stimuli allows choosing where to saccade and the convergence time decreases.

5   Robot Results

Once the proposed approach has been validated on a simulated model of the camera, we tested it in the UJI humanoid torso, namely *Tombatossals*. The robotic head (see Fig. 1) mounts two cameras with a resolution of 1024×768 pixels that can acquire color images at 30 Hz. The cameras move by means of

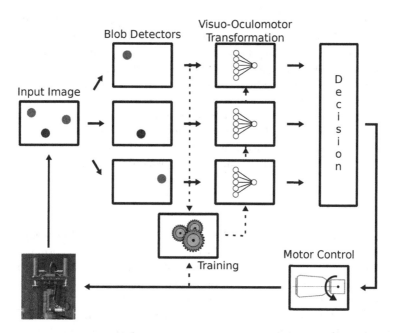

Fig. 6. Sketch of the control system. The image acquired by the camera is processed simultaneously by three blob detectors, each one tuned on a different color (red, green and blue). The visual position of the detected visual stimuli is then sent to three RBFs. Each RBF computes the mean and the standard deviation of the saccade required to gaze to the stimulus. The exploration behavior chooses and performs the movement with higher variance. The visual position of the selected stimulus after the saccade is used to train simultaneously the three networks.

a common tilt and two independent pan motors. In this work we used only the left camera.

We train the network using three visual stimuli to reduce the convergence time (see Section 4.3). The stimuli were composed by colored blobs (red, green and blue) that were displayed on a computer monitor.

The control system is sketched in figure 5. It is composed by several modules that run in parallel and that are wrapped into ROS (Robot Operating System) nodes [22]. The images acquired by the camera are processed by three blob detectors, each one tuned to detect a particular color. The outputs of the blob detectors are sent to three instances of the same neural network. Each network converts the visual location of the stimulus into a saccadic movement (mean and standard deviation). A decision making module that is configured to explore the visual space, produces a movement to the stimulus with higher variance. Once the movement is completed, the target is detected again by the blob detector. The visual positions of the stimulus, before and after the movement are sent, together with the performed movement, to the training module. The training model is in charge of comparing the outcome of the network with the desired output, in order to updated the weights and the covariance matrix as described by equation (3). In this experiment we used the linear approximation technique because it does not require any previous knowledge about the visuo-oculomotor transformation. Finally, weights and covariance matrices are sent to every RBF.

The training step is iterated 350 times. After 200 interactions of the learning process system brought the stimulus in the center of the image with an error smaller than 7 pixels. Then, we recorded a dataset of 200 pairs of *visual position - ground true movement* that we used to test the network. The error of the network was calculated as Euclidean distance between the ground truth movement and the output of the network. The resulting error was of $0.094° \pm 0.071°$.

6   Conclusions

In this work, we have presented a new bio-inspired algorithm to speeds-up the learning of saccade control. This control requires to learn the transformation that converts the visual location of the target into a gaze direction. The environment surrounding the agent provides several visual stimuli, so that the robot can choose to explore the one with higher uncertainty. In order to keep trace of the uncertainty, the neural network is trained using a Kalman filter. Using a probabilistic representation of the sensorimotor transformation is consistent with new neuroscience models [11].

Simulated and experimental results were conducted to validate the proposed model. Indeed, the achieved results shows that, in the best case, the system can halve the number of iterations required to correctly learn the transformation. The proposed methodology is general and can be applied with different learning strategy such as supervised learning and feedback error learning.

A more elaborated image processing is required to use the model with natural images. Future work will thus focus on detecting the most salient visual features and implementing a feature matching algorithm able to provide the estimated detection accuracy.

References

1. Antonelli, M., Grzyb, B., Castelló, V., del Pobil, A.: Augmenting the reachable space in the nao humanoid robot. In: Workshops at the Twenty-Sixth AAAI Conference on Artificial Intelligence (2012)
2. Antonelli, M., Chinellato, E., Del Pobil, A.P.: On-line learning of the visuomotor transformations on a humanoid robot. In: Lee, S., Cho, H., Yoon, K.-J., Lee, J. (eds.) Intelligent Autonomous Systems 12. AISC, vol. 193, pp. 853–861. Springer, Heidelberg (2012)
3. Castet, E., Masson, G.S.: Motion perception during saccadic eye movements. Nature Neuroscience 3(2), 177–183 (2000)
4. Chao, F., Lee, M., Lee, J.: A developmental algorithm for ocular-motor coordination. Robotics and Autonomous Systems 58(3), 239–248 (2010)
5. Chen-Harris, H., Joiner, W., Ethier, V., Zee, D., Shadmehr, R.: Adaptive control of saccades via internal feedback. The Journal of Neuroscience 28(11), 2804 (2008)

6. Chinellato, E., Antonelli, M., Grzyb, B., del Pobil, A.: Implicit sensorimotor mapping of the peripersonal space by gazing and reaching. IEEE Transactions on Autonomous Mental Development 3, 45–53 (2011)
7. Chinellato, E., Antontelli, M., del Pobil, A.P.: A pilot study on saccadic adaptation experiments with robots. In: Prescott, T.J., Lepora, N.F., Mura, A., Verschure, P.F.M.J. (eds.) Living Machines 2012. LNCS, vol. 7375, pp. 83–94. Springer,
   Heidelberg (2012)
8. Chinellato, E., Grzyb, B.J., Marzocchi, N., Bosco, A., Fattori, P., del Pobil, A.P.: The dorso-medial visual stream: From neural activation to sensorimotor interaction. Neurocomputing 74(8), 1203–1212 (2011)
9. Collins, T., Doré-Mazars, K., Lappe, M.: Motor space structures perceptual space: Evidence from human saccadic adaptation. Brain Research 1172, 32–39 (2007)
10. Deubel, H.: Separate adaptive mechanisms for the control of reactive and volitional saccadic eye movements. Vision Research 35(23-24), 3529–3540 (1995)
11. Fiser, J., Berkes, P., Orbán, G., Lengyel, M.: Statistically optimal perception and learning: from behavior to neural representations: Perceptual learning, motor learning, and automaticity. Trends in Cognitive Sciences 14(3), 119 (2010)
12. Forssén, P.: Learning saccadic gaze control via motion prediciton. In: Fourth Canadian Conference on Computer and Robot Vision (CRV), pp. 44–54 (2007)
13. Haykin, S.S., et al.: Kalman filtering and neural networks. Wiley Online Library (2001)
14. Hoffmann, H., Schenck, W., Möller, R.: Learning visuomotor transformations for gaze-control and grasping. Biological Cybernetics 93(2), 119–130 (2005)
15. Jordan, M., Rumelhart, D.: Forward models: Supervised learning with a distal teacher. Cognitive Science: A Multidisciplinary Journal 16(3), 307–354 (1992)
16. Kawato, M.: Feedback-error-learning neural network for supervised motor learning. Advanced Neural Computers 6(3), 365–372 (1990)
17. Marjanovic, M., Scassellati, B., Williamson, M.: Self-taught visually guided pointing for a humanoid robot. In: From Animals to Animats 4: Proc. Fourth Intl. Conf. Simulation of Adaptive Behavior, pp. 35–44 (1996)
18. McLaughlin, S.: Parametric adjustment in saccadic eye movements. Attention, Perception, & Psychophysics 2(8), 359–362 (1967)
19. Park, J., Sandberg, I.W.: Universal approximation using radial-basis-function networks. Neural Computation 3(2), 246–257 (1991)
20. Pouget, A., Sejnowski, T.J.: Spatial transformations in the parietal cortex using basis functions. Journal of Cognitive Neuroscience 9(2), 222–237 (1997)
21. Pouget, A., Snyder, L.: Computational approaches to sensorimotor transformations. Nature Neuroscience 3, 1192–1198 (2000)
22. Quigley, M., et al.: Ros: an open-source robot operating system. In: ICRA Workshop on Open Source Software (2009)
23. Schenck, W., Möller, R.: Learning strategies for saccade control. Künstliche Intelligenz (3/06), 19–22 (2006)
24. Schnier, F., Zimmermann, E., Lappe, M.: Adaptation and mislocalization fields for saccadic outward adaptation in humans. Journal of Eye Movement Research 3(3), 1–18 (2010)
25. Shibata, T., Vijayakumar, S., Conradt, J., Schaal, S.: Biomimetic oculomotor control. Adapt. Behav. 9(3-4), 189–207 (2001)
26. Sun, G., Scassellati, B.: A fast and efficient model for learning to reach. International Journal of Humanoid Robotics 2(4), 391–414 (2005)

Sensory Augmentation with Distal Touch: The Tactile Helmet Project

Craig Bertram[1], Mathew H. Evans[1], Mahmood Javaid[2], Tom Stafford[1], and Tony Prescott[1]

[1] Sheffield Centre for Robotics (ScentRo), University of Sheffield, UK
{c.bertram,mat.evans,t.stafford,t.j.prescott}@sheffield.ac.uk
[2] Wellcome Trust Biomedical Hub, Mood Disorders Centre, Department of Psychology, Univeristy of Exeter, UK
m.javaid@exeter.ac.uk

Abstract. The Tactile Helmet is designed to augment a wearer's senses with a long range sense of touch. Tactile specialist animals such as rats and mice are capable of rapidly acquiring detailed information about their environment from their whiskers by using task-sensitive strategies. Providing similar information about the nearby environment, in tactile form, to a human operator could prove invaluable for search and rescue operations, or for partially-sighted people. Two key aspects of the Tactile Helmet are sensory augmentation, and active sensing. A haptic display is used to provide the user with ultrasonic range information. This can be interpreted in addition to, rather than instead of, visual or auditory information. Active sensing systems "are purposive and information-seeking sensory systems, involving task specific control of the sensory apparatus" [1]. The integration of an accelerometer allows the device to actively gate the delivery of sensory information to the user, depending on their movement. Here we describe the hardware, sensory transduction and characterisation of the Tactile Helmet device, before outlining potential use cases and benefits of the system.

Keywords: Tactile devices and display, Human-computer interaction, Perception and psychophysics, Tactile and hand based interfaces, Mixed/augmented reality, Accessibility technologies, People with disabilities, Sensors and actuators, Sensor applications and deployments, Arduino.

1   Introduction

Touch is a richly informative sensory modality that can provide information about both the identity and location of objects. However, touch sensing is fundamentally local in humans, and distal environmental information must be gathered by other senses. In certain environments and situations a person's ability to acquire information about the distal environment, and objects within it, using vision or audition may be impaired, such as when a fire-fighter is searching a smoke-filled building. Such environments are usually noisy, often dark, and visually confusing. In such circumstances a rescue worker's vision and hearing may already be stretched trying to

N.F. Lepora et al. (Eds.): Living Machines 2013, LNAI 8064, pp. 24–35, 2013.

make out shapes and structures through the smoke, or to listen out for the cries of trapped people above the background noise. The inspiration for this project comes from our past research on tactile sensing in rodents, whose facial whiskers give early and rapid warning of potential hazards or nearby objects of interest through a purely haptic channel, and the active direction and focusing of the whiskers to provide further information. We have recently shown that robots can be effectively controlled using information about their environment from arrays of whisker-like sensors [2]. Here we propose to use artificial distance detectors to provide people with a similar controllable sense of distal space.

Various technologies have been developed to provide information about the world to assist navigation and exploration, usually through sensory substitution. Sensory substitution involves presenting the characteristics of one sensory modality – the 'substituting modality' – in the form of another sensory modality – the 'substituted modality' Examples of sensory substitutions include presenting luminance (a characteristic of visual perception) in the form of a grid of vibrating tactile elements [3] or as a auditory landscape [4]. The translation from a characteristic of one modality to another in sensory substitution devices is usually fixed, and consequently the function of these devices is likewise inflexible. Another approach to assistive technology is sensory augmentation, where the device extends perceptual capabilities. This is contrasted with sensory substitution, where the sensory experience provided by the device is reducible to an existing modality [5]. As a sensory augmentation device is not designed to merely translate a scene, different patterns of activity can be used to communicate different types of information from moment to moment, allowing for greater functional flexibility compared to sensory substitution devices. In sensory augmentation, the aim is to reproduce the function of a sensory modality as a way of interacting with the world, rather than translate the form of one modality into the form of another [6]. Sensory augmentation can also include the creation of an additional 'sense' by presenting information about the aspects of the world that are outside human perceptual capabilities, such as a perception of magnetic north [7]. As this sense would be novel, its presentation cannot be a translation of its form into another sense and thus devices doing so cannot be considered to perform sensory substitution (although they might be more specifically called sensory enhancement devices [8]). Extra senses provided by sensory augmentation can be interpreted and used to guide behavior [8] and aid those with impaired senses, such as providing the visually impaired with spatial sensory information thus aiding movement [7].

The beneficiaries of sensory augmentation are not limited to the sensorily impaired. Personnel working in environments that temporarily restrict their sensory capabilities may also benefit from sensory assistance. Fire-fighters in smoke-filled buildings are often unable to visually locate important objects such as people, doorways, furniture, etc. Although details vary between countries, the universal best-practice method employed by fire-fighters traversing smoke-filled environments is primarily haptic, and involves maintaining contact with a wall or guide-rope and exploring the interior of the room by moving with an extended hand or tool (e.g. [9, 10]). This practice has been used for most of the past century and has not so far been supplanted by more technologically sophisticated approaches.

Attempts have been made to provide the kind of augmented spatial awareness that may be useful in such circumstances. For instance, the Haptic Radar [11], is a modular electronic device that allows users to perceive real-time spatial information from multiple sources using haptic stimuli. Each module of this wearable "haptic radar" acts as a narrow-field range detector capable of sensing obstacles, measuring their approximate distance from the user and transducing this information as a vibro-tactile cue on the skin directly beneath the module. The system has been shown to be effective and intuitive in allowing the wearer to avoid collisions, and the authors discuss the possibility of presenting multiple layers of information at multiple resolutions. However, the functioning of the system is insensitive to task or condition, and consequently there is a risk that the information is not specific or informative enough when it is needed or desired. At times, such a system could also overload the user with excessive or irrelevant information that could be distracting or make it harder to achieve goals. An 'augmented white cane' device has also been demonstrated recently [12]. Ultrasonic sensors and vibro-tactile elements were mounted on a hand-held device, allowing the user to infer the spatial configuration of an environment by sweeping the device like a search light. The downside to this approach is that the user only receives information from a small portion of the environment at any one time, and that it restricts the use of the user's hand for other tasks. An objective of the Tactile Helmet project is to provide a system that overcomes these limitations of inflexibility and impracticality, or at least mitigates their worst effects, by providing a wearable assistive device capable of providing useful information through context sensitive function.

Fig. 1. The Tactile Helmet outside (left) showing the ultrasound sensors, inside (centre) showing actuators mounted around the headband, and a close-up of an actuator (right)

The Tactile Helmet device (shown in Figure 1) is a wearable sensory augmentation device, comprising an array of ultrasonic sensors mounted on the outside of a firefighter's helmet to detect objects in the nearby environment, and a tactile display composed of actuators – vibro-tactile elements to physically engage the user's head. The translation of information from the sensory array to the tactile display is controlled by a context sensitive algorithm. In the current implementation, the tactile display communicates object distance and warns the wearer of imminent collisions when moving. The function of the device is undergoing development: expanding its functioning, but with the aim of optimising the delivery of useful information to the wearer (see section 2.2).

In addition to drawing inspiration from the vibrissal system of rodents, who navigate extremely efficiently using distal tactile signals obtained through their whiskers, we consider that there are further good reasons to employ a head-mounted tactile display. First, experiments have demonstrated that while people find it difficult to process additional information through an already busy channel, such as sight or audition, they will often still have 'bandwidth' for information signals provided through the surface of the skin [13]. Second, there is also evidence that people may process touch stimuli preferentially (compared to signals in other modalities) when attention is divided [14]. Finally, we previously compared the sensitivity threshold and response times at five candidate locations: the hand, the outside of the thigh, the temples, the forehead and the back of the head [15]. We found comparable levels of sensitivity in the fingertips and temples, but quicker speed of response for stimulation sites on the head. In other words, the skin around the forehead and temples provides a sensitive and rapidly responsive site to transmit tactile information, which we take advantage of with the Tactile Helmet device.

A key component of the philosophy behind the Tactile Helmet device is the idea of active sensing, whereby the information presented to the user about an object is selected based on task-related demands. Previous approaches [3, 16] have proposed fixed-function devices, for example translating light levels and presenting them as 'haptic images' to the user, allowing the user to 'see' the environment through the sense of touch. In contrast, the display of the Tactile Helmet device could potentially communicate information about a range of properties: rate of change of position of the object with respect to the user; the physical attributes of the object; the time since first detection of the object by the sensory array; and the current situation or objectives of the user. Importantly, the Tactile Helmet device aims to communicate only the information most relevant to the present situation and will be flexibly controlled to do so. In this way the device should provide rich, task relevant information when it is most required without being unduly distracting.

This flexible approach to functionality will be achieved through another important feature of this device: the low bandwidth nature of the haptic display. In previous research, some laboratories have tested pixellated haptic displays, with many closely packed elements, in order to translate a visual scene into a tactile scene with reasonable resolution [16] – this is sensory substitution. In contrast, the present device uses only a handful of display elements, and seeks to use changes in activity (e.g. frequency or amplitude modulation of the tactile stimulation from the haptuators, or patterns of stimulation) to communicate information. This approach has two benefits – first, it can take advantage of hyperacuity to allow users to interpret simultaneous variable low-resolution stimulation at a higher perceived resolution. Second, as the system is not designed solely to translate a scene different patterns of activity can be used to communicate different types of information from moment to moment. This flexibility of function is a strong benefit of sensory augmentation over substitution devices. The possibility also exists that several overlapping channels of activity could be presented simultaneously, although more complex signals would necessarily require more time to master, and potentially take more cognitive load to process.

The remainder of this paper describes the hardware and software components of the Tactile Helmet device, before characterising the sensory field of the system and detailing some preliminary experimental work designed to explore the utility of the device in an exploration task. Finally, we will consider some future directions for the project, including improvements and experimental evaluations.

2   Methods

2.1   Hardware Overview

Figure 2 shows the configuration of present embodiment of the complete unit, divided into two main parts: computational and control aspects (main figure) and transducers and environmental sensors fitted to a firefighter's helmet (figure inset). The helmet shell is fitted with an array of eight Sensors – a ring of ultrasonic range finding sensors[1], which scan the local environment of the wearer, and communicate information about the environment through four Actuators[2]. Transformation of the ultrasound sensor signal to the actuator output is governed by measurements from an intertial management unit (IMU[3]). The transformation is performed by the Computation Unit (a small netbook running Matlab in a Windows 7 environment). The computation unit also controls the system as a whole – reading sensor data, executing algorithms, sending actuation commands to the actuators. Sensor and IMU data is sent to the computation unit via a Sensor-Actuator Bridge[4] on an I2C bus (4-line serial bus with 7-bit addressing). The bridge also acts as an interface between the control unit and the Actuator Driver Boards (in-house design including PIC microcontroller (dsPIC30F2011), amplifier, and other necessary circuitry). The PIC receives the actuation commands via a UART bus, interprets the commands, prepares the output waveform and drives the amplifiers. The amplifiers then send their output to the actuators. All sensors, and the IMU, are connected to the Sensor-Actuator Bridge via the above mentioned I2C bus (a single four-core umbilical). Power to the amplifiers is provided from a 12V battery via separate cabling. From the amplifiers, the actuator signal travels on separate lines for each actuator.

---

[1]   Eight Devantech SRF08, frequency: 40 kHz, range: up to 6 m; robot-electronics.co.uk

[2]   TactileLabs Haptuators, model TL002-14-A, frequency range: 50-500 Hz, peak voltage: 3.0 V, acceleration at 3.0V, 125 Hz, 15 g load: 3.0 G; tactilelabs.com

[3]   9-dof SensorStick; sparkfun.com

[4]   Arduino Mega (arduino.cc) board with an ATmega microcontroller running at 16MHz.

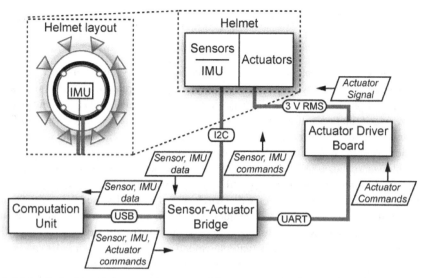

Fig. 2. Main: A diagram of the control process of the current embodiment of the Tactile Helmet device. Dashed inset: A schematic representation of the current embodiment of the helmet, corresponding to the dashed region 'Helmet' in the main figure. The helmet (yellow ellipse) is fitted with eight ultrasound sensors (gradated blue triangles) and an IMU, connected to a shared I2C bus (connection shown in blue) – the Sensor-Actuator Bridge (main figure). Inside the helmet is an adjustable headband (black circle) fitted with four actuators (small red circles) connected to the Actuator Driver Board (main figure), which is in turn is connected to the Sensor-Actuator Bridge (connection shown in red). The Sensor-Actuator Bridge then connects to the Computation Unit (connection shown in green).

2.2   Context Sensitive Sensory Transformation Algorithms

The helmet is able to perform context sensitive selection of one of two general modes of function dependent on acceleration measurements from the IMU. At walking speeds or higher, the helmet functions in 'explore' mode as a unidirectional proximity warning system. The signal amplitude of each of the four actuators is calculated from the pairwise average of the signal from two adjacent sensors to the actuator, which was initially scaled according to a cubic function (see Equation 1):

$$a = \left(1 - \frac{x}{m}\right)^3 \qquad (1)$$

Where a is the amplitude of the signal sent to the actuator, x is the pairwise average of the distance measurement from the two sensors adjacent to the actuator, and m is the maximum distance to which the sensors are set to measure, which we chose to be 200 cm. Following pilot testing, the transformation was changed to a piecewise function where $a = 1$ if $x < 100$, and according to an adjusted version of Equation 1 function (see Equation 2) if $x \geq 100$.

$$a = 2\left(1 - \frac{x}{m}\right)^3 \tag{2}$$

The intention of the cubic scaling was to account for the Weber-Fenchner law, whereby as stimulus intensity increases, a greater absolute increase in stimulus intensity is necessary to produce the same increase in perceived intensity [17, 18]. Further, frequency of stimulation follows a stepwise function such that objects closer than 50 cm trigger a shift to a lower frequency (80 Hz) 'warning' signal to indicate immanent collision from the normal stimulation frequency (150 Hz).

At slower speeds, the helmet currently gates the activity of the actuators such that they are not active when the user is still. An alternative mode for this situation is being developed, where the helmet will switch to a forward focused 'scanning' mode, intended to provide more detailed spatial information about the area on which the user is focusing. The rear two actuators continue to function as collision sensors according to equation 1, but the front two actuators respond such that $a = 0$ if $x \geq 100$ and calculated according to equation 3 if $x < 100$, to produce fine resolution of distance at shorter ranges (note that the warning signal is not present for these actuators in this mode).

$$a = \left(1 - \frac{x}{0.5m}\right)^3 \tag{3}$$

Further, the frequency of the signal now also varies to indicate the difference between the signals of the two sensors to which it responds, with lower frequencies indicating a relatively smaller signal, and thus an object relatively closer to the more lateral sensor, and higher frequencies indicating the same for the medial sensor, thus providing the user with more spatial information. This IMU-based switching between exploration and scanning encourages the wearer to actively engage in their environment which may result in the user acquiring richer environmental information, analogous to exploration in active touch sensing [1].

3   Device Characterization

A protocol was developed to measure the sensory range and extent of the helmet device to determine whether there are any significant blind spots or overlaps between sensors. The helmet was placed on a level surface atop a tripod, with the sensors 1m from the ground. A modification was made to sensing code such that the computation unit (the netbook) would output a sound if a particular sensor detected an object within its range. At approximately regular angular intervals around each sensor, a rubber balloon inflated to around 15 cm in diameter and held at 1 m off the ground was incrementally moved away from the helmet until it was out of range. By recording the closest locations that did not elicit a sound from the netbook at steps of 10 degrees, it was possible to map out the extent of each sensor. The process was repeated with the balloon at heights of 145, 135, 80 and 60 cm to get an estimate of the vertical extent of the range of the sensor. Figure 3 displays a map of sensitivity of the helmet based on this signal characterization, where the shape represents the area where the balloon would be detected, and is constructed from averaging across four sensors and orienting it according the location of the sensors on the helmet.

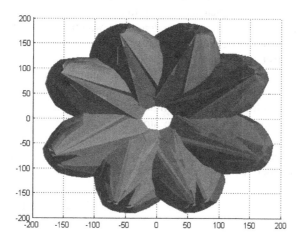

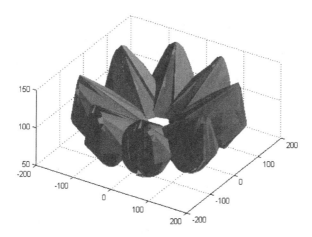

Fig. 3. Sensory extent of the Tactile Helmet device. (above) From above, showing the horizontal extent of the sensory field. Sensors overlap to ensure that there are no blind spots. (below) From the side, showing the vertical extent of the sensory field.

4 Preliminary Experiments

After construction and characterisation of the helmet, a preliminary experiment was conducted to determine whether the distal spatial information could be successfully interpreted and used. A 4 m long corridor was constructed with exits on the left and right side 1 m before the end. Participants were tested in both "helmet on" condition, where the helmet worked as described above, and "helmet off" condition, where the actuators were disabled. On each trial a randomly selected exit was narrowed so that it was too small to exit through. Participants were blindfolded instructed to navigate along the corridor "using all means available to them" (including their hands in both conditions) until they found the exits, to determine which of the exits was passable, and to exit the corridor. Participants were able to navigate and successfully locate the exit with little contact or collision with the wall. Video inspection of the trials suggests that, although participants used their hands in both conditions, they were relying on their hands much less in the helmet on condition than when no information was available from the helmet Some preliminary data from the experiment is shown in Figure 4. Although participants were slower to complete the task in the helmet on condition, experience with the helmet was limited, and performance is likely to improve with practice. Within each condition, performance increased (i.e. time to exit through a gap decreased) across trials, suggesting a learning effect.

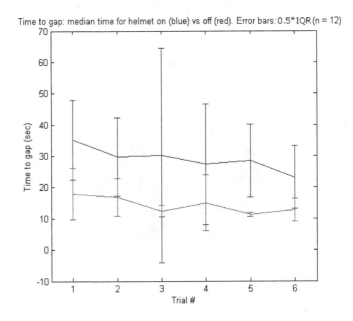

Fig. 4. Time taken for participants to navigate along a corridor, find, and then exit through a gap

5   Conclusion

This paper presents a proof of concept of the Tactile Helmet device. The device is contrasted with previous devices in several ways: it provides sensory augmentation, rather than sensory substitution by providing the functionality of a novel sense of distal spatial awareness rather than translating the form of existing distal senses; it makes use of the spare 'bandwidth' of tactile stimulation to convey information, which may be of particular use in confusing visual and auditory environments such as those which may be encountered by fire-fighters; it is head mounted, which provides a balance of sensitivity and response time to tactile stimulation whilst also leaving the hands-free for other tasks including direct haptic exploration of surfaces; and finally the device is developed within a philosophy of active sensing, where the behaviour of the system changes depending on the task and the behaviour of the user. Preliminary experiments have demonstrated that the helmet can be used in the absence of vision to assist users in navigating along and exiting a constructed corridor.

The on-going goal of the project is to work towards an ideal transformation of the distance information provided by the sensors – more generally into a sense of space, but more specifically into a task related account of the environment, and objects and affordances within it. As 'ideal' is probably best defined as most useful for a user, this transformation would likely be high bandwidth, highly specific, low latency, and low in intrusiveness. At present, the device represents a first step towards these goals.

Initial psychophysics experiments [15] suggested that the best option for scaling for the transformation of distance to vibro-tactile haptuator signal was a hyperbolic function, as this would counteract the effects of the Weber-Fechner law. However, when this signal was used in pilot navigational tasks, participants found that stimulus amplitude at long distances was too low, and too hard to detect to be of use. As a compromise, scaling for distances outside of arms range was selected for the navigation experiments described here. Further psychophysics experiments will provide information for improving the scaling further.

The intrusiveness of the transformation is minimised by controlling the flow of information according to thresholds of rotational and/or linear acceleration. However, this is currently quite imprecise due to the nature of the IMU and also quite coarse – the device is switched on or off, or between states depending on a fixed acceleration threshold. Future implementations could use the IMU more intelligently, for example by dynamically compensating for the potential risk of increased speed by providing more intense stimulation or initiating stimulation sooner in a proportional manner.

The bandwidth and specificity have only been briefly touched on with the current incarnation of the device. The function of the helmet is basic, and consequently the function hasn't been tailored to one or more specific functions or to make use of a high bandwidth signal. The near-term goal of the project, however, is to develop a 'language' of signals that is able to communicate information with more depth than simply scaling as a function of range, while potentially communicating information about several aspects of the environment simultaneously. This could include 'special case' signals to indicate the presence of particular salient features in the world, for example trip or collision hazards, or task affordances. As with the general functioning of the helmet, the detection of particular special cases could be tailored to particular situations or tasks.

A long-term goal is to develop a system to coordinate information between multiple helmets about the environment and other users, which could be used to aid navigation and provide additional support. This could take the form of constructing a map of a building by synthesising information from multiple helmets as users move through it,, or providing information about the location and status of other device users, both to each other, and to a centralised command unit that could support coordinated tracking.

The current direction of development for the Tactile Helmet device has been as a navigation aid to fire-fighters and other emergency personnel. However, as noted earlier, sensory substitution and augmentation devices have also shown potential for use in aiding those with restricted senses, such as providing the visually impaired. For this kind of day-to-day usage, developing a lightweight version of the Tactile Helmet device is a priority. This could be achieved by replacing the safety helmet that the device is currently mounted in with a lightweight cap or headband. Further, once a set of functions has been settled upon, the computer notebook and related hardware that is required to run flexible calculation software could be replaced with a self-contained piece of specifically constructed hardware, which would also reduce weight.

Finally, as well as a technological development, the Tactile Helmet device also represents an opportunity to investigate the nature of sensory perception and the integration of additional senses provided by sensory augmentation. A 'ladder of integration' has been proposed [15], which suggests that a series of tests could be used to determine the extent to which an extra sense provided by sensory augmentation has become integrated into a users 'cognitive core'. The Tactile Helmet device already satisfies the lower steps of the ladder – the stimulation provided by the device is detectable and users can respond to it i.e. use it to navigate. There is also evidence that it reaches the middle steps – users of the device are able to use the device to explore their environment without instruction, and participants in preliminary experiments made spontaneous statements about the nature of the task that framed the sensation from the device in distal terms, e.g. a participant who felt "a wall had appeared from nowhere", rather than reporting a sudden onset of tactile stimulation. Further experimentation could explore the nature of perception of the device, how perception and performance of tasks using the device change over time, and whether a change in the nature of the percept is related to changes in task performance. After satisfying the middle steps, investigating whether the perception provided by the device is subject to sensory illusions would indicate whether the novel, artificial sense has been truly integrated; an effective combination of man, machine, and code. Although this kind of work is removed from the practical applications of the technology, it has the possibility to be greatly informative. A sensory augmentation device that is interpreted in an integrated, heuristic manner would have a smaller cognitive load than one that required attention and intentional processing. Implementing a framework to measure this aspect of using a sensory augmentation device, rather than focusing on performance rates could help improve product development for the field as a whole.

Acknowledgements This research was funded by an EPSRC HEIF Proof-of-Concept award, and an EPSRC/UoS KTA, "A wearable active sensing device using tactile displays".

References

1. Prescott, T.J., Diamond, M.E., Wing, A.M.: Active touch sensing. Phil. Trans. R. Soc. B 366, 2989–2995 (2011)
2. Prescott, T., Pearson, M., Mitchinson, B., Sullivan, J.C., Pipe, A.: Whisking with robots: From rat vibrissae to biomimetic technology for active touch. IEEE Robotics & Automation Magazine 16, 42–50 (2009)
3. Bach-Y-Rita, P., Collins, C.C., Saunders, F.A., White, B., Scadden, L.: Vision Substitution by Tactile Image Projection 221, 963–964 (1969), Published online: March 08, 1969, doi:10.1038/221963a0
4. Durette, B., Louveton, N., Alleysson, D., Hérault, J.: Visuo-auditory sensory substitution for mobility assistance: testing TheVIBE. Presented at the Workshop on Computer Vision Applications for the Visually Impaired (October 2008)
5. Auvray, M., Myin, E.: Perception with compensatory devices: from sensory substitution to sensorimotor extension. Cogn. Sci. 33, 1036–1058 (2009)
6. McGann, M.: Perceptual Modalities:Modes of Presentation or Modes of Interaction? Journal of Consciousness Studies 17, 72–94 (2010)
7. Kärcher, S.M., Fenzlaff, S., Hartmann, D., Nagel, S.K., König, P.: Sensory augmentation for the blind. Front Hum. Neurosci. 6, 37 (2012)
8. Nagel, S., Carl, C., Kringe, T., Martin, R., Konig, P.: Beyond sensory substitution— Learning the sixth sense. Journal of Neural Engineering (2005)
9. International Association of Fire Chiefs: National Fire Protection Agency: Fundamentals of Fire Fighter Skills. Jones & Bartlett Learning (2004)
10. Denef, S., Ramirez, L., Dyrks, T., Stevens, G.: Handy navigation in ever-changing spaces: an ethnographic study of firefighting practices. In: Proceedings of the 7th ACM Conference on Designing Interactive Systems, pp. 184–192. ACM, New York (2008)
11. Cassinelli, A., Reynolds, C.: Augmenting spatial awareness with Haptic Radar. In: Proceedings of the 10th IEEE International Symposium on Wearable Computers, pp. 61–64. IEEE (2006)
12. Gallo, S., Chapuis, D., Santos-Carreras, L., Kim, Y., Retornaz, P., Bleuler, H., Gassert, R.: Augmented white cane with multimodal haptic feedback. In: 2010 3rd IEEE RAS and EMBS International Conference on Biomedical Robotics and Biomechatronics (BioRob), pp. 149–155 (2010)
13. Sklar, A.E., Sarter, N.B.: Good vibrations: tactile feedback in support of attention allocation and human-automation coordination in event-driven domains. Hum. Factors 41, 543–552 (1999)
14. Hanson, J.V.M., Whitaker, D., Heron, J.: Preferential processing of tactile events under conditions of divided attention. Neuroreport 20, 1392–1396 (2009)
15. Stafford, T., Javaid, M., Mitchinson, B., Galloway, A., Prescott, T.J.: Integrating mented Senses into Active Perception: a Framework. Presented at the Royal Society meeting on Active Touch Sensing, Kavllie Royal Society International Centre, Buckingsham-shire (2011)
16. Bach-y-Rita, P., Kaczmarek, K.A., Tyler, M.E., Garcia-Lara, J.: Form perception with a 49-point electrotactile stimulus array on the tongue: a technical note. J. Rehabil. Res. Dev. 35, 427–430 (1998)
17. Dehaene, S.: The neural basis of the Weber-Fechner law: a logarithmic mental number line. Trends Cogn. Sci. 7, 145–147 (2003)
18. Hecht, S.: The visual discrimination of intensity and the Weber-Fechner law. J. Gen. Physiol. 7, 235–267 (1924)

Benefits of Dolphin Inspired Sonar for Underwater Object Identification

Yan Paihas[1], Chris Capus[1], Keith Brown[2], and David Lane[2]

[1] Hydrason Solutions Ltd,
K.E.Brown@hw.ac.uk,
http://osl.eps.hw.ac.uk/
[2] Heriot-Watt University, Edinburgh, Scotland, UK

Abstract. The sonar of dolphins has developed over many years of evolution and has achieved excellent performance levels. With this inspiration, wideband acoustic methods for underwater sensing are being developed. In this paper we explore what we expect to gain from the wide bio-inspired beampattern of such a sonar. The system employed here (the BioSonar) uses wideband sensors based on dolphin sonar, covering a frequency band from around 30kHz to 150kHz and having a frequency dependent beamwidth considerably larger than that of conventional imaging sonars. We highlight the benefits of the transducers' beamwidth, indicating how these properties may be exploited to give improved sonar performance.

Keywords: Bio-inspired, dolphins, sonar.

1 Dolphins' Sonar and the BioSonar System

Dolphins' sonar is the result of millions of years of evolution. Dolphin echolocation systems are known for their excellent performance and have been studied for decades [1]. Dolphins significantly outperform man-made sonars for many applications. They show excellent capabilities for object detection and identification, especially in complex environments such as very shallow water and cluttered locations. As an example, reference [2] studied the capability of dolphins to correctly identify the contents of aluminium flasks suspended in mid-water. The US Navy, as part of their marine mammal programme, have successfully trained dolphins for complicated tasks such as mine detection and harbour inspection.

In this paper we look at some aspects of the performance of a bio-inspired sonar (BioSonar) that uses signals and beam patterns that were developed after analysing the dolphins' sonar systems. Some of the aspects of the designs are quite different from conventional high performance sonar systems. Here we look at how some of these design choices can aid improved object detection and identification.

1.1 Signals for the BioSonar

Based on observations of dolphins' echolocation clicks, Houser [3] introduced a taxonomy for the variety of clicks emitted by dolphins based on their frequency

N.F. Lepora et al. (Eds.): Living Machines 2013, LNAI 8064, pp. 36–46, 2013.

content. This particular click taxonomy identified several click types based on distributions in two main spectral regions, one at low frequency (<70 kHz) and the other at higher frequency (>70 kHz). Capus [4] developed bio-inspired signals from a knowledge of dolphin sonar systems and from observations of signals and strategies used by bottlenose dolphins performing object recognition tasks. Dolphins use the equivalent of a forward-looking sonar. The BioSonar has also been found to be effective running in a side-looking configuration. This permits the BioSonar to be operated in conjunction with existing sidescan sonars, giving cross-verification between the systems.

For any given array configuration, sensor beamwidths are inherently frequency dependent, narrower at the high frequency end. The -3 dB BioSonar beamwidths vary from around 8° to 40° between the high and low frequency extremes of the available bandwidth. These match beamwidths found in recent studies [5] and [6]. In these studies it was also observed that dolphins were able to alter the beamwidths of their signals, which may support different ways of using their sonars. The BioSonar system is based on the dolphin's sonar model, with proven effectiveness in object identification and tracking activities [7]. The operational frequency band used in the BioSonar is 30-130 kHz with bio-inspired signals that match Hauser's taxonomy. Further details of the BioSonar are given by Brown et al. [7].

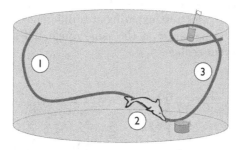

Fig. 1. Dolphin's behaviour during the free-swimming experiment. **(1)** detection and localisation of the target, **(2)** target identification and **(3)** return to the boat.

Figure 1 shows a sketch of a dolphin's behaviour during a free-swimming experiment. During phases 1 and 3, the dolphin is respectively searching for the target on the seafloor and the boat at the surface. The signals emitted during these longer range detection phases are typically low frequency. When the dolphin finds the target (phase 2), it starts an interrogation strategy, *pinging* the target more frequently. The variety of clicks used during the interrogation is much greater than during the other phases, and the click pattern contains all of the click types described by Houser in the taxonomy. It is during this phase that the dolphin is trying to perform the object identification and will be wanting to receive a number of returns on the object of interest to improve the identification. The structure of these echolocation signals and that of bats is very similar. It is this structure that has inspired the signals used in this study.

2   Characteristics of the BioSonar

The horizontal beamwidth of the BioSonar is considerably wider than conventional sidescan - from around 0.4° for a sidescan system to 6°-8° for the current narrowest beam version of the BioSonar. This means that the BioSonar acoustic shadow of a standing target mixes with the reverberation from its surroundings and unlike sidescan images acoustic shadow does not show up. However, as opposed to sidescan imagery, the most valuable target information acquired by the BioSonar lies in the highlights.

The target highlights are in general higher than the background, but a valuable feature, noticeable in the BioSonar image, is the consistency of the target echo over consecutive pings. For SNR above 3 dB the target echo is very clear. For classification, by looking at the high resolution time domain echo, we can distinguish the full echo structure and identify different contributions within it. Figure 2 shows difference in SNR for the same target in two test areas used for these studies. Note that the target is an aluminium cylindrical bottle (around 70 cm heigh and 30 cm diameter). The seafloor of first test area in Fig. 2 *(left)* is made of mud mixed with crushed shells and then highly reflective. In the second test area in Fig. 2 *(right)* the seabed is made of fine sand and then less reflective than the first one [8]. In this test experiment a difference of 8 dB in SNR has been observed for the same target at the same range and the same orientation but lying on two different seabeds.

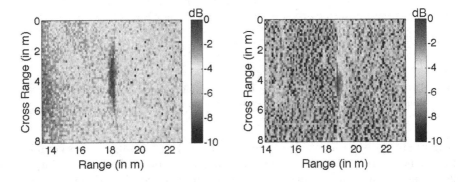

Fig. 2. BioSonar images of objects normalised to the maximum target echo in two test areas. Note that the range for both targets is around 18 metres.

Under low SNR conditions we use an enhanced FFT algorithm, which provides clean and consistent spectra from the target responses. On a straight pass, the angle of view varies slowly and the major features of the spectra (such as peaks and notches for example) can be tracked from one ping to the next. This consistency between consecutive pings is an important factor for discrimination and for false alarm reduction. For classification, the ping-to-ping consistency ensures a clustering of echo responses in the feature space.

Sonar Beamwidth. The beamwidth of the sonar system is a joint function of the transmit and receive beam patterns. For the current BioSonar system two receive settings are available. For the narrower receive setting, the joint - 3dB beamwidth of the system is around 6° at 120 kHz. At the lower frequency end (around 40 kHz), the joint beamwidth of the sonar is around 18°. We can reasonably assume that the beamwidth variation across the frequency band is near linear.

For missions following a classic lawnmower pattern, the sonar beamwidth impacts directly on the number of hits per pass. Since the beamwidth is frequency dependent, this also varies with frequency. For views of targets acquired at a range of 15 to 20 m, we can expect to record between 8 and 20 hits per pass depending on the frequency band. The number of hits per pass is an important consideration from a statistical point of view. If we consider the recorded echoes as a stochastic process (this hypothesis is especially valid for seabed returns and cluttered environments), probability dictates that over a one hour mission (around 10^6 pings) a not insignificant number of returns may give responses similar in some way to a target echo we are interested in.

In the following paragraphs we aim to demonstrate the importance of the wide beamwidth of the BioSonar and the importance of the echo consistency between hits. For the sake of the discussion, let us simplify the problem and let us consider that the echo amplitude is the result of a random binary process. We assume that the random variable is similar to an unbiased coin (output value: 0 or 1). Consider a sampling interval of 10 cm and suppose that we are looking for a specific target with highlights as shown in figure 3 *i.e.* with the specific sonar response: 1010000001. The target of interest covers 10 samples which corresponding to a 1 m extent.

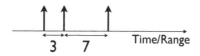

Fig. 3. Example of target model

Note that if this echo model appears over simplistic it matches however the dynamic one can expect in the feature domain. Over 1 m range data we can reasonably extract around 10 significant features. The reader can refer to [9] for feature extraction in wideband sonar data for object recognition.

In our model the seabed response will follow the random variable described earlier and the target response will be deterministic. In order to compute the false alarm rate we just need to calculate the chance of observing an echo from the seabed matching the specific target: $\left(\frac{1}{2}\right)^{10} \approx 9.77 \times 10^{-4}$. So, the probability of such an event is relatively low, but considered in the context of an hour-long mission with 10 Hz ping rate and 60 metres range covered by each ping, we can estimate that this particular event will occur on average more than 2000 times per mission. Assuming a transit of 3 knots, an hour-long mission will cover around 0.5 km^2 of terrain. Using the information on only one ping will then result in more than 4000 false alarms per km^2.

Trials data have shown repeatedly that target echoes change slowly over small angular changes during a transit. So we can assume that the target echo will retain some consistency over the number of hits per pass. From our estimate of our minimum eight hits per pass, the probability of getting the same return as this specific target over 6 consecutive pings during the one hour mission can be computed at approximately 3.7×10^{-12}. To put this number in perspective: if the BioSonar system was inspecting the totality of the surface of Earth (oceans and lands) we would have in average 1.5 false alarm for the totality of the globe. So this shows that It is the target echo consistency provided by the BioSonar and its wide beam pattern that holds the key to reducing false alarms. Figure 4 displays an example of ping-to-ping spectral consistency for a test object labelled ID 1 over five consecutive pings.

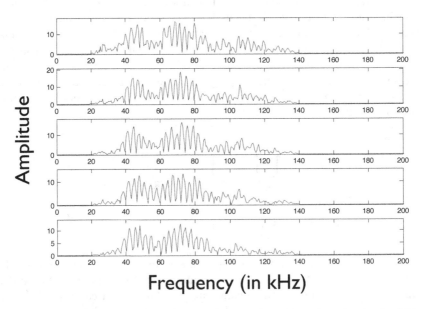

Fig. 4. Spectral consistency for Object ID 1 responses over five consecutive pings

Signal-to-noise ratio is traditionally defined as the power ratio between the signal and the background noise. This is equivalent to the square amplitude ratio between these two quantities. With wideband sonar systems such as the BioSonar the natural approach for analysis, feature extraction, etc. is to work directly in the frequency domain. Considering all the target hits during one pass, there will be greater consistency in the central pings in terms of energy distribution in the spectra. For this reason the traditional approach to SNR using an unmodified echo amplitude ratio is less appropriate, since in our case it is better seen as a continuous function of frequency: $SNR(f)$. We know from [9,10] that target resonances are characterised by interferences in the spectra. So features used for identification will in most cases take these interferences into account. Characterising the SNR directly at the frequency level allows us to estimate the features' distributions and the performance of the system.

3   Experiments

One of the main purposes of this work has been to investigate interaction between objects in the wideband echoes. Dolphins have shown excellent abilities when trying to find objects in cluttered environments and here we try to understand the performance of a bio-inspired sonar. When two targets are close together or when one particular target of interest is adjacent to other clutter objects, we expect the echoes to interact with one other, strongly influencing the overall response.

An important observation is that if we compare images for a single target and for two closely spaced targets, figures 5 (*left*) and 6 (*left*), the separation task is almost impossible using the sidescan sonar. In the intensity imagery, at this resolution, the echo responses merge, appearing as one object.

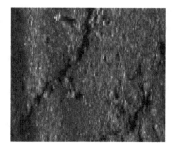 

Fig. 5. Target in clutter: (*left*) Sidescan snapshot, (*right*) BioSonar image snapshot

Looking at the BioSonar image on the other hand, we see that many of the echoes have two clear specular contributions giving rise to a double arc structure, see figure 6 (*right*). A parallel observation is that the enhanced Fourier transform of the double target characteristically contains higher frequency oscillations than

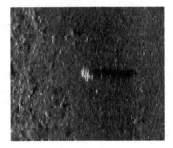 

Fig. 6. Closely spaced target: (*left*) Sidescan snapshot, (*right*) BioSonar image snapshot

are seen in the enhanced Fourier transform of the typical single target response. Following on from these observations we can utilise a spatial domain timing technique to distinguish the double targets from the single targets. The spatial timing transform is defined by the following equation [11]:

$$\mathrm{TD}[s(t)] = \mathrm{FT}^{-1}\left[|\mathrm{FT}(s(t))|^2\right] \tag{1}$$

where $s(t)$ is the echo in the time domain, and FT the Fourier transform. The spatial timing domain response is a measure of the timing density between the individual echo contributions within the full target echo and is consequently directly related to the configuration of the objects on the seafloor. In fact, with only the insight of the energy density in the spatial timing domain for a target response, it is possible to infer whether the inspected target is a single or double cylinder from the echo interaction .

The Masking Phenomena. The techniques outlined above can be used to draw out the capability of the wideband system in distinguishing between a single and double target. It is also important to emphasise the natural limitations to this capability. We can do this by extending the masking phenomenon introduced by Kanizsa for image analysis and interpretation [12].

In image analysis the masking phenomenon occurs when a certain pattern in the image overwhelms the interpretation of the image itself, and thus masks some part of the information. Figures 7 and 8 display two examples of masking.

Figure 7 illustrates a masking phenomenon by texture embedding. The base of the triangle is lost to our perception, as it becomes part of the texture defined by the horizontal parallel lines. In this case the information *triangle* is also lost (even though the triangle is still present in the image).

Figure 8 presents an example of masking by concealment. The hexagon (on the left of the image) is still present in the figure to the right, but it is concealed by the dominance of the parallelograms and entirely disappears from the image interpretation.

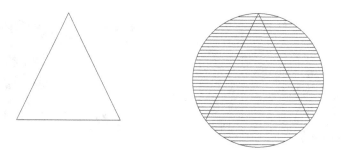

Fig. 7. Example of masking by embedding texture

Fig. 8. Example of masking by concealment

Sonar images are different from optical images, the fundamental difference being that sonar produces range images. This introduces a range dependency in sonar interpretation and results in a change of geometry of the observed scene. Consequently, we can define at least three types of masking in sonar: masking by Reverberation Level, masking by equal range and masking by shadowing. These individual phenomena are well known in the sonar literature, but the particular design of the BioSonar (especially the wider horizontal beamwidth) influences the expression of the masking effects.

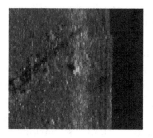 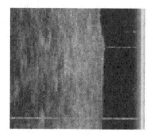

Fig. 9. Example of masking by reverberation level. *(left)* Sidescan image, *(right)* BioSonar image.

 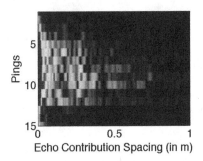

Fig. 10. Example of masking by equal range. *(left)* Sidescan image of two closely spaced targets. *(right)* Spatial timing transform (cf. Eq. 1) of BioSonar data of the same object.

An example of masking by reverberation level (RL) is presented in figure 9. By virtue of its wider beamwidth, the BioSonar integrates more seafloor area at equivalent range than the sidescan. For high grazing angles the RL is relatively much higher and in the case of figure 9 *(right)* just overcomes the target echo in the BioSonar image.

Figure 10 displays a sidescan image of two closely spaced objects alongside the wideband spatial timing transform. The figure presents a rather special case where the two targets are directly in line with the vehicle trajectory. In this particular example it is possible to infer from the splitting of the target shadow in the sidescan image that the target is in fact made up of two objects. In this special case, the spatial timing domain is of little value. The reason behind this is that because of the almost perfect alignment of the two targets with the AUV trajectory, the ranges between the BioSonar and each target are almost identical. The echo contributions of the individual targets overlap considerably and it is difficult to distinguish them using the spatial timing information.

The third masking effect discussed here is masking by shadowing. In this case the two objects lie on a line orthogonal to the sonar trajectory and the closer object casts an acoustic shadow over the second one. The second object is not strongly insonified and a much reduced backscatter echo contribution is recorded. Figure 11 displays an example of partial masking by shadowing. Figure 11 *(left)* displays the high resolution BioSonar image of a double target. By highlighting only the tail of the joint target echo (figure 11 *(right)*) we can bring out the response of the second identical object. The echo amplitude in this case is around one-tenth that of the closer object. It is important to note that these masking phenomena occur in very particular situations. Overcoming the masking effects can be easily achieved by acquiring multiple views of each target.

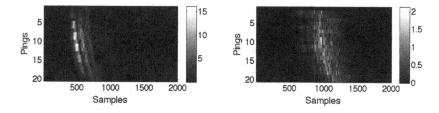

Fig. 11. Example of masking by Shadowing. (*left*) high resolution BioSonar image, (*right*) highlighted tail of the target echo.

4   Conclusions

A sonar system that was inspired by that of a dolphin has been tested to see its effectiveness for object identification and detection. It was found that when used in a sidescan configuration the system was effective at separating very closely spaced objects. This performance arises both from the frequencies used to excite the objects and from the wider beam patterns which ensure that there will always be several pings on each target. Consistency of response across pings allows object identification to be performed with an overall low false alarm rate. This corresponds to the observed behaviour of dolphins during their target identification phase where they attempt to get more information about the object. Multi-look evaluation also helps to overcome a variety of masking issues so detection of objects in cluttered environments becomes easier. Again this is what dolphins do during their identification phase and swim around the object. Overall moving from sonars that are image based using very narrow beam widths to bio-inspired wider beamwidths and not relying on imagery for the processing offers significant advantages for object detection and identification.

References

1. Au, W.: The Sonar of Dolphins. Springer (1993)
2. Moore, P., Roitblat, H., Penner, R., Nachtigall, P.: Recognizing successive dolphin echoes with an integrator gateway network. Neural Networks 4, 701–709 (1991)
3. Houser, D.S., Helweg, D.A., Moore, P.W.B.: Classification of dolphin echolocation clicks by energy and frequency distributions. J. Acoust. Soc. Am. 19 (1999)
4. Capus, C., Pailhas, Y., Brown, K.E., Lane, D.M., Moore, P., Houser, D.: Bio-inspired wideband sonar signals based on observations of the bottlenose dolphin (Tursiops truncatus). J. Acoust. Soc. Am. 121(1), 594–604 (2007)
5. Moore, P.W., Dankiewicz, L.A., Houser, D.S.: Beamwidth control and angular target detection in an echolocating bottlenose dolphin (Tursiops truncatus). J. Acoust. Soc. A 124(5), 3324–3332 (2008)
6. Starkhammar, J., Moore, P.W., Talmadge, L., Houser, D.S.: Frequency-dependent variation in the two-dimensional beam pattern of an echolocating dolphin. Biology Letters (2011)

7. Brown, K., Capus, C., Pailhas, Y., Petillot, Y., Lane, D.: The Application of Bioin-spired Sonar to Cable Tracking on the Seafloor. EURASIP Journal on Advances in Signal Processing (2011), doi:0.1155/2011/484619

8. APL-UW, High-Frequency Ocean Environmental Acoustic Models Handbook (October 1994)

9. Pailhas, Y., Capus, C., Brown, K., Moore, P.: Analysis and classification of broadband echoes using bio-inspired dolphin pulses. J. Acoust. Soc. Am. 127(6), 3809–3820 (2010)

10. Pailhas, Y.: Sonar Systems for Object Recognition. PhD thesis, Ocean Systems Lab, School of EPS, Heriot-Watt University (2013)

11. Pailhas, Y., Capus, C., Brown, K., Petillot, Y.: Design of artificial landmarks for underwater SLAM. IET Radar, Sonar & Navigation 7(1), 10–18 (2013)

12. Kanizsa, G.: Grammatica del Vedere. Il Mulino, Bologna (1980)

Time to Change: Deciding When to Switch Action Plans during a Social Interaction

Eris Chinellato[1], Dimitri Ognibene[1], Luisa Sartori[2], and Yiannis Demiris[1]

[1] Department of Electrical and Electronic Engineering, Imperial College London, UK
[2] Department of General Psychology, University of Padova, Italy
e.chinellato@imperial.ac.uk

Abstract. Building on the extensive cognitive science literature on the subject, this paper introduces a model of the brain mechanisms underlying social interactions in humans and other primates. The fundamental components of the model are the "Action Observation" and "Action Planning" Systems, dedicated respectively to interpreting/recognizing the partner's movements and to plan actions suited to achieve certain goals. We have implemented a version of the model including reaching and grasping actions, and tuned on real experimental data coming from human psychophysical studies. The system is able to automatically detect the switching point in which the Action Planning System takes control over the Action Observation System, overriding the automatic imitation behaviour with a complementary social response. With such computational implementation we aim at validating the model and also at endowing an artificial agent with the ability of performing meaningful complementary responses to observed actions in social scenarios.

Keywords: Social interaction, motor simulation, action observation, action planning, motor primitives.

1 Introduction

While observing a partner executing an action, it appears that the observer's motor system is pre-activating for the execution of the same action [11,21,12]. In most cases, the motor signal is never released, and such activation remains mostly unconscious (but measurable by neurophysiological experimental techniques [18,16,14]). The consequences of this phenomenon for the development of the motor system and the acquisition of social skills are nevertheless fundamental. In fact, this well recognised cognitive mechanism constitutes a typical "mirror" effect, and elicits phenomena such as understanding and interpretation of a partner's actions [20,11]. Such resonance represents also a natural substrate for imitation behaviors [13]. When required though, the motor system will stop resonating, for preparing a complementary action response [18,19].

Thus, two strictly interrelated processes, which have been called Action Observation System (AOS), and Action Planning System (APS), are contextually active during social interactions [3]. The APS is the neural system which,

N.F. Lepora et al. (Eds.): Living Machines 2013, LNAI 8064, pp. 47–58, 2013.

using proprioception and sensory input regarding the surrounding environment, and according to the subject personal objectives and motivations, plans and monitors the execution of all sort of actions. The AOS is instead in charge of following the actions of a partner, mainly by matching them to a subject own motor repertoire.

In this work we investigate the nature of the link between AOS and APS, exploring their behavior in different environmental and social conditions. We model both Action Observation and Action Planning systems with competitive structures, in which candidate actions composed by pairs of inverse and forward models are dynamically evaluated and compared, in the first case referred to observed actions, in the second to planned ones. We have been testing interaction mechanisms between the systems which can explain the effects described in the literature, while constituting the base for the generation of skilled complementary responses in human-robot social setups. With our model, we are able to reproduce the mirroring to complementary switching effect observed in human studies. Such skill can been applied to actual interaction data in which human movements are tracked and interpreted in real-time by a robotic system, in order to interpret the subject actions and prepare an appropriate response.

2   Modeling AOS and APS

Our model is based on the competition among candidates composed by pairs of inverse and forward models. Similar frameworks, based on the concepts of competition, simulation and hierarchy, have been successfully applied to various behavioural tasks in the past [10,22], but the introduction of a dual competitive system is completely novel, and allows to explain a number of neuroscience findings difficult to justify otherwise.

Three inverse/forward model pairs in a competitive framework can be observed in Fig.1. An inverse model computes the motor plan required to achieve a target state considering the current state. A forward model estimates the next state, given the current state and the motor plan. Coupling inverse and forward models allows the motor system to perform a feed-forward control in order to anticipate the evolution of an action and its effects on the environment. In a competitive system, such as that of Fig.1, the accuracies of various inverse/forward model pairs in predicting the next state can be compared in order to assess what pairs constitute the most suitable representations of the ongoing phenomenon. More precisely, in the case of APS, the most accurate model pair is the one best suited to achieve the goal state from the initial state. The motor plan devised by the most accurate inverse model is forwarded to the motor cortex. If a motor signal is finally released, such plan is thus employed by the subject in order to actually achieve the goal state. In the AOS case, the most accurate model pair is the one which is best at capturing the action performed by a partner. It is important to clarify that we use our own motor system, i.e., the models we have learnt by pursuing our own goals, in order to represent and interpret environment and movements of the partner [8,4]. We believe that, in many conditions and as a

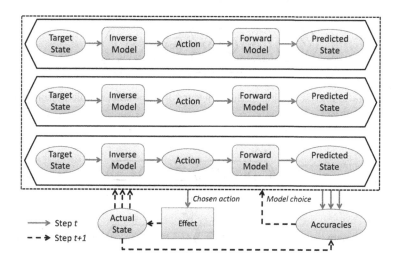

Fig. 1. Competitive framework among a number of candidate Inverse/Forward model pairs concurring for prediction and control of executed and observed actions

default behavior, the output of the most accurate inverse model is again sent to the motor cortex, and constitutes the repeatedly registered automatic imitation or motor resonance [13]. In most cases, the release signal for this motor plan is never released, and the activation remains mostly unconscious.

Action Primitives

The inverse models in the above described framework represent motor primitives which are likely maintained in the premotor cortex of humans and other primates. A higher level, goal-based representation of such motor primitives is properly coded by the posterior parietal cortex [15]. This more abstract representation is instantiated with specific boundary conditions corresponding to the current state of the environment. So, e.g., the general primitive *reaching* is instantiated with the final objective position of the end effector, probably but not necessarily corresponding to the presence of an object. The same general primitive can also generate different inverse models in the premotor cortex, such as when multiple potential goal objects are available for a reaching action.

Primitives are organised hierarchically, and can be merged in time and space to form more complex ones. For example, a grasping action is composed by a transport and a grip components, which in turn are formed by more basic primitives (such as move arm forward, adduct a finger, and so on). The complete set of primitives of different complexity levels represent the whole motor repertoire available to a subject (see e.g. the very simplified representation in Fig.2).

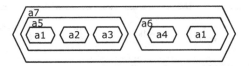

Fig. 2. Schematic representation of the structure of a possible, simple motor repertoire available to a subject

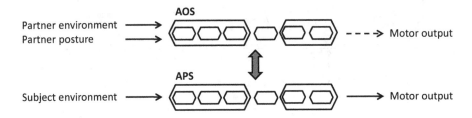

Fig. 3. General framework of the AOS/APS model

AOS and APS Interacting

AOS and APS are composed by matching primitives of the type described above. According to our model, the motor cortex can receive activation signals from either of the two systems. The influence of each system on the planned motor activity depends on both environmental and social variables, such as presence of objects in the common workspace, availability of gaze information, instruction to perform a collaborative task. All these aspects constitute triggers for switching between the two possible modalities. AOS and the typical mirror effects it elicits represent the default behaviour of the compound system, which by default resonates with the partner's actions. On the other hand, following suitable cues, APS can take charge over AOS, and the candidates which are directly related to the real spatial configuration of the observer prevail. Monitoring of the partner's movements is still be performed by AOS, in a decoupled way, and can directly affect on-line action execution if necessary.

The described experimental setup below, testbed for our model, constitutes also a typical example of the way the two systems work and interact.

3   Experimental Setup

As a first approximation to the objectives of our model, we have implemented a version of it which plans either Whole Hand or Precision grips on objects according to the environmental and social context and the movements of a visible partner. To test the behaviour of our model we have taken the data of a real psychophysical experiment designed to analyse what motor response subjects are preparing during different stages of a social interaction [18].

Fig. 4. Sequence from the social interaction video shown to the subjects of the experiment. The rightmost frame shows the moment in which subjects begin to interpret the movements of the demonstrator as requiring a complementary response which overcome the default resonating behaviour. Adapted from Sartori *et al.* [18].

The subjects of the experiment observe a video showing an actor performing a sequence of movements, representing a social interaction. Fig. 4 shows a sequence of the video of one of the conditions of the experimental protocol, in which the last frame represents a change in social requirement from the demonstrator. In the condition shown in Fig. 4, the actor/experimenter pours coffee from a thermos into three cups placed at reachable distance for her, and finally move her hand towards a fourth cup which is out of reach for her, and closer to the observer. It was shown in [18] that the motor system of the observer pre-activates for a whole hand grip, similar to that required to grasp the thermos, while the experimenter is pouring coffee into the three close cups. It thus appears that the observer motor system is resonating with the partner's at this stage. When the actor extends her hand to reach the far cup (last frame), the observer motor pre-activation changes, switching to a precision grip, suitable to grasp the fourth cup. This apparently happens so that the observer can prepare for a complementary movement – approach the cup to the partner – which represents a natural social response. It is worth noting that, in an experimental condition not shown here, pre-activation does not change if the experimenter move the hand back towards herself, without approaching the far cup. Thus, soon after a qualitative change in the nature of the social interaction, subjects switch their motor plan from mirroring to complementary.

Fig. 5 depicts the trajectory of the demonstrator hand, approximately from her own viewpoint, during the whole video. Relevant time-steps are highlighted by labeled empty circles, representing: (1) movement start; (2) thermos grasp (first frame of Fig. 4); (3) (4) (5) coffee pouring into the three close cups (second to fourth frames of Fig. 4); (6) movement end (last frame of Fig. 4). The filled round marker represents the point in time at which human experiments showed that subjects had switched from the default resonating behavior (AOS dominance) to the preparation of a complementary social response (APS dominance) [18]. The filled square marker shows the same switch point as detected by our model, as explained in Section 5.

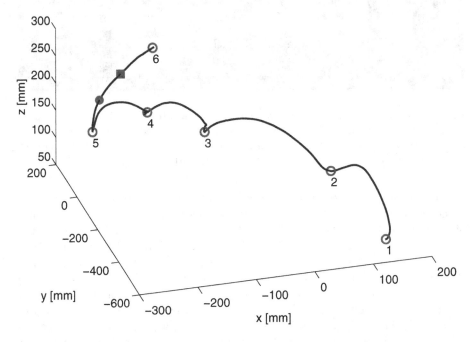

Fig. 5. Trajectory in 3D of the demonstrator hand during the whole video. The filled round marker is the point in time at which subjects were found to have switched from resonating to complementary motor response. The filled square marker shows the same switching point as detected by our model.

4   Implementation

The model implementation thus far consists of three main components. The first component is a representation of the motor primitives available to the subject, the second component is the competitive structure, common to both AOS and APS, and the third component is the associative memory relating action observation to action production, which takes also into account contextual information.

4.1   Motor Primitives

Motor primitives for reaching and grasping, respectively, are implemented with two artificial neural networks. We represent a human arm with three degrees of freedom, two for the shoulder (one for flexion/extension, the other for abduction/adduction) and the third for the elbow extension. The hand is represented with two degrees of freedom, one for the *first dorsal interosseus* (FDI) muscle, serving index finger flexion/extension, and thus participating in all grasping actions, and the other for the *abductor digit minimi* (ADM) muscle, serving little finger abduction, and hence only participating in Whole Hand grips.

The neural networks (standard feed-forward back-propagation ANNs) are trained simulating autonomous exploration of the environment, mapping proprioceptive information on joint angles with egocentric position of the target position. Training of the arm reaching network is done by providing egocentric end effector position in input and arm joint values in output. It have thus three inputs and three outputs. The hand grasping network receives in input the size of the object and outputs the state (yes/no) of the FDI and ADM muscles required to grasp that object.

For these experiments we have omitted the processing of visual information and its mapping to joint space representation. In fact, all of the above can be done in a biological plausible way, similarly to how it is performed in the Posterior Parietal Cortex, without the need to employ Cartesian coordinates [5].

4.2   Competition

Competition among candidates is performed according to the principles described in Sec. 2, separately for AOS and APS, and concurrently for different muscles. This means that, in each system, competition among reaching candidates and competition among grasping candidates, although clearly related, are performed in parallel.

A fundamental point, for both its practical and theoretical implications, is the choice of the candidates that are competing at any one moment. For APS, the list of candidates is obtained from the set of objects visible to the observer. Out of reach, or even unachievable objects such as those visible on a video, are valid candidates for the grasping competition, as indicated by studies showing neural pre-activation upon simple visualization of graspable objects [9]. It has not been as consistently shown how this principle applies to reaching primitives, and we propose that the mechanism is only slightly different. In our framework, no pre-activation is possible, and thus no candidates available, for reaching actions that are impossible or unattainable in the current conditions (e.g. objects on a video), but reaching candidates are generated for those objects that are just out of reach for the subject, implying a tentative reaching or an "ask" signal. AOS candidates are generated similarly to APS's, but considering the partner's reference frame in terms of object position and graspability. There is a critical issue here though, which is that, at the motor primitives level, the observer represents the world of the partner in his own egocentric coordinates. In other words, the above means that the subject takes the perspective of the partner and represent her world as he were in her position. This perspective taking mechanism, requiring a transformation from allocentric to egocentric coordinates by the observer, seems to be performed by a network connecting parietal areas and Premotor Cortex [14]. The important consequence of such transformation is that the subject can employ his own motor primitives to evaluate the movements of the partner.

To exemplify the generation of candidates and their competition, let us consider the case of the experimental setup described in Sec. 3. For what concerns grasping, all visible objects are represented in terms of the types of grasp they afford, and visual to motor transformation for grasping at this stage is performed as described in [7,6]. Candidates are thus Precision or Whole Hand Grips, which will dominate according to what object is or is expected to be the action target (PG for spoon, WHG for mug). The representation is the same for subject and observer. For what concerns reaching, objects are represented both relatively to the subject point of view (in the APS) and to the partner's point of view (in the AOS). Candidates are thus the location of the potential targets in the two different egocentric spaces. Objects farther than the maximum arm extension are labeled as out of reach.

4.3   Social Associative Memory

The relation between AOS and APS is coordinated by a social associative memory which matches certain actions to their natural social response, irrespective of who is actually performing the action. If action B (e.g. take) usually follows action A (e.g. give), the observation of a partner executing A elicits the pre-planning of B by the observer. On the other hand, If the subject executes A, he expects to see the partner performing B in response. A different response would either be classified as an anomaly to discard, or instead constitute an important new relation worth to be memorised. In any case, it is the comparison between predicted and observed stimuli, both on the personal and the social side, which drives the use and plastic modification of action components and their relations.

A social memory of this kind could be stored in the Hippocampus, but its management according to the contextual states of environment and interaction is most likely performed by the Pre-Frontal Cortex [16]. Indeed, the medial Pre-Frontal Cortex (mPFC) has been observed to be more responsive during observation of social movements than individual movements [1].

In this work, we taught our system that when a partner extends her hand towards a target that is not reachable for her (e.g. the furthest, out of reach object), she most likely expects our reaction in terms of handing her that same object. This makes the subject move its focus from the spoon (mirroring effect) to the cup closer to himself (complementary response). This switching generates an automatic change in grasp planning [9,7], immediately reflected by different activations of ADM and FDI muscles.

5   Results

The implementation of reaching primitives described above, although simple, allows us to obtain interesting results that clearly indicate the direction to follow in the next model development steps. Detection of the switching point between the dominance phases of AOS and APS by the model occurs when the actor/partner performs a reaching actions toward a target (the far cup) which is not reachable for her. Such behaviour is interpreted as an "ask" stimulus, eliciting a social complementary response of type "give" on the same target object, i.e.

move the far cup closer to her, so that she can complete the action. No complex reasoning is required for this response, as common ask/give scenarios associating an action with a typical response are learnt by the associative memory so that, upon observation of an "ask" stimulus, a "give" response on the same target is automatic.

The filled square marker in Fig. 5 shows the switching point as detected by our model, while the filled round marker is the actual switching point observed in human studies. It can be observed how, even though the model is able to detect the change in the nature of the interaction before the end of the movement, the human subjects are much faster in this task. Almost identical results were obtained in a dual protocol in which precision grasping was substituting whole hand grasping [18].

We believe there are three reasons for such discrepancy. The first is probably the limited accuracy of the model, probably in the correspondence between the proper and the observed kinematics parameter, which we are currently improving. The second limit is also something that can be dealt with, and is the quality of the sensory information provided to the model (e.g., very importantly, information on the wrist position, instead of the more revealing end effector). The third factor, which cannot really be overcome, is the natural social abilities of humans, which are able to take contextually into account a number of different aspects (posture, voice, gaze, sounds) which cannot be all included in the model.

6   Employing the AOS/APS Model Framework in Human-Robot Interactions

A final, long-term goal of this work is to endow an artificial system, as a humanoid robot, with more advanced social skills when engaged in interactions with human partners. In a previous work complementary to this, and also aimed at achieving the above described skills, we implemented a system for dynamic attention allocation able to actively control gaze movements during a visual action recognition task [17]. Similarly to what described for the reaching prediction in the cognitive science setup described above, the system is able to predict the goal position of the partner hand while it moves towards one of a number of visible targets. At the same time, robot gaze is controlled with the purpose of optimizing the gathering of information relevant for the task. An example of gazing behaviour by the robot, on a relatively cluttered virtual environment, can be observed in Fig. 6(a). Such skills have also been applied to actual interaction data, in which human movements are tracked and interpreted in real-time by the system, as shown in Fig. 6(b).

We are now extending the action prediction abilities shown by the system in the above described experiments for interpreting the possible social meaning of the partner's actions in order to prepare an appropriate response. This is done by introducing the *social associative memory* introduced in Sec. 4 to the system. Additionally, two contextual variables need to be taken into account: reachability and object identity. The first allows to discriminate what objects can

(a) Virtual experimental setup   (b) Real world experimental setup

Fig. 6. Virtual and real experimental setups of robot gazing behaviour. In the virtual setup stimuli are represented by different coloured shapes. The red cylinder represents the robot gaze direction, see [17] for details. In the real setup stimuli are cube blocks.

be acted upon by either the subject and the partner, and the second introduces the environmental variability necessary to elicit different types of actions allowing to create a relatively complex instance of social memory.

Robotic implementation can represent a valuable testbed for the AOS/APS social interaction model, and at this stage we are able to advance some hypotheses of the effects we expect to observe by applying the model to real world interactions. First of all, the system good performance in action prediction (see [17]) should allow for a fast and reliable detection of the switching point between the AOS dominated resonance phase and the APS controlled social response. Second, we expect to observe a further improvement in such performance, consistently with the additional confidence the system can achieve in certain classes of social interactions, by practicing them. Finally, we plan to show with the robot a number of effects typically observed social interaction studies on human subjects, such as automatic imitation [13], cross-modal priming [12], interference [2] and familiarity effects [11].

7   Conclusions

A novel framework for modeling social interactions according to insights provided by cognitive science studies was presented in this work. It was shown how the framework is consistent with most findings, and how its implementation allowed us to replicate some effects observed in actual human studies. Methods and expected outcomes of applying the model to an existing robot system designed for endowing a humanoid robot with social skills were discussed. We are currently working on improving the performance of the model on various experimental neuroscience tasks, both on saved data and in real time. We are developing in parallel new skills for our robotic system thanks to the integration of different social frameworks. With all of the above, we aim at exploring further the nature of the relation between the AOS and APS systems, and the way they modulate their activity in order to generate final motor programs.

Acknowledgments. This research has received funding from the European Union Seventh Framework Programme FP7/2007-2013 – Challenge 2: Cognitive Systems, Interaction, Robotics – under grant agreement No [270490]- [EFAA], and from the University of Padova, under program Bando Giovani Studiosi 2011, L. n.240/2010.

References

1. Becchio, C., Manera, V., Sartori, L., Cavallo, A., Castiello, U.: Grasping intentions: from thought experiments to empirical evidence. Front Hum. Neurosci. 6, 117 (2012)
2. Blakemore, S.-J., Frith, C.: The role of motor contagion in the prediction of action. Neuropsychologia 43(2), 260–267 (2005)
3. Buccino, G., Binkofski, F., Fink, G.R., Fadiga, L., Fogassi, L., Gallese, V., Seitz, R.J., Zilles, K., Rizzolatti, G., Freund, H.J.: Action observation activates premotor and parietal areas in a somatotopic manner: an fmri study. Eur. J. Neurosci. 13(2), 400–404 (2001)
4. Cattaneo, L., Barchiesi, G., Tabarelli, D., Arfeller, C., Sato, M., Glenberg, A.M.: One's motor performance predictably modulates the understanding of others' actions through adaptation of premotor visuo-motor neurons. Soc. Cogn. Affect Neurosci. 6(3), 301–310 (2011)
5. Chinellato, E., Antonelli, M., Grzyb, B.J., del Pobil, A.P.: Implicit sensorimotor mapping of the peripersonal space by gazing and reaching. IEEE Transactions on Autonomous Mental Development (2011) (in press)
6. Chinellato, E., del Pobil, A.P.: Neural coding in the dorsal visual stream. In: Asada, M., Hallam, J.C.T., Meyer, J.-A., Tani, J. (eds.) SAB 2008. LNCS (LNAI), vol. 5040, pp. 230–239. Springer, Heidelberg (2008)
7. Chinellato, E., del Pobil, A.P.: The neuroscience of vision-based grasping: a functional review for computational modeling and bio-inspired robotics. Journal of Integrative Neuroscience 8(2), 223–254 (2009)
8. Costantini, M., Committeri, G., Sinigaglia, C.: Ready both to your and to my hands: mapping the action space of others. PLoS One 6(4), e17923 (2011)
9. Culham, J.C., Valyear, K.F.: Human parietal cortex in action 16(2), 205–212 (2006)
10. Demiris, Y., Khadhouri, B.: Hierarchical attentive multiple models for execution and recognition of actions. Robotics and Autonomous Systems 54, 361–369 (2006)
11. Fabbri-Destro, M., Rizzolatti, G.: Mirror neurons and mirror systems in monkeys and humans. Physiology (Bethesda) 23, 171–179 (2008)
12. Fadiga, L., Craighero, L., Olivier, E.: Human motor cortex excitability during the perception of others' action. 15(2), 213–218 (2005)
13. Heyes, C.: Automatic imitation. Psychol. Bull. 137(3), 463–483 (2011)
14. Jackson, P.L., Meltzoff, A.N., Decety, J.: Neural circuits involved in imitation and perspective-taking. Neuroimage 31(1), 429–439 (2006)
15. Jastorff, J., Begliomini, C., Fabbri-Destro, M., Rizzolatti, G., Orban, G.A.: Coding observed motor acts: different organizational principles in the parietal and premotor cortex of humans. J. Neurophysiol. 104(1), 128–140 (2010)

16. Mukamel, R., Ekstrom, A.D., Kaplan, J., Iacoboni, M., Fried, I.: Single-neuron responses in humans during execution and observation of actions. Curr. Biol. 20(8), 750–756 (2010)

17. Ognibene, D., Chinellato, E., Sarabia, M., Demiris, Y.: Towards contextual action recognition and target localization with active allocation of attention. In: Prescott, T.J., Lepora, N.F., Mura, A., Verschure, P.F.M.J. (eds.) Living Machines 2012. LNCS, vol. 7375, pp. 192–203. Springer, Heidelberg (2012)

18. Sartori, L., Bucchioni, G., Castiello, U.: When emulation becomes reciprocity. Soc. Cogn. Affect Neurosci. (May 2012)

19. Sartori, L., Cavallo, A., Bucchioni, G., Castiello, U.: From simulation to reciprocity: the case of complementary actions. Soc. Neurosci. 7(2), 146–158 (2012)

20. Spunt, R.P., Lieberman, M.D.: The busy social brain: evidence for automaticity and control in the neural systems supporting social cognition and action understanding. Psychol. Sci. 24(1), 80–86 (2013)

21. Wilson, M., Knoblich, G.: The case for motor involvement in perceiving con-specifics. Psychol. Bull. 131(3), 460–473 (2005)

22. Wolpert, D.M., Doya, K., Kawato, M.: A unifying computational framework for motor control and social interaction. Philos. Trans. R. Soc. Lond. B. Biol. Sci. 358(1431), 593–602 (2003)

Stable Heteroclinic Channels
for Slip Control of a Peristaltic Crawling Robot

Kathryn A. Daltorio[1], Andrew D. Horchler[1], Kendrick M. Shaw[2],
Hillel J. Chiel[2], and Roger D. Quinn[1]

[1] Department of Mechanical Engineering, Case Western Reserve University,
10900 Euclid Ave, Cleveland, Ohio 44106-7222
rdq@case.edu
[2] Department of Biology, Department of Neurosciences,
and The Department of Biomedical Engineering, Case Western Reserve University,
10900 Euclid Ave, Cleveland, Ohio 44106

Abstract. Stable Heteroclinic Channels (SHCs) are continuous dynamical systems capable of generating rhythmic output of varying period in response to sensory inputs or noise. This feature can be used to control state transitions smoothly. We demonstrate this type of controller in a dynamic simulation of a worm-like robot crawling through a pipe with a narrowing in radius. Our SHC controller allows for improved adaptation to a change in pipe diameter with more rapid movement and less energy loss. In an example narrowing pipe, this controller loses 40% less energy to slip compared to the best-fit sine wave controller.

Keywords: stable heteroclinic channels, biologically-inspired control, worm-like robots, peristalsis.

1   Introduction

Coordinated oscillators can generate life-like motion in biologically-inspired robots. These oscillators are often based on limit cycles, systems of equations that stabilize into a regular periodic output. Changing the relative phase of an oscillator can change the locomotory gait [1]. These controllers may be similar to the way animals control their bodies because groups of neurons connect to generate repeating cycles, often referred to as central pattern generators (CPGs). See [2] for a review of controllers inspired by this concept. One of our previous papers was based on the Wilson-Cowan model to adjust the speed and spatial resolution of traveling waves [3]. We added feedback to limit radial expansion in [4]. However, we want variable dwell times at phases in the oscillation cycle in response to environmental feedback, which led us to use a different mathematical oscillator: stable heteroclinic channels.

Stable heteroclinic channels (SHCs) are a framework for continuous dynamic oscillation that can produce regular intervals of steady output [5–7]. In other words, the system can "pause" near defined equilibrium points. This allows a designer to treat the equilibrium points as a sequence and yet the dynamic equations provide

N.F. Lepora et al. (Eds.): Living Machines 2013, LNAI 8064, pp. 59–70, 2013.
© Springer-Verlag Berlin Heidelberg 2013

smooth continuous transitions. The transitions can be controlled by feedback terms. In this way, SHC pattern-generating networks can be varied to produce behavior ranging from the smooth cyclic outputs found in limit-cycle CPGs to the more sensory-gated state-like behavior found in finite state machines [8].

The underlying dynamical structure is based on the principle of "winnerless competition" [6–7]. Conceptually, an SHC consists of a sequence of saddle points, which are equilibrium points that attract trajectories in certain directions (referred to as a stable manifold), and repel trajectories in others (an unstable manifold). By setting up the unstable manifold of one saddle to lead directly into the stable manifold of another saddle, the SHC manifold is formed and sequences of transitions between the saddles can be generated [9]. Here we shall refer to the saddle equilibrium points that form the overall SHC manifold as the "nodes" of the SHC. Paths in phase space connecting two different equilibrium points are called heteroclinic connections as opposed to homoclinic connections, which loop back to the same equilibrium point. Stable (attracting) cycles of heteroclinic channels were first found in turbulent flow equations [5]. Here we use Lotka-Volterra equations [10–13] but these systems of saddle equilibria can be produced by a variety of models ranging from a chain of piecewise-continuous linear saddles [8] to pulse-coupled oscillator networks [14] to networks of Hodgkin-Huxley neurons [15]. The Guckenheimer-Holmes system [5], which has a mathematical structure that closely resembles the Lotka-Volterra system, is also commonly used [16–18]. In other work, we have employed SHCs in a neuromechanical model of feeding behavior in the sea slug [19] and in a PWM speed controller of a wheeled millipede-inspired robot [20]. In this work, we use SHCs to control a simulation of worm-like locomotion.

2   Simulating Peristaltic Locomotion

Lack of traction is a primary failure mode of soft-bodied peristaltic robots. Therefore, their controllers should be designed to mitigate slip. There are two complementary components to slip reduction. First, the anchoring segments should be controlled to generate sufficient traction. Second, the remaining segments should be controlled to advance in such a way that little traction is required. Thus, each segment should be in one of two control states: anchoring or non-anchoring. The shape of the segments must be a continuous function of time, so it makes sense to choose a continuous mathematical system, SHCs, to represent the actuation.

We will consider a simulated worm robot with 12 actuated segments, each with one degree of freedom: length. The segment length is coupled to height such that as length decreases, the height increases until the segment touches the ground. We model a line of 13 point masses with a segment defined between each pair of masses. We assume that segments have stiffness that resists changes in length and a lesser stiffness that resists differences in adjacent segment heights. We assume radial symmetry (e.g., no gravity induced deformations) in a pipe environment. If the robot is in contact with the pipe in multiple places, these deformations will have effects not only locally for the contacting segment, but also in other segments. An actuator at each segment applies force to extend and retract the segment. Given actuation forces, our simulation determines Coulomb friction ground (pipe) contact forces and node locations. Segment lengths and segment heights are derived from node locations. We assume Coulomb friction coefficients of 0.12 (static) and 0.1 (sliding) to solve for the friction state of the segments (slipping forward, slipping backward, static friction, or not-in-contact). The details of the simulation solver and model will be found in [21]. This simulation is an abstraction of our recent worm-like robot [22].

3   Implementation of SHC Control System

The segments of the worm-like robot in our simulation are controlled by actuation forces. Positive forces increase segment length and decrease segment height. Segments should extend, then retract, and then anchor to the pipe wall farther ahead. Thus there are two control objectives—traction control for anchoring segments and wavelength control for advancing segments—that have been implemented as two control modes: the anchor mode and the progress mode with paired extension and retraction. Each segment is controlled by one of the two modes. Unlike finite state machines, smooth mode transitions are desired. Limit cycles commonly used in CPGs produce smooth outputs akin to sine waves in which the amplitude is at the maximum value for only an instant. Since we want the contacting segments to stay firmly anchored as the other segments progress, we could threshold this output. But a better option may be to control the actuation forces with SHCs.

We use competitive Lotka-Volterra equations [13] for our controller in the form of the following system of stochastic differential equations:

$$\mathrm{d}a_i^t = a_i^t \left[\alpha - \sum_{j=1}^n \rho_{ij} a_j^t \right] \mathrm{d}t + \eta_i^t \mathrm{d}W_i^t \tag{1}$$

where a_i^t is the level of activation of SHC node i as a function of time, t. a_i^t varies between 0 (when the segment is extending or retracting) and 1 (when the segment is anchoring). $\alpha = 200$ is the instantaneous activity growth rate, η_i^t is a parameter controlling noise levels over time, W_i^t is an n-dimensional Wiener process (like standard Brownian motion). Since our simulated worm robot has 12 segments, we use a 12-node SHC and specify ρ, the connection matrix, as

$$\rho = \alpha \begin{bmatrix} 1 & \gamma & \gamma & \gamma & \gamma & 0 & 0 & \zeta & \gamma & \gamma & \gamma & 0 \\ 0 & 1 & \gamma & \gamma & \gamma & \gamma & 0 & 0 & \zeta & \gamma & \gamma & \gamma \\ \gamma & 0 & 1 & \gamma & \gamma & \gamma & \gamma & 0 & 0 & \zeta & \gamma & \gamma \\ \gamma & \gamma & 0 & 1 & \gamma & \gamma & \gamma & \gamma & 0 & 0 & \zeta & \gamma \\ \gamma & \gamma & \gamma & 0 & 1 & \gamma & \gamma & \gamma & \gamma & 0 & 0 & \zeta \\ \zeta & \gamma & \gamma & \gamma & 0 & 1 & \gamma & \gamma & \gamma & \gamma & 0 & 0 \\ 0 & \zeta & \gamma & \gamma & \gamma & 0 & 1 & \gamma & \gamma & \gamma & \gamma & 0 \\ 0 & 0 & \zeta & \gamma & \gamma & \gamma & 0 & 1 & \gamma & \gamma & \gamma & \gamma \\ \gamma & 0 & 0 & \zeta & \gamma & \gamma & \gamma & 0 & 1 & \gamma & \gamma & \gamma \\ \gamma & \gamma & 0 & 0 & \zeta & \gamma & \gamma & \gamma & 0 & 1 & \gamma & \gamma \\ \gamma & \gamma & \gamma & 0 & 0 & \zeta & \gamma & \gamma & \gamma & 0 & 1 & \gamma \\ \gamma & \gamma & \gamma & \gamma & 0 & 0 & \zeta & \gamma & \gamma & \gamma & 0 & 1 \end{bmatrix}^T \tag{2}$$

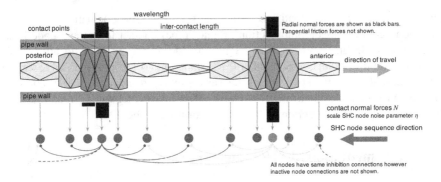

Fig. 1. Each of the 12 body segments is associated with a node of the SHC. SHC nodes inhibit their nearby neighbors, except the immediately posterior node (see the connection matrix (2)). This asymmetry causes the active SHC node to shift rearward as the body moves forward. The segments associated with inactive nodes extend and retract to maintain the inter-contact length. The segments associated with active nodes anchor to the pipe wall by regulating normal forces. The normal forces (gray bars) can scale the SHC node noise. The magnitude of the noise shortens or prolongs the period of the active node, (see Fig. 2), thus keeping it aligned with each new contacting segment (see Fig. 3).

where $\zeta = 0.025$ and $\gamma = 3$. Note that the matrix is circulant, the diagonals wrap around, and thus each node has identical connections to other nodes based on relative location.

We integrate (1) numerically using the Euler-Maruyama method, resulting in the following system of difference equations:

$$\Delta a_i^{t+\Delta t} = a_i^t \left[\alpha - \sum_{j=1}^{n} \rho_{ij} a_j^t \right] \Delta t + \eta_i^t \sqrt{\Delta t}\, g_i^t \tag{3}$$

where $\Delta t \approx 0.001$ sec. is the integration timestep and $g_i^t \sim \mathcal{N}(0,1)$ is a normally distributed random variable with zero mean and unit variance. The state variable a_i is augmented by Δa at each timestep. The timestep size, Δt, is approximate because smaller timesteps are used when multiple contacts must be resolved (as described in [21]). Wiener process noise is time-invariant so different size time-steps can be used. Also the state variable a_i is prevented from exceeding boundary conditions as the noisy forcing might perturb the state, a_i, outside of the valid regime ($0 \leq a_i \leq 1$). Scaling the normal variates by the square root of the timestep, $\sqrt{\Delta t}$, makes the stochastic term act as a time-scale-invariant Wiener process noise. As a consequence, it is a good model for continuous Levy walks such as Brownian motion or zero-mean memory-less white noise. The noise term is explored further below. The parameter values of the connection matrix, ρ, were chosen using a Matlab toolbox we have developed for the design, analysis, and simulation of SHC networks [23].

Equation (1) defines an SHC manifold with as many dimensions as segments. The activity in each dimension (or state) grows exponentially in a stochastic manner. When properly tuned, the activation value is quite small for most state and close to 1 for the other nodes. We will call nodes with more than 90% of activation "active." Active SHC nodes are colored red in Fig. 1 for one simulation timestep. The stochastic growth is limited by inhibition from other nodes. Each node inhibits the activation of most of its nearby neighbors to the degree of its activation (wrapping around the ends of the simulated worm as shown by the dashed lines in Fig. 1). However, the node immediately posterior to a given node is not inhibited, and thus the activity of that node slowly grows until it is large enough to inhibit the node that was originally most active. This is a class of Lotka-Volterra population competition dynamics equations in which each segment population activity inhibits ("preys on") the activity of other nodes, except for the node immediately posterior in the cycle. Eventually, the activity in this subsequent node grows because the current active node does not inhibit it. It becomes the active node and further inhibits the previous active node. This creates the traveling wave that will drive the sequence of actuator extensions and retractions in order to contact and follow the walls of the pipe. There are two waves per body because the inhibition only strongly affects four neighboring nodes ahead and behind it, so there are uninhibited nodes six segments (half the body length) away.

Segments corresponding to active SHC nodes are controlled by the anchoring control mode to firmly expand against the pipe. When a segment's SHC node is inactive, a weighted sum of the neighboring activities is integrated to determine the control force (a progress extension/retraction mode of the actuator force controller).

In pipe narrowings, segments without feedback touch the ground early, and thus cannot extend or retract properly. This requires either slippage or more traction on the anchoring segments. However, a narrowing in a pipe should be an opportunity to get more horizontal traction per contact pressure because of the change in ground angle. This can be accomplished by using the SHC framework to allow segments to compete to be anchors. This keeps the SHC state in phase with the segment contact state.

Instead of a fixed noise magnitude, $\eta = \varepsilon$. The noise magnitude in each direction is scaled based on contact normal forces:

$$\eta_i^t = \varepsilon + 10^{-6} \left| N_i^{t-\Delta t} \right|^{1.5} \tag{4}$$

where $\varepsilon = 10e-12$. The effect of the Gaussian noise on the mean first passage time for linear (and linearized) saddle systems has been characterized by [24]. Figure 2 shows how adding more noise terminates the activity of a node more quickly.

This gives an SHC node a competitive advantage in the competition to become active and hence anchor its associated segment. This makes sense because the ability to generate large contact normal forces is a mechanical advantage for an anchor point. The result is that the contact point is in phase with the active SHC node. There are two advantages to having the feedback in the stochastic term in (3) to modulate the period of the SHC. First, it suggests that this method will handle noisy sensor data well. Second, by isolating the control parameters in the stochastic term, we can use tools developed for characterizing similar stochastic differential equations.

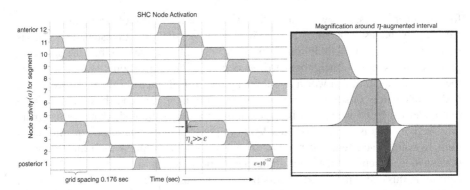

Fig. 2. The SHC activation sequence with the noise magnitude, $\eta = \varepsilon = 10e-12$ (see (4)) results in SHC nodes being active ($a > 0.9$) for a mean period of 0.176 seconds. Changing the noise magnitude η alters this mean period [24]. A temporary increase in the noise magnitude η prior to the activation of the subsequent node can precipitate an early transition from the prior activate state. In this figure, for illustrative purposes, the noise magnitude η is equal to $\varepsilon = 10e-12$ at all times except during the brief interval indicated by the green bar when η is set to 1. Note that the vertical lines are evenly spaced at the average activation duration. The mean phase of the overall system is unaffected by an early transition of one of the states.

4   Comparison of Three Controllers

We will compare three different controllers of increasing complexity. In controller (A), the actuation forces are clipped sine waves with a different phase offset for each segment. (B) The actuation force is controlled by the SHC but without feedback on the noise term, i.e. equation (4) is not used but rather set to a constant. (C) The SHC has timing feedback from normal contact forces generated by the segment with the pipe as described in equation (4).

The best clipped sine wave controller (A) was a regression fit to the output of the SHC controller after settling into regular motion in a constant 38 mm radius pipe. The wave has mean at 10% of the sine wave amplitude and a maximum value clipped at 70% of the amplitude:

$$\text{Controller A: } F_i^t = \min\left(50, 70\sin\left(50\frac{2\pi}{n}i + 0.95(2\pi)t\right) + 7.5\right) \tag{5}$$

Here, F_i^t (in Newtons) represents the force of the actuator for segment i at time t, n is the total number of segments, 12. The negative values, corresponding to segment retraction and anchoring, were unclipped. See Fig. 4 for an example wave at a single segment.

In both controllers (B) and (C), the control force is not a sine wave but instead determined based on previous force values and the SHC node activities. When SHC node activity crosses the threshold 0.9, the actuator force at that segment, F_i, is decremented by an initial amount. Then, if any slip is detected we further decrease the actuator force proportional to the magnitude of the slip. In the absence of slip, generally shortly after contact, the actuator force is controlled to maintain a chosen actuator force. Upon exiting the anchoring mode, the maximum of the current force or the onset force was augmented by the initial decrement amount.

For the nonanchoring segments under (B) and (C), we paired the extension and retraction within a single wavelength so that the segments of similar lengths (e.g., the segments before and after the contact point) would have matched length rates-of-change to maintain the inter-contact length. Nominally, there are six segments in a wavelength: one contacting and two extending paired to two retracting. That leaves one segment unpaired in the middle. The middle segment should be at the maximum extension suitable for the segment, resulting in a larger inter-contact length than an unclipped sine wave. Having a larger vs. smaller inter-contact length is better for getting through pipe narrowings. If the inter-contact length is too small, the stride length can go to zero, stalling progress. Also, this middle segment acts as a buffer if the forward anchoring segment becomes inactive at a different time than the posterior anchoring segment. The SHC inhibition cycle specified by connection matrix (2) discourages less than five or more than seven segments between anchoring segments, but if even outside those bounds, this weighted integration would be robust.

Controller (C) is different from (B) in that we provide feedback to the SHC used for timing. Controller (B) used a fixed noise magnitude, $\eta = \varepsilon$. In controller (C), the noise magnitude in each direction is scaled based on contact normal forces as described in equation (4).

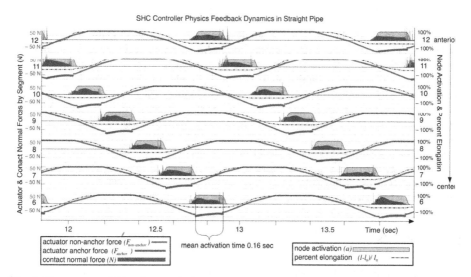

Fig. 3. Key state variables for steady locomotion in a straight pipe with SHC feedback from equation (4). Contact forces (normal force shown here) and segment shape (percent elongation plotted here) are determined from actuator forces. In turn, the contact forces trigger the growth of SHC node activity, keeping node activation and contact forces in phase.

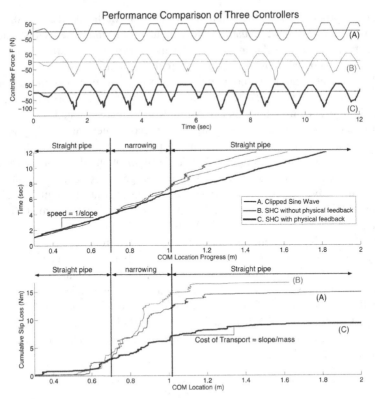

Fig. 4. Comparison of three different controllers in a pipe narrowing. **(Top)** The force, F_6^t, applied by the actuator of a single segment, for each of the three controllers. Controller (A) forces are unaffected by the narrowing. Controller (B) forces have modified amplitudes and Controller (C) forces have modified amplitudes and periods. This latter adaptation gives (C) better performance as is shown in the next two panels. **(Center)** The mapping between time and center-of-mass (COM) progress. The reciprocal of the slope is the speed. Controller (C) has the most consistent and fastest speed through the narrowing. **(Bottom)** Progress vs. cumulative slip loss. The slope times the mass is the Cost of Transport (COT). Each controller can be efficiently tuned for a straight fixed-diameter pipe. However, when the pipe changes diameter, controller (C) performs the best. The location of the pipe narrowing is labeled on the x-axis, but the narrowing effect extends beyond the two lines because the anterior segments enter the narrowing before the COM and the posterior segments leave the narrowing after the COM.

5   Results

The initial conditions in the simulation are such that the simulated worm robot is at rest and not touching the walls. As expected, the simulated worm robot does not start to progress until traction is gained. An ill-designed controller could result in no forward progress, backward motion, or constant-velocity floating without touching the walls. An effective controller will lead to the simulated worm robot settling into forward motion with small friction forces.

We tested the three controllers discussed in Section 4 in a simulated pipe with a constant 0.38 m radius at first ($x < 0.7$ m), then a smooth narrowing and widening in radius, followed by a long constant 0.38 m radius pipe (after $x > 1.0$ m). All three controllers were able to navigate this environment. First, we will consider behavior in the straight portion of the pipe. The values of the key parameters after settling into motion in the long straight (constant-radius) pipe are plotted for the (C) controller in Fig. 3. In the straight pipe, all three controllers develop similar normal forces and segment deformations. The (B) controller has longer and less regular activation times because it lacks feedback to terminate the SHC node activity when the next segment makes contact. The (A) control was fit to the (C) output forces in the 6th segment and phase-offset for each segment. Thus (A) and (C) have the same actuator force ranges. Second, we will consider the behavior in the pipe narrowing, Fig. 4. In addition to the desired behavior in straight pipes, each of these controllers allows the simulated worm robot to pass through a 30% decrease in the pipe radius. The passage with the (C) controller is diagrammed in Fig. 5.

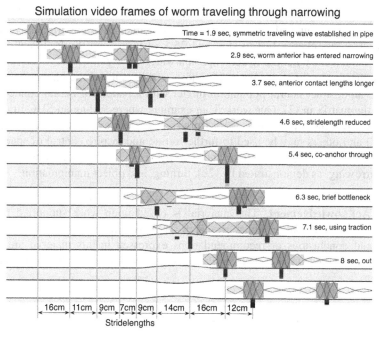

Fig. 5. The simulated worm robot with Controller (C) shown when the third node from the posterior (third segment from left) is in the middle of its active period. The contact forces are dark bars (normal component in black; positive and negative tangential components in blue and pink, respectively). Vertical lines indicate the contact points of this segment (their spacing is the stride length). Note that the stride length starts out at 16 cm, and becomes smaller as the contact length increases. Simulation times and brief descriptions are provided at right.

Next we consider energy efficiency. We assume that the irrecoverable energy loss is the sum of the work done against friction. We refer to this as the Cumulative Slip Loss in Fig 4. This is a quantity we can numerically integrate from our simulation by measuring the slip over the average friction force for each segment and each timestep. Note that the slope of the Cumulative Slip Loss plotted against the progress of the Center of Mass (COM) is proportional to the cost of transport, a common efficiency parameter. Since the total mass for each simulated run is the same, we can see that controller (C) results in lower (better) cost of transport and less energy required.

6   Discussion

In this paper we present an example of locomotion in which timing and coordination is critical. Uncoordinated motion wastes energy, slows progress and may not even result in forward progress, and can get the body in a state that is increasingly difficult to maneuver out of (e.g., getting stuck with contact points lodged too close together in a narrowing). Sufficient coordination for a specific task can be accomplished without feedback as in the robot Softworm [22] or our sine wave controller (A). However, with feedback, the simulated worm robot can do the same task more efficiently (in our case with 40% less energy lost to slip, and 25% of maximum the frictional cost of transport) and has the potential to accomplish a wider variety of goals.

The stochastic dynamic oscillator we chose, stable heteroclinic channels, can be considered an abstract neural population model with a pattern of neural connectivity related to the matrix in (2). Our work is an example where adding an SHC improves performance. But there is more to learn from biology. For example, the experiments by [25–27] and others may be used to further refine and inspire control of speed and dynamic adjustment of wave number and spacing, as well as explore new behaviors such as burrowing, as demonstrated by [28], turning, and object manipulation.

Funding Acknowledgement. This material is based upon work supported by the National Science Foundation under Research Grant No. IIS-1065489. Any opinion, findings, and conclusions or recommendations expressed in this material are those of the authors(s) and do not necessarily reflect the views of the National Science Foundation.

References

1. Ijspeert, A.J., Crespi, A., Ryczko, D., Cabelguen, J.-M.: From swimming to walking with a salamander robot driven by a spinal cord model. Science 315, 1416–1420 (2007)
2. Ijspeert, A.J.: Central pattern generators for locomotion control in animals and robots: a review. Neural Networks 21, 642–653 (2008)
3. Boxerbaum, A.S., Horchler, A.D., Shaw, K.M., Chiel, H.J., Quinn, R.D.: A controller for continuous wave peristaltic locomotion. In: Proc. IEEE Int. Conf. on Robotics and Automation, pp. 197–202 (2011)

4. Boxerbaum, A.S., Daltorio, K.A., Chiel, H.J., Quinn, R.D.: A Soft-Body Controller With Ubiquitous Sensor Feedback. In: Proc. 1st Int. Conf. on Living Machines (2012)
5. Guckenheimer, J., Holmes, P.: Structurally stable heteroclinic cycles. Math. Proc. Camb. Phil. Soc. 103, 189–192 (1988)
6. Afraimovich, V.S., Zhigulin, V.P., Rabinovich, M.I.: On the origin of reproducible sequential activity in neural circuits. Chaos: An Interdisciplinary Journal of Nonlinear Science 14, 1123–1129 (2004)
7. Laurent, G., Stopfer, M., Friedrich, W., Rabinovich, M.I., Volkovskii, A., Abarbanel, H.D.I.: Odor Encoding as an Active, Dynamical Process: Experiments, Computations and Theory. Ann. Rev. Neuro. 24, 263–297 (2001)
8. Shaw, K.M., Park, Y.-M., Chiel, H.J., Thomas, P.J.: Phase Resetting in an Asymptotically Phaseless System: On the Phase Response of Limit Cycles Verging on a Heteroclinic Orbit. SIAM Journal on Applied Dynamical Systems 11, 350–391 (2012)
9. Rabinovich, M., Huerta, R., Varona, P., Afraimovich, V.S.: Transient cognitive dynamics, metastability, and decision making. PLoS Computational Biology 4, e1000072 (2008)
10. Gause, G.F.: Experimental Studies on the Struggle for Existence. J. Exp. Biol. 9, 389–402 (1932)
11. Gilpin, M.E.: Limit Cycles in Competition Communities. The American Naturalist 109, 51–60 (1975)
12. May, R., Leonard, W.: Nonlinear aspects of competition between three species. SIAM Journal of Applied Mathematics 29, 243–253 (1975)
13. Afraimovich, V., Tristan, I., Huerta, R., Rabinovich, M.I.: Winnerless competition principle and prediction of the transient dynamics in a Lotka-Volterra model. Chaos: An Interdisciplinary Journal of Nonlinear Science 18, 043103 (2008)
14. Neves, F.S., Timme, M.: Controlled perturbation-induced switching in pulse-coupled oscillator networks. J. Phys. A: Math. Theor. 42, 345103 (2009)
15. Nowotny, T., Rabinovich, M.I.: Dynamical Origin of Independent Spiking and Bursting Activity in Neural Microcircuits. Phys. Rev. Lett. 98, 128106 (2007)
16. Ashwin, P., Karabacak, O.: Robust Heteroclinic Behaviour, Synchronization, and Ratcheting of Coupled Oscillators. Dynamics, Games and Science II 2, 125–140 (2011)
17. McInnes, C., Brown, B.: A dynamical systems approach to micro-spacecraft autonomy. Proc. 20th AAS/AIAA Space Flight Mechanics Meeting 136, 1199–1218 (2010)
18. Li, D., Cross, M.C., Zhou, C., Zheng, Z.: Quasiperiodic, periodic, and slowing-down states of coupled heteroclinic cycles. Phys. Rev. E 85, 016215 (2012)
19. Shaw, K.M., Cullins, M.J., Lu, H., McManus, J.M., Thomas, P.J., Chiel, H.J.: Investigating localized sensitivity in the feeding patterns of Aplysia californica. In: Front. Behav. Neurosci. Conference Abstract: 10th Int. Congress of Neuroethology (2012)
20. Webster, V.A., Lonsberry, A.J., Horchler, A.D., Shaw, K.M., Chiel, H.J., Quinn, R.D.: A Segmental Mobile Robot with Active Tensegrity Bending and Noise-driven Oscillators. In: Proc. IEEE/ASME Int. Conf. on Advanced Intelligent Mechatronics, July 9-12 (2013)
21. Daltorio, K.A., Boxerbaum, A.S., Horchler, A.D., Shaw, K.M., Chiel, H.J., Quinn, R.D.: Efficient worm-like locomotion: slip and control of soft-bodied peristaltic robots. Bioinspiration & Biomimetics (2013) (in review)
22. Boxerbaum, A.S., Shaw, K.M., Chiel, H.J., Quinn, R.D.: Continuous wave peristaltic motion in a robot. International Journal of Robotics Research 31, 302–318 (2012)
23. Horchler, A.D.: SHCTools: Matlab ToolBox for Simulation, Analysis, and Design of Stable Heteroclinic Channel Networks, Version 1.1. GitHub (2013), http://github.com/horchler/SHCTools (accessed May 2013)

24. Stone, E., Holmes, P.: Random Perturbations of Heteroclinic Attractors. SIAM Journal of Applied Mathematics 50, 726–743 (1990)
25. Moore, A.R.: Muscle tension and reflexes in earthworm. The Journal of General Physiology 5, 327–333 (1923)
26. Gray, B.Y.J., Lissmann, H.W.: Studies in Animal Locomotion VII: Locomotory Reflexes in the Earthworm. Journal of Experimental Biology 15, 518–521 (1938)
27. Quillin, K.: Ontogenetic scaling of hydrostatic skeletons: geometric, static stress and dynamic stress scaling of the earthworm lumbricus terrestris. The Journal of Experimental Biology 201, 1871–1883 (1998)
28. Omori, H., Murakami, T., Nagai, H., Nakamura, T.: Kubota.: Planetary Subsurface Explorer Robot with Propulsion Units for Peristaltic Crawling. In: IEEE Int. Conf. on Robotics and Automation, pp. 649–654 (2011)

Design for a Darwinian Brain:
Part 1. Philosophy and Neuroscience

Chrisantha Fernando

Dept. of Electronic Engineering and Computer Science
Queen Mary University of London
ctf20@eecs.qmul.ac.uk

Abstract. Fodor and Pylyshyn in their 1988 paper denounced the claims of the connectionists, claims that continue to percolate through neuroscience. In they proposed that a physical symbol system was necessary for open-ended cognition. What is a physical symbol system, and how can one be implemented in the brain? A way to understand them is by comparison of thought to chemistry. Both have systematicity, productivity and compositionality, elements lacking in most computational neuroscience models. To remedy this woeful situation, I examine cognitive architectures capable of open-ended cognition, and think how to implement them in a neuronal substrate. I motivate a cognitive architecture that evolves physical symbol systems in the brain. In Part 2 of this paper pair develops this architecture and proposes a possible neuronal implementation.

1 Introduction

This is the first of a two-part paper showing that physical symbol system with the properties defined by Fodor and Pylyshyn could plausibly exist in the brain. I will explain what a physical symbol system is and what properties to expect from it, by showing how biochemical systems in cells can be interpreted as physical symbol systems. Having described the substrate independent properties of a physical symbol system, I will then describe two possible implementations of physical symbol systems in the brain, one based on activity patterns, and the other based on connectivity patterns. Finally I consider cognitive architectures that are physical symbol systems, and discuss their current algorithmic limitations, in order to motivate our own architecture described in Part 2.

The Hungarian chemist Tibor Ganti described the concept of fluid automata, or chemical machines [1] that are made of molecules. Molecules are objects composed of atoms that have specific structural relations to other atoms in the molecule. A molecule is assembled according to a combinatorial syntax, i.e. a set of chemical structural constraints such as valance, charge, etc that determine how atoms can legally join to make a molecule. Combinatorial semantics determines how a molecule with a particular structure will react (behave) in a given environment. Semantic content in terms of a chemical symbol system equates

N.F. Lepora et al. (Eds.): Living Machines 2013, LNAI 8064, pp. 71–82, 2013.

to its chemical function, i.e. its reactivity. The function of a molecule is a function of the semantic content of its parts, translated; the reactivity of a benzene ring is modified by its side-groups such as methyl groups. In short, the chemical symbols and their structural properties cause the system behaviour.

This physical symbol system composed of a vat of chemical molecules operates in parallel rather than in series. It is constrained by kinetics. Its function (reactivity) is subjected to non-encoded influences such as temperature that influences the thermodynamic equilibrium position of chemical reactions. These aspects do not tend to intrude in normal conversations about physical symbol systems, but they are germane to processing taking place in physical symbol systems made of chemicals. A pleasing example of a symbolically specified computation in chemistry is a chemical clock. The two autocatalytic cycles of the Belousov-Zhabotinsky (BZ) reaction constitute a fluid automaton that implements a chemical clock. Whilst it is the symbolic organisation of the molecules that specifies the reaction network topology, it is at the level of the analog (dynamical systems) operations of the assembled reaction network that the clock like phenomenon of the BZ reaction exists. Continuous behavior results from the ensemble properties of a multitude of discrete symbolic chemical operations. The tremendous success of chemistry is to have robustly linked the chemical system level with the macroscopic level.

The fact the chemical molecules are physical symbol systems gives the macroscopic properties of chemistry the same abstract characteristics found in any physical system system. These are now examined. For example, chemistry has productivity. Productivity means that a system can encode indefinitely many propositions, that there is recursive assembly of parts of the representation to produce the proposition, and that an unbounded number of non-atomic representations is entailed [2]. If we replace the term proposition with adaptation, then in biological evolution, the number of possible adaptations is unlimited in the same sense, being the developmental result of a symbolic genetic and epigenetic encoding. Whether it is helpful to replace proposition with adaptation is a moot point, but to me this seems natural because adaptations have a form of truth-value in the sense that they are the entities to which fitness applies. Similarly, the capacity for chemical reactivity is unlimited, i.e. there are many more possible chemical reactions than could be implemented in any realistically sized system. An unlimited number of molecules can be produced allowing an unlimited many chemical reactions, and this is made possible with only a finite set of distinct atom types. Therefore, an unbounded set of chemical structures must be composite molecules. In physical symbol systems terms it is also the case that an indefinite number of propositions can be entertained, or sentences spoken. This is known as the productivity of thought and language. If neural symbol systems exist, they must have the capacity of being combined in unlimited ways. The next section deals with how this could potentially work. Finally, there is reason to believe that no non-human animal has the capacity for productive thought [3].

The second property of chemistry is that it is systematic, i.e. the capacity for atoms to be combined in certain ways to produce some molecules is intrinsically connected to their ability to produce others. This is neatly shown by how chemists learn chemistry. There are several heuristics that chemists can learn to help predict properties of molecules and how they will react based on their structure. Only a chemist lacking in insight will attempt to rote learn reactions. Similarly with systematicity in language. The ability to produce and understand a sentence is intrinsically connected with the ability to produce and understand other sentences. Languages are not learned by learning a phrasebook, they have syntax which allows the systematic combination of entities.

The third property of chemistry is that an atom of the same type makes roughly the same contribution to each molecule in which it occurs. For example, the contribution of a hydrogen atom to a water molecule influences many reactivity properties of that molecule. Hydrogen atoms suck electrons and this is a property of the hydrogen atom itself, whatever in most cases it binds to. This means there is systematicity in reactivity (semantics) as well as in structure (syntax). This is known as compositionality in physical symbol system speak. In the same way, lexical items in sentences have approximately the same contribution to each expression in which they occur. This approximate nature suggests that there is a more fundamental set of atoms in language than words themselves, e.g. linguistic constructions.

Why was the idea of a physical symbol system even entertained in chemistry? Why did people come to believe in discrete atoms coming together systematically to form molecules, instead of for example some strange distributed connectionist model of chemistry? Lavoisier discovered a systematic relationship between chemical reactions, i.e. the conservation of mass. Proust discovered the law of definite proportions, i.e. that compounds can be broken down to produce constituents in fixed proportions. Dalton extended this law of multiple proportions that explained that when two elements come together to form different compounds (notably the oxides of metals), they would come together in different small integer proportions. Are there any analogous reasons to believe in physical symbol systems in the brain, based on observation of human thought and language?

Open-ended cognition appears to require the ability to learn a new symbol system, not just to implement an already evolved one. Children can learn and manipulate explicit rules [4] and this implies the existence of a neuronal symbol system capable of forming structured representations and learning rules for operating on those representations. For example, Gary Marcus has shown that 7 month old infants can distinguish between sound patterns of the form ABA versus ABB, where A and B can consist of different sounds e.g. "foo". "baa" etc. Crucially, these children can generalize this discrimination capacity to new sounds that they have never heard before, as long as they are of the form ABA or ABB. Marcus claims that performance in this task requires that the child must extract "abstract algebra-like rules that represent relationships between placeholders (variables), such as the first item X is the same as the third item Y, or more generally that item I is the same as item J" Several attempts have been made to explain the performance of these children without a PSS (e.g. using connectionist models) but Marcus has criticised these as smuggling in symbolic rules in one way or another by design. For Marcus it seems that the system itself must discover the general rule. In summary, the problem with a large set of connectionist learning devices is that a regularity learned in one component of the solution representation is not applied/generalized effectively to another part. Marcus calls this the problem of training independence, a solution to which is one of the fundamental requirements for a learning system to be described as symbolic or rule based.

Concerning the discussion on whether connectionist models without explicit symbols nor rules could be sufficient for open-ended cognition it is informative to quote the discussion of limitations of their own connectionist/computational neuroscience model given by Rougier et al (2005) [5]. Their model was able to form representations of categories of class in the Wisconsin Card Sorting Task. The authors in the main paper claim to produce complex representations without symbols. However, they moderate their claims in the Supplementary Material as follows. "Our model does not directly address this capacity for generativity... it was never asked to perform entirely new tasks... people can recombine familiar behaviors in novel ways. Such recombination, when built upon a rich vocabulary of primitives, may support a broad range of behavior and thus contribute to generatively.". The paper acknowledges the simplicity of the algorithm implemented, i.e. "random sampling + delayed replacement" and acknowledges the need for mechanisms that can undertake more sophisticated forms of search. "Generativity also highlights the centrality of search processes to find and activate the appropriate combination of representations for a given task, a point that has long been recognized by symbolic models of cognition... Our model implemented a relatively simple form of search based on stabilizing and destabilizing PFC representations as a function of task performance, which amounts to random sampling with delayed replacement... An important focus of future research will be to identify neural mechanisms that implement more sophisticated forms of search.".

Another paper of interest is that of Tani and Nolfi [6]. It self-organizes the connections between neural module experts on the basis of minimising of prediction error of the expert predictors at the lower levels. The higher level then predicts the gate opening dynamics of the lower level predictors and propagates this upwards. Two levels are modelled. It is claimed that such systems are an alternative to explicit symbolic models of complex environments. However, the system is very severely theoretically limited. The reason is that for a high dimensional space of models, there would be a combinatorial explosion of possible higher-order models which could be composed from combinations of lower order models. This problem has also been faced by Goran Gordon in a set of papers about hierarchical curiosity loops [7]. A generative process for constructing novel experts is needed. Each module effectively then becomes an atom in a physical symbol system, with its inputs and outputs specifying its connections to other atoms, i.e. permitting arbitrary variable binding. Incidentally, the use of prediction error also becomes infeasible for open-ended systems because the space of trivial predictions is also too large. Therefore, neither of these two connectionist papers propose how complex representations can emerge without explicit symbols. The former because of reasons the authors admit themselves, and the later due to its non-extensibility to high dimensional systems.

In summary, the following is a concise definition of a symbol system adapted from Harnad to emphasise the chemical aspects [8]. A symbol system contains a set of arbitrary atoms (or physical tokens) that are manipulated on the basis of explicit rules that are likewise physical tokens or strings (or more complex structures, e.g. graphs or trees) consisting of such physical tokens. The explicit rules of chemistry generate reactions from the structure of atoms and molecules (plus some implicit effects, e.g. temperature). The rule-governed symbol-token manipulation is based purely on the shape of the symbol tokens (not their meaning), i.e., it is purely syntactic, and consists of rulefully combining and recombining symbol tokens, in chemical reactions. There are primitive atomic symbol tokens and composite symbol-token strings (molecules). The entire system and all its parts – the atomic tokens, the composite tokens, the syntactic manipulations both actual and possible and the rules - are all "semantically interpretable". The syntax can be systematically assigned a meaning e.g., as standing for objects or as describing states of affairs. For example, semantic interpretation in chemistry means that the chemical system exhibits chemical reactivity. In biochemistry this is extended to a higher level. In the chemicals inside cells, it is fair to say that properties of those chemicals, e.g. the concentration, or their configuration, actually stands for states of affairs in the environment outside the cell, for example the conformation of a signalling molecule may represent the glucose concentration outside the cell. In the same way a neural symbol system exhibits behavior such as the child's capacity to distinguish ABA from ABB in grammar learning tasks.

This analogy has helped me to understand what Fodor and Pylyshyn were talking about in their famous paper denouncing connectionism [2]. From this basis I developed what I believe may be two plausible candidates for a neuronal symbol system. The first has been described in detail in two previous publications (Fernando 2011; Fernando 2011). In short it consists of arbitrary physical tokens (spatiotemporal patterns of spikes) arranged into molecules or symbol structures (groups of spatiotemporal patterns of spikes passing through the same network). These tokens undergo explicit rule-governed symbol-token manipulation (reactions involving transformation of the spike patterns), and a process of stochastic optimization can learn the explicit rule sets. Thus the proposed neural representation of an atomic symbol token is a spatiotemporal patterns of spikes. Neurons detect a particular spatiotemporal spike pattern if the axonal delays from the pre-synaptic neuron to the detector neuron are properly matched to the spatiotemporal pattern such that depolarization reaches the detector neuron body simultaneously. This implementation of neural symbol-tokens (atoms) uses the concept of polychronous computing and a modification of the concept of wavefront computing [9] [10]. The construction of molecular symbol structures from atomic symbol-tokens requires binding of atomic symbol-tokens together such that they can be subsequently manipulated (reacted) as a function of the structure of the molecule. In my framework, compositional neural symbolic structures exist as temporally ordered sequences of symbols along chains of neurons. A symbol-structure can be described as a sequential combination of spike patterns, a feature that has been described in real neuronal networks as a "cortical song" [11]. A great many such short chains of combined spatiotemporal spike patterns may exist in the brain. Each chain being akin to a register in a computer, blackboard or tape that can store symbol-tokens of the appropriate size. A single symbol-token could be read by a detector neuron with the appropriate axonal delay pattern when interfacing with the chain. A parallel symbol system in the brain consisting of a population of such chains, each capable of storing a set of symbol token molecules and operating on these molecules in parallel is conceivable. Interaction between (and within) such chains constitutes the operations of symbol-manipulation. Returning to the chemical metaphor, such interactions can be thought of as chemical reactions between molecules contained on separate chains, and rearrangements within a molecule expressed on the same chain. Whilst in a sense a chain can be thought of as a tape in a Turing machine (due to the serial nature of the strings), it also can be thought of as a single molecule in a chemical system (due to the existence of multiple parallel chains). This constitutes the core representational substrate on which symbol manipulation will act.

Receiving input from the chain and writing activity back into the chain is done by a detector neuron with specific input and output delays in relation to the chain. A detector neuron only fires when the correct pattern of input is detected. In effect the neuron is a classifier. Where the classifier fails to fire, the same pattern enters the chain as leaves the chain. This is because the spatiotemporal organization of these patterns does not match the spatiotemporal tuning curve of the detector neuron. Only when the spatiotemporal spike pattern matches does the detector neuron fire. Once fired, the output of the detector neuron is injected back to the neurons of the chain. If the output of the detector neuron slightly precedes the normal passage of the untransformed pattern through the chain, then the refractory period of the output neurons of the chain prevents interference by the original untransformed pattern, which is thereby replaced by the new pattern specified by the detector neuron. Such a detector neuron we will now call a classifier neuron because it is a simple context free re-write rule with a condition (detection) and an action pole of the type seen in Learning Classifier Systems (LCS) [12] [13].

It can be seen that such classifier neurons are selective filters, i.e. the classifier neuron is only activated if the spatiotemporal pattern is sufficiently matched with the axonal delays afferent upon the neuron. The above classifier implements an implicit rule. An implicit rule is a rule that operates on atomic or molecular symbol structures without being specified (encoded/determined/controlled) by a symbol structure itself. There is no way that a change in the symbol system, i.e. the set of symbols in the population of chains, could modify this implicit matching rule. The implicit rule is specified external to the symbol system. Whenever the symbol passes along this chain, it will be replaced by the new symbol, irrespective of the presence of other symbols in the system.

In a symbol system (as in chemistry), symbols are manipulated (partly) on the basis of explicit rules . This means that the operations or reactivity of symbols depends on/is controlled by/is causally influenced by their syntactic and semantic relationship to other symbols within the symbol-structure and between symbol structures. If inhibitory gating neurons are combined with classifier neurons then it becomes possible to implement explicit rules within our framework. If an inhibitory gating unit must receive a particular pattern of spikes in order for it to become active, and if that unit disinhibits a classifier neuron that is sensitive to another symbol token then it is possible to implement a context-sensitive re-write rule. The rule is called context sensitive because the conversion of X to Y depends on the relation of X to another contextual symbol T. A set of context-sensitive re-write rules is capable of generating a grammar of spike-patterns. Consider starting the system off with a single symbol-token S. Probabalistic application of the rules to the initial symbol S would result in the systematic production of spike patterns consisting of grammatically correct context-sensitive spike pattern based sentences. A major implementation issue in real neuronal tissue would be the fidelity of transmission of spatiotemporal spike patterns. The information capacity of such a channel may fall off with decreasing fidelity of copying in that channel in a manner analogous to Eigen's error catastrophe in genetic evolution [14]. Finally, we have shown how it is possible to undertake stochastic optimization in the above system by copying with mutation of neuronal classifiers [15]. This mechanism depends on spike-time dependent plasiticity (STDP) by which one neuron infers the re-write rule implemented by another neuron, with errors. This permits a population of re-write rules to be evolved. The details of this mechanism are beyond the scope of this paper but can be found elsewhere [15].

What kinds of behavior require a physical symbol system, if any? We have understood something about how chemical and genetic systems generate reactions and cellular behavior. But what about how neural representations produce organismal behavior? Whilst a gene is a functional unit of evolution, we can ask, what are the possible functional units of thought at the algorithmic level, i.e. what are the entities in the brain that encode behavioral traits whose probability of transmission during learning is correlated with reward (or value functions of reward) [16] [17]. It may well be the case that there exist units of thought that do not have compositionality, systematicity and productivity at all. For example, a solitary feed-forward neural network that implements a forward model which predicts the next state of the finger as a function of the previous joint angles and the motor command given to the finger, does not satisfy the definition of a symbol system discussed above. Such models of the world may be complex, e.g. inverse models of the motor commands required to reach a desired goal state, but may lack the properties of a physical symbol system, e.g. [6] and [5]. The units may contribute to many domains from the linguistic to the sensorimotor, or occupy a level of specialized niches, e.g. units in the cerebellum, or units in the primary visual cortex. By analogy with units of evolution [18] they may be called units of thought, although their individuality may be more a convenience for measurement and description than a strict reality, for example it is possible that they are physically overlapping. I define them as the entities in the brain whose survival depends upon value, whatever form such value takes, e.g. extrinsic or intrinsic values [19]. How far can an animal get in cognition without physical symbol systems? In short the mystery is this; what are the kinds of representation upon which search must act in order to produce cognition, and how do certain classes of representation map to certain classes of cognition? A concrete proposal is given in Part 2. The short answer is that it is open-ended generative processes that require physical symbol systems, of the type for example that Rougier et al say their connectionist model cannot do [5].

The second question is what kinds of learning algorithm are good for generative search? In answering this question I have found it important to notice that both thought and evolution are intelligent, and undertake knowledge-based search as defined by Newell [20]. I propose that both open-ended thought and open-ended evolution share similar mechanisms and that a close examination of these two processes in context is helpful. Traditionally thought has been considered to be directed and rational [21], whereas evolution by natural selection has been considered random and blind [22]. Neither view in their extreme forms is correct. A necessary feature of creative thought is stochastic search [23] [24] and evolution by natural selection whilst it may be a blind watchmaker is not a stupid one, it can learn to direct its guesses about what variants are likely to be fit in the next generation [25]. Apparently a grandmaster, maybe Reti, was once asked "How many moves do you look ahead in chess?" to which he answered"One, the right one!" We now know that evolution makes moves (mutants) in the same way, not by explicitly looking ahead, but by recognizing what kinds of moves tended to be good in the past in certain situations and biasing mutation accordingly [26] [27][28]. In fact, this principle has been exploited in the most sophisticated black-box optimization algorithms such as covariance matrix adaptation evolution strategies (CMA-ES) that rivals the most sophisticated reinforcement learning algorithms [29]. How do the mechanisms of cognition differ from the mechanisms of genetics in discovering how to make new adaptive ideas/organisms effectively? It seems that processes of systematic inference are in place in both adaptive systems, e.g. bacterial genomes are organized into Operons that through linkage disequilibrium tend to bias variability between generations along certain dimensions, i.e. crossover is a structure sensitive operation, a notion that was coined to describe the Classical cognitive paradigm [2]. Structure sensitive operations are the symbolic equivalent of methods to bias variation in continuous systems such as CMA-ES.

Both evolution and thought are generative processes with systematicity and compositionality because both depend on structure sensitive operations on representations [2] that far exceed the complexity of CMA-ES. We are able to generate educated guesses about hidden causes, to make up explanations for which there is as yet insufficient data to use deduction. This is evidenced by children's ability to learn language [30][31][32]. In fact there is considerable evidence much of which I find convincing, that no non-human animal can entertain models of the world that require the hypothesis of unlimited kinds of hidden causes [3]. Genetic and epigenetic generative mechanisms seem rather inflexible in comparison to those of thought. From close examination of these abilities can we have some idea what units of thought must be capable of representing about the world, i.e. how they limit the kinds of concept we can entertain? Perhaps we have a chance by looking at an example of constrained creativity from Gestalt psychology in the form of insight problems [33]. The 9-dot problem is a task in which you must cover each dot in a 3 x 3 matrix of dots with just 4 straight lines without removing your pen from the paper. Through the application of introspection we seem to solve this problem by generating a set of possible solutions, inventing some kind of fitness function for the solutions such as how many dots they cover, and mutating the better solutions preferentially in a systematic way. We may not be able to verbalize the scoring function nor how the initial set of solutions was chosen or varied. Often we will reach an impasse, a kind of sub-optimal equilibrium in thinking about this problem. The impasse may be diffused by suddenly discovering the correct solution that arises in a punctuated event called the Ahha moment. How are the solutions to the above problem represented, how many distinct solutions are represented at once in the brain, how are they varied over time, how do solutions interact with each other, how does a population of solutions determine behavior, and how are the scoring functions represented in the brain, and how many there are? We have proposed elsewhere that the solution of such problems involves a Darwinian search in the brain [34]. It may be the case that careful modelling of the above kinds of thinking process can uncover the kinds of representation and the search algorithm used to search in the space of such representations.

In cognitive science it is useful to put your money where your mouth is and actually try to build a robot to do something. Part 2 shows how we propose to build a robot that has open-ended creativity and curiosity that exhibits productivity, compositionality, and systematicity of thought. The most advanced attempts to build robots capable of open-ended creativity are briefly discussed. The general principle in all that work has been to learn predictive models of self and world. These are typically implemented as supervised learning devices such as neural networks. The networks are not physical symbol systems. However, some approaches attempt to combine predictors compositionally. The approach that I have decided to take in the remainder of Part 1 is to examine the limitations of these approaches in producing productive behavior in robots.

The approach taken by most groups in trying to achieve open-ended cognition is to learn predictive models of the world that become as rich as possible, and to guide action selection on the basis of how well certain models are being learned. The idea is that as we grow up we try to model the world in greater and greater richness and depth. An early and beautiful approach to this problem was by A.M. Uttley who invented a conditional probability computer and connected it to a Meccano robot that was controlled with a hard-wired Braitenberg type controller [35][36] . This computer learned to associate sensory states with motor actions when this external controller was controlling the robot, and could learn the sensorimotor rules, e.g. turn left if left sensor dark and right sensor light, and eventually substitute this hard-wired controller. Note that this was an entirely associationist form of model and would fall under what Fodor and Pylyshyn call Connectionism. More recent approaches learn state(t)-motor(t)-next state(t+1) triplets through experience, generating a network of anticipations [37]. Once such a cognitive map is learned it can be used for model-based reinforcement learning (dynamic programming) to backpropogate rewards associated with goal sensory states to earlier states and thus to permit planning of high-value trajectories. The capacity to simulate sense-action trajectories is a critical function of a cognitive map. The sophistication of mental simulation is a significant constraint on intelligence and creativity. Note that there is no physical symbol system involved in the above algorithm for cognitive map building. How then might a physical symbol system help in learning models of the world?

An obvious limitation of the above kind of cognitive map formation is that it may fail when there are unobserved causes. One solution is the generation of classifiers that infer the existence of hidden-causes [38] and then utilize this in further modelling of the world. By this means an unlimited conceptual world opens. This is where physical symbol systems may well enter the picture. Once a space of higher-level classifiers exists, the combinatorial explosion of possible conditions and predictions requires an effective method for sparse search in this higher-order space. The space of higher-order predictions is also highly structured. In summary, the units of thought in TGNG are s(t)-m(t)-s(t+1) triplets that store transition probabilities. There is no need for compositionality and systematicity of such units because they can be entirely instructed by supervised learning methods (at least in low dimensional sensorimotor systems). Efficient sparse search algorithms would however be needed when hypothesising hidden states for which there is insufficient sensory evidence to undertake supervised learning. So, by thinking about the problem from this robotic perspective, I have identified a domain in which stochastic processes and representations with compositionality and systematicity would be expected to bring efficiency, i.e. during search in a structured space of hidden causes.

Interestingly, a modification of the simple cognitive maps above that potentially adds compositionality and systematicity has been proposed in order to scale reinforcement learning to "domains involving a large space of possible world states or a large set of possible actions" [39]. The solution has been to chunk actions into higher-level actions consisting of sequences of primitive actions, called "options" [40]. Options can be hierarchically organized policies. Each abstract action specifies a policy (a map from states to actions) that once started from in initiation state, executes a sequence of actions (which may also be options) until a termination state is reached. They work by reducing the size of the effective search space by allowing structuring of search by modular recombination (concatenation) of options. The critical question is how the correct set of useful options can be chosen in the first place. Typically options are hand-designed, with options specifying sub-goals to be achieved. In none of the models are options stochastically explored. The options framework is an admirable first step towards achieving a truly systematic and compositional physical symbol system for describing actions. It is related to the notion of action grammar developed by Patricia Greenfield (Greenfield 1991) which also seeks to identify the behavioural correlates of neural symbol systems.

To conclude, for any reasonable sized sensory and motor dimensionality, and for reasonable time-periods, exhaustive search in predictive model space is not possible. The computation is exponential in time and space. To solve this problem, units of thought (e.g. predictors and actors) should exhibit compositionality and systematicity, of the kind for example that we see in the options framework. The system should have the ability to form hierarchical descriptions of action and hierarchical predictive models. Part 2 presents our fluid option framework or fluid action grammar in which option-like symbol structures are co-evolved in the brain along with their fitness functions.

References

1. Ganti, T.: The Principles of Life. Oxford University Press, Oxford (2003)
2. Fodor, J., Pylyshyn, Z.: Connectionism and cognitive architecture: A critical analysis. Cognition 28, 3–71 (1988)
3. Penn, D., Holyoak, K., Povinelli, D.: Darwin's mistake: Explaining the discontinuity between human and nonhuman minds. Behavioral and Brain Sciences 31(2), 109–130 (2008)
4. Marcus, G.: The Algebraic Mind: Integrating Connectionism and Cognitive Science. MIT Press (2001)
5. Rougier, N., Noelle, D., Braver, T., Cohen, D., O'Reilly, R.: Prefrontal cortex and flexible cognitive control: Rules without symbols. Proc. Natl. Acad. Sci. U. S. A. 102, 7338–7343 (2005)
6. Tani, J., Nolfi, S.: Learning to perceive the world as articulated: an approach for hierarchical learning in sensory-motor systems. Neural Networks 12, 1131–1141 (1999)
7. Gordon, G.: Hierarchical exhaustive construction of autonomously learning networks (2011)
8. Harnad, S.: The symbol grounding problem. Physica D 42, 335–346 (1990)
9. Izhikevich, E.M.: Polychronization: computation with spikes. Neural Computation 18(2), 245–282 (2006)
10. Izhikevich, E.M., Hoppensteadt, F.: Polychronous wavefront computations. International Journal of Bifurcation and Chaos 19, 1733–1739 (2009)
11. Ikegaya, Y., et al.: Synfire chains and cortical songs: Temporal modules of cortical activity. Science 304, 559–564 (2004)
12. Holland, J., Reitman, J.: Cognitive systems based on adaptive algorithms. ACM SIGART Bulletin 63, 43–49 (1977)
13. Wilson, R.: Function approximation with a classifier system (2001)
14. Eigen, M.: Selforganization of matter and the evolution of biological macromolecules. Naturwissenschaften 58(10), 465–523 (1971)

15. Fernando, C.: Symbol manipulation and rule learning in spiking neuronal networks. Journal of Theoretical Biology 275, 29–41 (2011)
16. Price, G.: Selection and covariance. Nature 227, 520–521 (1970)
17. Price, G.: The nature of selection. Journal of Theoretical Biology 175(3), 389–396 (1995)
18. Maynard Smith, J.: The problems of biology. Oxford University Press, Oxford (1986)
19. Stout, A., Konidaris, G., Barto, A.: Intrinsically motivated reinforcement learning: A promising framework for developmental robotics (2005)
20. Newell, A.: Unified Theories of Cognition. Harvard University Press (1990)
21. Newell, A., Simon, H.A.: Human problem solving. Prentice-Hall, Englewood Cliffs (1972)
22. Dawkins, R.: The Selfish Gene. Oxford University Press, Oxford (1976)
23. Perkins, D.: Insight in minds and genes. In: Sternberg, R., Davidson, J. (eds.) The Nature of Insight, MIT Press, Cambridge (1995)
24. Simonton, D.: Foresight in insight? a darwinian answer. In: Sternberg, R., Davidson, J. (eds.) The Nature of Insight. MIT Press, Cambridge (1995)
25. Pigliucci, M.: Is evolvability evolvable? Nature Reviews Genetics 9, 75–82 (2008)
26. Kashtan, N., Alon, U.: Spontaneous evolution of modularity and network motifs. Proc. Natl. Acad. Sci. U. S. A. 102(39), 13773–13778 (2005)
27. Izquierdo, E., Fernando, C.: The evolution of evolvability in gene transcription networks (2008)
28. Parter, M., Kashtan, N., Alon, U.: Facilitated variation: How evolution learns from past environments to generalize to new environments. PLoS Computational Biology 4(11), e1000206 (2008)
29. Stulp, F., Sigaud, O.: Path integral policy improvement with covariance matrix adaptation (2012)
30. Goldberg, A.: Constructions: A Construction Grammar Approach to Argument Structure. University of Chicago Press., Chicago (1995)
31. Steels, L.: Experiments on the emergence of human communication. Trends in Cognitive Sciences 10(8), 347–349 (2006)
32. Steels, L., De Beule, J.: Unify and merge in fluid construction grammar (2006)
33. Sternberg, R., Davidson, J.: The Nature of Insight. MIT Press, Cambridge (1995)
34. Szathmary, E., Fernando, C.: Concluding remarks. In: The Major Transitions Revisited, MIT Press, Cambridge (2009)
35. Andrew, A.: Machines that learn. The New Scientist, 1388–1391 (1958)
36. Uttley, A.: Conditional probability computing in a nervous system. In: Mechanisation of Thought Processes, vol. 1, pp. 121–147. H.M. Stationery Office, London (1959)
37. Butz, M., Shirinov, E., Reif, K.: Self-organizing sensorimotor maps plus internal motivations yield animal-like behavior. Adaptive Behavior 18(3-4) (2010)
38. Nessler, B., Pfeiffer, M., Maass, M.: Stdp enables spiking neurons to detect hidden causes of their inputs (2010)
39. Botvinick, M., Niv, Y., Barto, A.: Hierarchically organized behavior and its neural foundations: A reinforcement learning perspective. Cognition 113(3), 262–280 (2009)
40. Sutton, R., Precup, D., Singh, S.: Between mdps and semi-mdps: A framework for temporal abstraction in reinforcement learning. Artificial Intelligence 112, 181–211 (1999)

Design for a Darwinian Brain: Part 2. Cognitive Architecture

Chrisantha Fernando, Vera Vasas, and Alexander W. Churchill

Dept. of Electronic Engineering and Computer Science
Queen Mary University of London
ctf20@eecs.qmul.ac.uk

Abstract. The accumulation of adaptations in an open-ended manner during lifetime learning is a holy grail in reinforcement learning, intrinsic motivation, artificial curiosity, and developmental robotics. We present a design for a cognitive architecture that is capable of specifying an unlimited range of behaviors. We then give examples of how it can stochastically explore an interesting space of adjacent possible behaviors. There are two main novelties; the first is a proper definition of the fitness of self-generated games such that interesting games are expected to evolve. The second is a modular and evolvable behavior language that has systematicity, productivity, and compositionality, i.e. it is a physical symbol system. A part of the architecture has already been implemented on a humanoid robot.

1 Introduction

The main objective of the developmental robotics field is to create open-ended, autonomous learning systems that continually adapt to their environment, as opposed to constructing robots that carry out particular, predefined tasks [1]. We present a design for a cognitive architecture capable of ontogenetic evolution of open-ended controllers and open-ended fitness functions (i.e. games). When implemented in a robot, the architecture is intended to generate playful interaction with the world, in which intuitively 'interesting' behaviours are discovered. The architecture in a sense provides a computational definition of what 'interesting' means. Eventually it can reasonably be expected to exhibit creative and curious playful behaviour when coupled with any situated and embodied device that has rich sensorimotor contingencies, i.e. where actions have sensory consequences. The overall cognitive architecture consists of an evolving population of game molecules, each with an associated population of actor molecules, and a long term memory of optimal game and atom molecules.

The fitness of an actor molecule in a population is determined by the game molecule associated with that population. The fitness of a game molecule is dependent on the fitness of actor molecules in its associated population. The fitness of a game molecule is a measure of the evolutionary progress of actor molecules associated with that game.

N.F. Lepora et al. (Eds.): Living Machines 2013, LNAI 8064, pp. 83–95, 2013.

The intuition behind this definition of actor molecule fitness is that we wish to select for actor molecules in the population that are good at its game. The intuition behind the fitness of a game molecule is that we wish to select for games that maximise the predicted rate of fitness increase in the actor population, i.e. we wish to select for games on which progress can be maximised. To do so one possibility is to exploit a simple result from evolutionary biology, Fisher's fundamental theorem of natural selection. The theorem states that 'the rate of increase in fitness of any organism at any time is equal to its genetic variance in fitness at that time' [2]. This applies under the following conditions: i. There must be no epistasis, that is, the fitness contributions of alleles in the units under selection must be additive and linear. ii. There must be no linkage disequilibrium, i.e. no non-random association of alleles at two or more loci in the genotype of the molecule under selection. iii. There must be no frequency dependent selection, i.e. the fitness of one actor molecule must be independent of its frequency in the population of actor molecules.

Whilst none of these conditions apply in any realistic and complex system; however, the deviation from the ideal is often sufficiently small that the approximation is valid. Violations of the three conditions above, e.g. epistasis, may cause the theorem to fail in predicting long term fitness changes, for example neutral evolution may be observed in which the population explores a fitness plateau where all solutions are of effectively equal fitness, until one solution happens to find a step to a higher fitness plateau. If such features dominate the system, then other more sophisticated predictors of evolutionary progress must be used to select for games on which progress can be made. For example, maximising the fitness entropy of offspring, or applying a Rechenberg type game fitness proportional to the proportion of offspring that are superior to the parent on the game, may be explored [3]. The key principle is that we are using measures that predict evolutionary progress to determine the fitness of a game.

Price's equation elaborates and generalises Fisher's fundamental theorem by including a term due to transmission bias, i.e. mutation, or environmental change. Fisher's equation is a special case of the first part of the Price equation and states that the rate of change of a trait (i.e. fitness itself) is proportional to the co-variance between fitness and that trait. In short, Fisher and Price reveal that by observing an instantaneous property of the population of actors, a game unit can determine whether progress in that game is expected in the future. Fisher and Price's insights permit an instantaneous measure of variance to stand in for a rate of change of fitness. Without realising this fundamental relation to Fisher and natural selection, the strategy of defining the fitness of a population of tests as the variance in another population of models has been proposed by Bongard and Lipson [4], however, they consider not the evolution of behaviours, but specifically of forward models or dynamic equations. The algorithm described here is considerably more general than evolving models of the world.

The specific method used here to approximate evolutionary progress in the actor population is to explicitly calculate the gradient of the normalised median fitness of the population of actors from the beginning to the end of 100 generations of actor evolution. Related methods [5] [6] [7] [8] use regression methods to

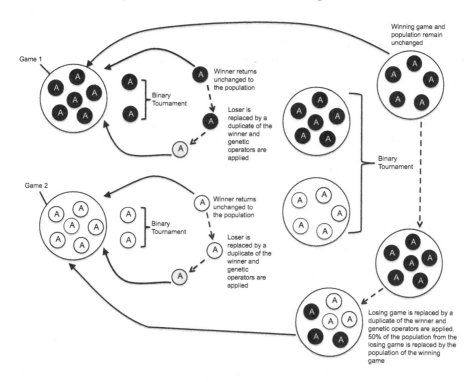

Fig. 1. The co-evolutionary algorithm. Each game has an associated population of actors in a separate island. After a certain number of binary tournaments between actors in each islands, a binary tournament takes place between games.

predict the rate of change of performance of a single actor over time. The high-level pseudo-code for the algorithm is outlined below, and illustrated in Figure 1. First initialise a population of game molecules. For each game molecule initialise a population of actor molecules. Iterate the following main loop. Evolve each actor population for 100 generations using a genetic algorithm, with fitness as defined by its associated game molecule. Determine the fitness of each game molecule based on the evolutionary dynamics in its actor population. Use fitness proportionate selection to replicate and mutate game molecules. Transfer the stable actor/game molecules at each game molecule generation to a long term memory store. Entities in the long term memory store are available for recombination in subsequent generations.

Some critical features of the encoding of actor and game molecules that permit open-ended evolution are now described. Let us first consider actor molecules. Specifically, the molecules are constructed such that they exhibit systematicity and compositionality [9] see Part 1 for a full discussion. In other words they are physical symbol systems analogous to atoms and molecules in chemistry. The function of an atom is systematically determined by its structure, and atoms can be composed into molecules whose function is determined by the arrangement and structure of atoms that compose the molecule. However, in contrast to all other cognitive architectures, e.g. CopyCat [10], SOAR [11], ACT-R [12], and CLARION [13], our system integrates connectionist and symbolic systems in a manner that preserves the full power of the physical symbol system as required by Fodor and Pylyshyn; see Gary Marcus' book "The Algebraic Mind" for a detailed criticism of previous attempts at hybrid cognitive architectures within the domain of symbolic connectionism [14]. There is no conflict in our framework with other machine learning or connectionist approaches because all the non-symbolic features of a wide range of supervised and unsupervised learning algorithms found in machine learning textbooks can be possessed inside an atom in our architecture.

2   Methods

2.1   Design of Atoms

Each atom is of the form shown in Figure 2. Each atom contains: a. A list of strings i that specify from where the input data will come into the atom, i.e. it labels the registers from which values are taken as input to the atom. These registers are found in a working memory that atoms can read from and write to asynchronously. b. A set of internal registers r that can store states. c. A transform function T that uses the input to calculate an output. Any transfer function is permitted, e.g. a feed-forward neuronal network, a logical function, a stochastic hill climbing algorithm, covariance matrix adaptation evolution strategy, Q-learning, in fact any function is allowed, and the greater the diversity of such functions the better. In later implementations users will be able to submit such atoms to an online web interface. A list of strings o that specifies to where in working memory the output data will be written when it is produced by c. e. A fitness as defined by the associated game molecule.

Thus, an atom is specified by a tuple i,r,T,o,f. The data inside an atom is encapsulated and not directly accessible by other atoms. Atoms send information to each other via the registers in working memory. This information in a register can for example be as simple as a signal for other atoms that observe this register to turn on or off, or as complex as a Turing complete computer program. However, in the examples provided, the most complex information passing that is needed is to pass a short list of floating point numbers.

2.2   Binding of Actor Atoms to Form Actor Molecules

The functional binding or linking of atoms to each other is via interactions mediated by a set of registers which may equate to the Basal Ganglia in the brain. The atoms correspond to neuronal groups in the cortex. These registers consist of a dictionary of key-value pairs, where the keys are the entries in the input and output tables of atoms that refer to the registers that contain the values. It is the values that are effectively passed between atoms. The linkage of

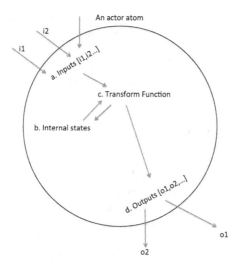

Fig. 2. General structure of an atom

atoms in this way creates actor molecules, which are defined as directed graphs of information flow between active atoms. These actor molecules are embedded within a larger network of potentially overlapping information flows between atoms. Sensory and motor grounding is achieved in the following way. There is a list of dictionary entries to which raw sensory data is written, and from which values are read to control raw motor output. Actor atoms can read from and write to these special registers. Game molecules are distinct from actor molecules only in that game molecules cannot contain any motor atoms, i.e. that control motors directly. Game molecules only output a scalar fitness value at the end.

2.3 Initialisation of Actor and Game Population

The actor population is initialised using innately specified behaviours loosely corresponding to reflexes. These reflexes may be tagged such that they can never be completely erased from the population of actions. The game population is initialised with a set of innate games that are designed to complement the set of innate 'reflexes'. For example, if a reflex actor molecule exists that moves the head to foveate on a face, then it may come with a matching game molecule that assigns fitness to it in proportion to the speed at which foveation was achieved over a trial.

2.4 Replication and Mutation of Molecules

Each molecule is capable of being replicated with a range of variation operators. These are of two types, intra-atomic mutations, and structural mutations to the molecule itself. The first type is specific to each atom, and the second type is general. Structural mutations include additions of atoms, removals of atoms, additions of links, removals of links. They are capable of producing recurrent molecular structures, also disjoint molecular structures are possible. The details of the mutation operators are available in the python software which is made available on a github repository here: `https://github.com/alexanderchurchill/darwinian-neurodynamics`.

2.5 Transfer of Stable Actor and Game Molecules to Long Term Memory

Combining diversity maintenance and the accumulation of adaptation is the evolutionary computation version of the stability plasticity dilemma. The solution used here is to take actor molecules and game molecules that have reached fixation in the population out of the population and into a long-term memory store that is immutable. In addition, a molecule similarity function is implemented which punishes molecules in the population in proportion to their similarity to molecules already in long-term memory. Over time, adaptive games and solutions accumulate in LTM, permitting the evolving populations to explore new games and solutions. Limiting the kinds of exploration that the agent engages in.

2.6 Games, Solutions, and Adjacent Possible Behaviours

Given the semi-formal operations defined above, it is now possible to hand-design a set of behavioural molecules and understand the behaviour that they will produce in the robot that they control. It is also possible to apply the variation operators to the specifications and see what kinds of adjacent possible behaviours result from random variation.

The algorithm is implemented in Python using the Naoqi API provided by Aldebaran robotics. This provides an ideal asynchronous modular toolbox that already contains a vast amount of high-level functionality. In effect, each python module or box in the choregraphe framework can be incorporated into an actor atom as defined above. For example, there may be a face recognition atom that takes visual input and outputs the presence or absence of a face in the current visual field, and the string value associated with that face, a sound localisation atom may return the angular displacement of the head that is required to rotate the head towards the predicted location of the sound. Such high-level actor atoms can also potentially be submitted by web-users. Conversely, low-level actor atoms also exist. Such a low-level actor atom may for example contain a simple 3 layer feed-forward neuronal network that maps directly from sensory input to motor output, thus implementing a pure reflex action. Atoms may contain any transfer function, e.g. a liquid state machine, an SVM etc, etc. In addition, atoms can contain unsupervised learning algorithms, e.g. EM algorithms, K-means clustering, Principle Component Analysis, Independent Component Analysis, mutual or predictive Information calculations, and these may be parameterized in different ways, with variation operators defined on the functions. Thus, the network of controllers that is produced is entirely general. Of-course, the more complex the atom the more brittle it will be to mutation. The research program generated by this paper is to discover the evolvable grammar of action used by a brain. Let us now consider some specific action molecules and their associated game molecules.

2.7   Reacting to Resistance

The action molecule described is capable of controlling the robot to hill climb in elbow joint angle space in order to maximize the prediction error of the angle of the elbow, see Figure 3(a). In other words, it tries to do actions that result in unpredictable elbow positions, for example, if an obstruction is encountered then it will explore those joint angles containing the obstruction preferentially because these are the joint positions that the forward model does not currently get right. Four atoms encode the closed-loop behaviour. The first atom takes the elbow angle and motor command as input, and uses these to update a forward model of the elbow angle. This requires storage of elbow angle and motor command for one time-step within the atom. The output that is written to working memory is the predicted elbow angle at the next time-step.

The second atom is then activated. This takes the predicted elbow angle and the actual elbow angle and calculates the squared error between these two inputs and writes to memory the output that is the prediction error of elbow angle. This atom then activates the third atom. The third atom in this chain molecule is a hill-climber atom that takes an input that it always treats as reward (in this case the input is prediction error of the elbow joint model. It tries to maximize reward by using hill-climbing on a parameter vector. If reward increases, the system keeps the current parameters and does further exploration, else the system reverts to the old parameters. Finally, these parameters are written to working memory that activates the final molecule to interpret these parameters as motor commands, which means to write them to the motor registers that are immediately executed. When this motor command to move the elbow is written to memory this reactivates the first atom in the molecule, thus causing the loop to iterate. After a fixed time period the loop is stopped, all atoms being inactivated. An example of a game associated with the above molecule is an atom that simply requires that the prediction error output of the second atom is maximized over some time period e.g. 10 seconds. Consider mutations of the above behavior molecule. The following one-step mutants are possible: 1. Bypass mutation of atom 1: Get input from shoulder angle not elbow angle, or from foot motor not elbow motor. 2. Bypass mutation of atom 2: Mutate transformation in atom 2, so that prediction error = Elbow angle -Predicted elbow angle, not the square of the later. 3. Bypass mutation of atom 3: Modify parameters of S.H.C. e.g. probability of accepting a worse set of parameters. Or try to minimize rather than maximize the input. 4. Bypass mutation of atom 4: Control another motor, e.g. the foot or the neck.

Many of the mutants will result in molecules that are non-functional, e.g. stochastic hill climbing will not be able to hill climb if one is trying to control

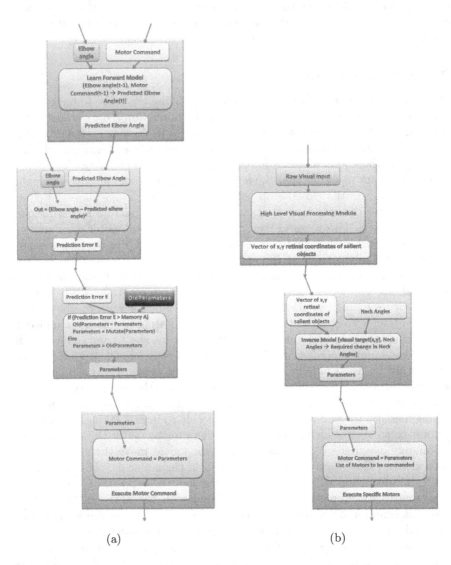

Fig. 3. Two example actor molecules. (a). A 4-atom actor molecule for optimal elbow exploration, with the property that it 'enjoys' exploring elbow joint positions associated with unexpected angle effects of motor commands to the elbow joint, i.e. reacting to resistance. (b) Actor molecule for orienting to a salient visual input by moving the neck.

the elbow angle by moving the foot. Molecules of this type will score poorly on the above game and will be unlikely to be selected on the basis of performance on that game. However, mutant 3 may result in interesting behavior where the agent tries to minimize not maximize prediction error. This would result again in poor performance on the original game however. For this variant to survive, a new game mutant is required that rewards minimization of the output of atom 2.

2.8   Orienting to Objects

Visual orienting to objects can be undertaken by a range of possible behavior molecules. Figure 3(b) shows one possible molecule. The first atom receives the raw sensory input and uses high level visual processing routines to extract a vector of x,y retinotopic coordinates of salient visual objects. For example the object recognition module in Naoqi by Aldebaran can be used. This output is read from working memory by an inverse model atom. This atom has learned a model mapping the desired visual location of the object to be foviated, and the current neck angle, to the desired neck angle required to achieve foviation. This output is then sent to a standard motor execution atom that takes the input, contains a list of motors to be commanded and writes to the motor register to execute the action.

A game that stabilizes the above molecule would be to minimize the distance between salient visual objects. Variants of the orienting to objects behavior that can be produced by mutation of the above molecule include, 1. Turning away from objects, achieved by replacing the inverse model with another model, and by mutating the game so as to maximize the visual distance between salient objects. 2. Orienting to different kinds of salient visual object, achieved by modifying the visual processing module in atom 1. 3. Moving the object in a specific way over the visual field, e.g. oscillating it in the visual field, achieved by modifying the inverse model in atom 2, and by modifying the game so as to reward high variance over time in the visual distance between salient objects.

2.9   Maximising Mutual Information between Effectors and Sensors

Figure 4 is an example of an unsupervised learning molecule that acts in order to detect sensorimotor contingencies, and stores combinations of effector-sensor pairs that have high mutual information. These pairs can be used by other systems to structure exploration at a later stage. This molecule detects one such SM contingency with the highest mutual information between the sensor and motor time windows. Notice, there is no motor control in this molecule, it does not influence behaviour at all, only observes behaviour. In a sense, it is a molecule of thought, not of action. It is of-course possible to have atoms with somewhat different information measures, for example predictive information may be used, or Granger causality. The cognitive architecture presented here is Catholic, an atom will be preserved if it is stable within the co-evolutionary dynamics of actions/thoughts and games, whatever that atom contains.

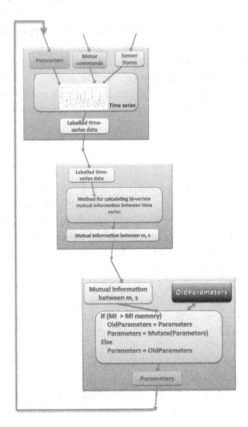

Fig. 4. Stochastic hill climbing in the space of sensor-motor pairs to find those with high mutual information between them

3   Results

An initial experiment was carried out using a fixed game of maximising the Euclidian distance of a NAO H25 humanoid robot from its starting position to its position after a 5 second trial. Actor molecules were initialised as a random sensor atom (S) connected to a transform atom (T), which in turn connected to a motor atom (M). S could utilise a maximum of 3 sensors, T transformed its inputs using either a linear weight matrix, Izhikevich spiking network or K-means clustering, while M used its inputs to send commands to up to 4 motors. A population of 10 individuals were evolved using a microbial Genetic Algorithm for 1000 binary tournaments (2000 evaluations). Each individual started a trial on its back in a predefined resting position. Parameters in S, T and A atoms had a small probability of mutation, such as the weights in the matrix or the number of clusters in a K-means atom. Additionally extra S-T-M groups could be generated with random parameters and added to the actor molecule.

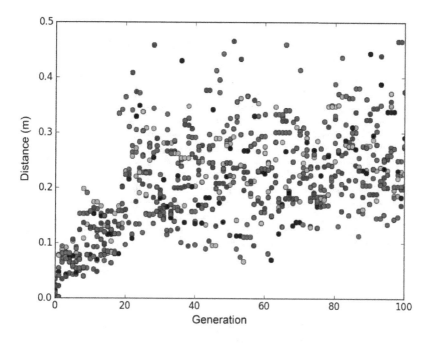

Fig. 5. The fitness in meters for each individual in the population at each 'generation' of 10 binary tournaments (20 evaluations)

This creates a disjoint graph, which can be thought of as a 'bag' of molecules. The fitness of individuals at every generation can be seen in Figure 5. We see that there is quick improvement at the beginning of the run, which soon stabilises to best distance scores of between 0.4 and 0.5 meters. The fitness scores are noisy, which is caused by randomness present in the molecules that changes the robots trajectory from trial to trial. The location of the centroids in the K-means atom and the initialisation of the of the parameters in the Izhikevich network reset at the start of each trial, which means that behaviour can differ slightly between trials. The best actor molecule found at the end of evolution can be seen in Figure 6. It consists of three disjoint S-T-M groups, one containing a K-means transform atom and the others using an Izhikevich network. One Izhikevich network controls the left hip, while the other controls the right. The left hip and right knee are also controlled by the K-means atom. This results in a series of rapid alternating step-like movement in the two legs of the robot and a sideways rocking motion, which propels the robots forwards, although often following a slightly curved trajectory.

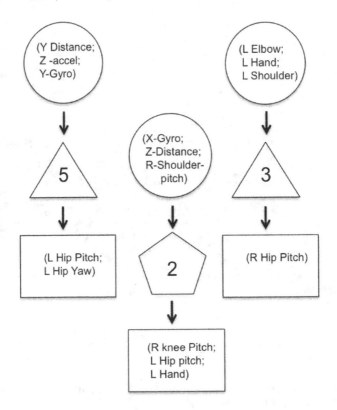

Fig. 6. The molecule of the best solution found. Circles represent sensor atoms with sensors described inside, triangles Izhikevich network atoms with the number of outputs inside, pentagons K-means atoms with the number of clusters described inside and squares motor atoms with the motors described inside.

4   Discussion and Conclusions

In this paper we have introduced an open ended cognitive architecture for the active learning of new behaviours through the co-evolution of actors and games. Both are made up of atomic units, and the way they are constructed can be interpreted as a "grammar of thought" for actors and a "grammar of action" for games. The best pairs of problems and solutions are stored in long term memory and can be drawn upon through recombination to help create and solve future tasks. A number of different possible games have been discussed and presented in the molecular structure of the proposed framework that allows a robot to learn through self exploration and environmental interaction. An initial experiment investigating the evolution of actor molecules on a fixed game showed that the system is able to produce interesting behaviour in the NAO robot. This system is in active development, and future work will be to include a large collection of atom types in a full game evolution algorithm, producing a wide range of different behaviour for a variety of real world robots.

The freedom present in our model to integrate any new algorithm from the Machine Learning literature, and the ability for other researchers to create new atoms and molecules through the open source code repository leads us to believe that the proposed architecture will present a spring board for the development of curious and playful robots.

Acknowledgements. Thanks to a John Templeton Foundation Grant "Bayes and Darwin" and a FP-7 FET OPEN Grant INSIGHT for funding this work.

References

1. Bellas, F., Duro, R., Faina, A., Souto, D.: Multilevel darwinist brain (mdb): Artificial evolution in a cognitive architecture for real robots. IEEE Transactions on Autonomous Mental Development 2(4), 340–354 (2010)
2. Fisher, R.: The Genetical Theory of Natural Selection. Clarendon Press (1930)
3. Rechenberg, I.: Evolutionstrategie 94. Frommann-Holzboog, Stuttgart (1994)
4. Bongard, J., Lipson, H.: Nonlinear system identification using coevolution of models and tests. IEEE Transactions on Evolutionary Computation 8(4), 361–384 (2005)
5. Baranes, A., Oudeyer, P.Y.: R-iac: Robust intrinsically motivated active learning. In: 2009 IEEE 8th International Conference on Development and Learning, Shanghai, China, pp. 1–6 (2009)
6. Baranes, A., Oudeyer, P.Y.: Intrinsically motivated goal exploration for active motor learning in robots: a case study (2010)
7. Baranes, A., Oudeyer, P.Y.: Active learning of inverse models with intrinsically motivated goal exploration in robots. Robotics and Autonomous Systems (2012)
8. Oudeyer, P.Y., Kaplan, F., Hafner, V.: Intrinsic motivation systems for autonomoys mental development. IEEE Transactions on Evolutionary Computation 11(2), 265–286 (2007)
9. Fodor, J., Pylyshyn, Z.: Connectionism and cognitive architecture: A critical analysis. Cognition 28, 3–71 (1988)
10. Hofstadter, D., Mitchell, M.: The copycat project: A model of mental fluidity and analogy-making. In: Fluid Concepts and Creative Analogies: Computer Models of the Fundamental Mechanisms of Thought, pp. 205–267. Basic Books, New York (1995)
11. Newell, A.: Unified Theories of Cognition. Harvard University Press (1990)
12. Anderson, J.: How Can the Human Mind Occur in the Physical Universe. Oxford University Press (2007)
13. Sun, R.: Duality of the Mind: A Bottom-up Approach Toward Cognition. Lawrence Erlbaum Associates, Mahwah (2002)
14. Marcus, G.: The Algebraic Mind: Integrating Connectionism and Cognitive Science. MIT Press (2001)

Virtual Modelling of a Real Exoskeleton Constrained to a Human Musculoskeletal Model

Francesco Ferrati, Roberto Bortoletto, and Enrico Pagello

Intelligent Autonomous Systems Laboratory
Department of Information Engineering (DEI)
University of Padua, Italy
{ferrati,bortolet,epv}@dei.unipd.it

Abstract. Exoskeletons represent one of the most important examples of human-oriented robotic devices. This paper describes an existing lower-limb exoskeleton designed to assist people with lower extremity paralysis or weakness during the movements of standing up and walking. Starting from the analysis of a real system developed about seven years ago, a virtual multibody model was realized in order to deeply understand how the device worked and find out some potential improvements in the actuators control and in the kinematic design. The virtual device was properly constrained to a human musculoskeletal model in order to simulate a real operating condition. The analysis of the simulation results suggested a kinematic modification of the system and a new dynamic model was developed in order to test the new design through the comparison of four different models.

Keywords: Exoskeleton, Bio-Inspired Robots, Lower Extremity Orthosis, Gait Cycle, OpenSim Simulation.

1 Introduction

An exoskeleton is a particular kind of robotic device designed to be worn by a subject in order to assist him during the execution of a motor task or improve his motion performances. Since the late 1960s, such devices have followed a strong evolutionary process and, after more than fifty years, some scientific results are gradually turned into real products mainly used for military [1–4] and clinical purpose [5, 6]. Considering their clinical applications, lower-limb exoskeletons represent multi-purpose medical devices, used to provide an efficient means of rehabilitation or walking assistance to people with lower extremity weakness or paralysis due to neurological disease or injury. From a structural point of view, they are mechatronic assistive devices externally applied to a subject and working in parallel with the musculoskeletal system [7]. A detailed historical review about lower limb exoskeletons can be found in [8], while [9] reports various projects around the world. Fig. 1 shows some of the latest devices.

N.F. Lepora et al. (Eds.): Living Machines 2013, LNAI 8064, pp. 96–107, 2013.
© Springer-Verlag Berlin Heidelberg 2013

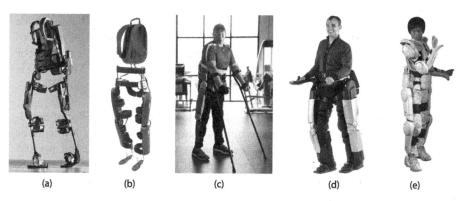

Fig. 1. (a) Ekso Suit; (b) ReWalk; (c) Vanderbilt Exoskeleton; (d) Rex; (e) HAL

Ekso Suit [10] by Ekso Bionics is one of the most advanced exoskeletons already on the market. It has three Degrees-of-Freedom (DOFs) per leg: two active joints at the hip and knee respectively, actuated by two brushless DC motors, and one passive joint at the ankle. The control system uses over fifteen sensors such as encoders and force sensing in order to assist body positioning and balancing. The stability of the subject is maintained using a pair of crutches, which also provide a command button to start the gait cycle when pushed. Thanks to its adaptable measures, the same device suits different anthropometric measures. Argo's *ReWalk* [11] is actuated by two motors per leg which control the hip and knee movement respectively. A remote controller at the wrist is used to select the operating mode and a pair of crutches is still used to maintain the user's stability. Two different versions of the system have been developed to be used in clinical rehabilitation centers or for everyday use. *Vanderbilt Exoskeleton* [12] is a project developed by the Center for Intelligent Mechatronics of the Vanderbilt University. It has two DOFs per leg actuated by two brushless DC motors both positioned at the thigh segment. It has a slim profile and no footplates or bulky backpack components are used. A lightweight Ankle-Foot-Orthosis (AFO) is used to stabilize the foot, but also in this case, the stability of the subject is maintained using a pair of crutches. It is able to automatically adjust the amount of assistance for users who still have some muscle control in their legs, and it is the only wearable exoskeleton that incorporates Functional Electrical Stimulation (FES) technology. A slightly different approach was used by Rex Bionics in developing the *Rex* [13]. Unlike other existing exoskeletons, the user selects the desired movement through the use of a joystick. Because the Rex doesn't require the use of any additional support, the user's hands are left free and the system can be used even by subjects with low strength in their upper limbs. Finally, one of the most famous exoskeletons is certainly the Cyberdyne's Robot Suit HAL (*Hybrid Assistive Limb*) [14]. Unlike the devices described above, HAL provides both a lower and/or an upper limb exoskeleton, in order to supply and/or improve human physical capability in various fields such as rehabilitation, physical training, heavy labour and rescue support. The most challenging aspect of HAL is the hybrid control strategy it uses. In fact, the bio-electrical input signals used to command the motors are directly acquired from a series of electromyography (EMG) surface electrodes through which it recognizes the user's motor command and starts the desired motion. Furthermore, if not good EMG signal is detectable, HAL can work through an alternative control strategy called *Robotic Autonomous Control* [15].

This paper analyzes an existing lower limb exoskeleton, designed and developed about seven years ago, with the aim of assisting people with lower-limbs paralysis or weakness in walking. After some years of inactivity, the present study wants to consider the system from scratch in order to improve its functions according to the current state-of-the-art. The rest of the paper is organized as follow: in section 2 a technical description of the considered exoskeleton and a kinematic and dynamic analysis are provided. The design phase of an accurate model of the real system is described in section 3, while section 4 describes the simulation approach used to make the system able to reproduce the walking motion. Section 5 provides a description of the obtained results, comparing the kinematic and dynamic behaviour of the four developed operating models. Finally, section 6 concludes the work and presents significant future developments.

2   Technical Specifications

The considered exoskeleton [16] was patented in 2006 by its inventor B. Ferrati and represents the result of a long time development period (the first patent dates back to 1982). Many versions of the system were realized over the years and the device studied here was successfully validated[1] on four subjects with spinal cord injury. Fig. 2 (a-b) show a snapshot of the exoskeleton whereas Fig. 2 (c) represents the kinematic scheme. The system consists of two actuated orthoses and a corset. Once worn, the system assists the user during the movements of standing up and walking. Considering the kinematic design, the system is characterized by a pseudo-anthropomorphic architecture. Although its design is close to the human's leg kinematics, the device includes only two flexion-extension DOFs per leg, constraining the whole motion in the sagittal plane. In fact, each orthosis consists of a series of three segments opportunely constrained via two purely rotary joints at the hip and knee. This design was dictated by the desire to make the system simple and robust. The aim of the project was not to perfectly reproduce human motion but to use a minimalistic design to provide a simple, low-cost and useful assistive device. Each orthosis is actuated by two ball screw DC motors opportunely linked to the thigh and shin segments in order to convert the linear motion to a rotation of the included joint.

The kinematic design allows the upper actuator to substitute the hip flexor muscle (*rectus femoris*) and the hip extensor (*hamstring muscles group*) while pulling and pushing respectively. On the other hand, the lower actuator acts as the knee flexor muscles (*hamstring, gastrocnemius*) while pulling and as the knee

---

[1] A video of an experimental session is available at: http://www.ferrati.com/

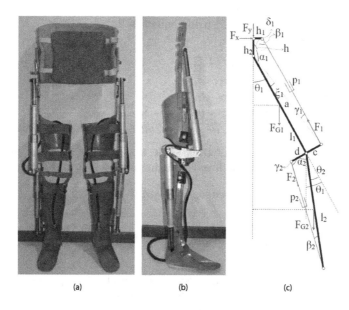

Fig. 2. (a) Front view of the real exoskeleton. (b) Right-side view of the real exoskeleton. (c) Kinematic scheme.

extensors (*quadriceps muscle group*) while pushing. The stability of the subject is maintained using a walker, which was also designed to contain the battery pack and the electronic components.

2.1 Kinematic and Dynamic Analysis

In order to study the physical behaviour of the real exoskeleton, a preliminary kinematic and dynamic analysis was made. Known the angle positions described by the hip and knee joints during a single gait cycle, the actuators forces necessary to generate locomotion were computed. Since the system was designed as an assistive device, only slow velocities are involved in the motion. Under this hypothesis, dynamic loads in the estimation are neglected.

The free body diagram of the system is represented in Fig. 2 (c). The l_1 segment represents the thigh whereas the l_2 represents the shin. θ_1 indicates the hip joint's position with reference to the vertical direction whereas θ_2 indicates the knee joint's position with reference to the thigh direction. Segments p_1 and p_2 indicate the variable lengths of the upper and lower actuators respectively. It is possible to calculate the length of the two actuators as function of the hip and knee angle positions. In particular, p_2 is computed by using equation 1.

$$\begin{cases} \alpha_2 = \frac{\pi}{2} - \theta_2 \\ p_2 = \sqrt{l_2^2 + d^2 - 2l_2 d cos\alpha_2} \end{cases} \tag{1}$$

On the other hand, p_1 is computed using equation 2.

$$\begin{cases} a = \sqrt{l_1^2 + c^2} \\ h = \sqrt{h_1^2 + h_2^2} \\ \xi_1 = arcsin\frac{c}{a} \\ \delta_1 = arcsin\frac{h_1}{h} \\ \alpha_1 = \pi - \theta_1 - \xi_1 - \delta_1 \\ p_1 = \sqrt{a^2 + h^2 - 2ahcos\alpha_1} \end{cases} \tag{2}$$

After computing the actuators lengths, some angles values for the lower and upper structures can be easily deduced by using equation 3 and 4 respectively.

$$\begin{cases} \beta_2 = arcsin\left(\frac{d}{p_2}sin\alpha_2\right) \\ \gamma_2 = \pi - \alpha_2 - \beta_2 \end{cases} \tag{3}$$

$$\begin{cases} \beta_1 = arccos\left(\frac{h^2 + p_1^2 - a^2}{2hp_1}\right) \\ \gamma_1 = \pi - \alpha_1 - \beta_1 \end{cases} \tag{4}$$

Considering the forces acting on the system, F_{G1} and F_{G2} represent the forces due to the gravity at the thigh's and shin's Center of Mass (COM), whereas F_1 and F_2 stand for the upper and lower actuators forces. m_1 and m_2 are the approximate weights of the thigh and shin segments applied in their COM respectively. The forces supplied by the shin and thigh actuators are computed as in 5 and 6 from a simple torque balance with respect to the knee and the hip center respectively.

$$F_2 = \frac{m_2 g \frac{l_2}{2} sin\left(\theta_1 - \theta_2\right)}{l_2 sin\beta_2} \tag{5}$$

$$F_1 = \frac{m_1 g \frac{l_1}{2} sin\theta_1 + m_2 g \left[l_1 sin\theta_1 + \frac{l_2}{2} sin\left(\theta_1 - \theta_2\right)\right]}{h sin\beta_1} \tag{6}$$

Equations 5 and 6 show how to obtain the upper and lower actuators forces as function of the θ_1 and θ_2 hip and knee angular positions.

3   Modelling

The implementation of a virtual model provides considerable advantages to the analysis of a device, such as unnecessary creation of a physical prototype of the system to test some variation in its configuration, early acquisition of more technical knowledge, reducing design time and costs and achievement of a better quality in the final product. In this perspective, virtual modelling is a useful tool both in the initial phase of functional design, in the mid-term evaluation of the alternatives and in the final kinematics and dynamics verifications. Starting from these considerations, all the physical features of the real system were imported in a simulator in order to study its dynamic behaviour and verify the effectiveness of some potential improvement without making any changes to the existing prototype. Considering the human-machine interaction involved in the use of an exoskeleton, the OpenSim[2] [17] simulator was chosen to implement

---

[2] Freely available at: https://simtk.org/home/opensim

and analyze the model from a biomechanical point of view. It is an open-source multibody dynamics engine developed by the Stanford University to implement mathematical models of the musculoskeletal system and create kinematic and dynamic simulations of human movement. By using the OpenSim built-in tools and a human musculoskeletal model provided within it, four different models were implemented, in order to study the behaviour of the system in different operating conditions and in particular the interactions that are generated between the human musculoskeletal structure and the exoskeleton.

– The first model represents the kinematic and dynamic characteristics of the existing exoskeleton. It is considered in order to examine the behavior of the device without the presence of a user.
– The second model differs from the previous one because of the adding of a human subject. This model uses an accurate musculoskeletal human model to reproduce the actual device operating condition, studying how the presence of a subject modifies the results previously obtained.
– The third model constitutes an improvement of the existing system. In fact, after the simulation of the real system model, a new multibody model was realized, adding an active DOF at the ankle joint in order to achieve a greater similarity in the gait kinematics (the real system doesn't have any DOF at the ankle).
– The fourth model adds the musculoskeletal model of a user to the modified virtual system.

3.1   Bodies, Joints, and Actuators

The first modelling step was to virtually reproduce all the mechanical parts using a CAD software. Starting from an accurate measurement of the actual components of the system and the identification of their materials, an accurate virtual reconstruction of each piece was processed. For each body the mass, the position of the COM, and the inertia matrix were computed. In particular, according to the convention used by OpenSim, the inertia matrix calculation was made with respect to a reference frame positioned at the origin of the absolute reference system of each part, and not respect to the COM. After the simulation of the actual system, a new virtual model was designed adding a flexion-extension ankle joint. The real system model and the modified one are shown in Fig. 3(a-b).

Within these models, each joint establishes a *parent-child* relation between two bodies. Three type of joints were used to define the relations between the components. A *Weld Joint* introduces no DOF and fuses two bodies together; this type of joint was used to define the relations between the corset body and the ground, between the corset rod and the corset body and between both the thigh and shin rods and their bootlegs. A *Pin Joint* introduces one coordinate about the common z-axis of the parent and child joints frames. *Pin joints* were used to define the hip and knee joints, and in the modified model for the ankle joint too. The geometry of the three joints is shown in Fig. 3(c). Finally, a *Custom*

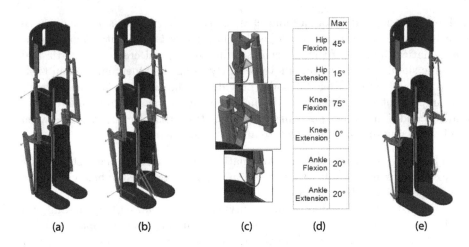

Fig. 3. CAD models of the real (a) and ankle-modified (b) systems. The reference frames for each DOF are given. Geometry (c) and range of motion (d) of the hip, knee and the ankle hinge joints. (e) Forces applied to the model of the real system.

Joint is used to specify 1-6 coordinates and a custom spatial transformation. It was used to define the musculoskeletal model's joints. The force of each actuator was represented by using a vector applied to the extreme points of each linear motor, as depicted in Fig. 3(d). This assumption allowed to use a simple point-to-point linear force model, which describes a force between two fixed points on two bodies. The force value is calculated as the product of an optimum strength value and a control value, while the associated sign is intended positive if it increases the distance between the extremities.

3.2   Constraints

Once the definition of the exoskeleton's virtual model was completed, the code for the human musculoskeletal multibody model was integrated as shown in Fig. 4. The aim was to get a complete model of the human-machine interaction in order to simulate with good approximation the system's actual operating configuration. To do this, a preexisting human musculoskeletal model provided by OpenSim was used. It comprises the physical model of both the lower limbs, the trunk and the head of a subject 175 cm tall and weigh 75 kg with 23 degree of freedoms and 91 muscles.

The human model was properly modified to fit the constraints imposed by the virtual device. In particular, the following changes were made.

– The joint between the human *torso* and the *ground* was replaced by a constraint between the *pelvis* and the exoskeleton's *corset* in order to reproduce the structural relationship between the trunk of the subject and the top of the system.

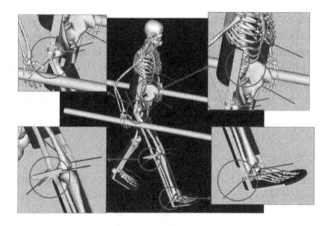

Fig. 4. Details of the *Coordinate Coupler* constraints within the ankle modified model.

- The DOFs were appropriately modified to respect the constraints imposed by the system on the human musculoskeletal structure.
- As the exoskeleton was designed to help people with lower-limbs paralysis to walk, the muscles of the lower extremities were disabled, delegating the actuation task to the exoskeleton's actuators only.
- To reproduce the kinematic constraints that occur when the device is worn, four-to-six constraints were added between the subject and the system's models. Since the use of the exoskeleton makes the movement of the subject's lower limbs equal to that of the device, a *Coordinate Coupler* type constraint was used to make the flexion-extension of the user's joints (hip, knee and ankle) dependent to the rotation of the system's corresponding DOFs.

4   Simulation

After the modelling phase, the motion behaviour of the four implemented models was analyzed. The kinematic and dynamic simulations were carried out using an OpenSim tool, called Computed Muscle Control (CMC). The CMC's purpose is to compute a set of muscle excitations (or more generally actuator controls) that will drive a dynamic model to track a set of desired kinematics [18]. In order to obtain a new effective force control for the real exoskeleton prototype, the gait cycle movement was considered and the actuators forces were computed. The data used in the simulations were obtained properly sampling the gait cycle data provided by OpenSim, which are widely validated in literature and freely provided with the software. A gait cycle of duration 4 s and a sampling time of 0.4 s was considered. Once the simulation is successfully completed, the CMC returns the experimental results for the effective joints' positions, velocities and accelerations, the generalized forces produced by the actuators to generate the motion, the actuators powers and speeds, the position errors of the joints and the actuation control values. In particular, one of the aims of the present work was to use the CMC tool to calculate the force of each actuator necessary to move the generalized coordinates along the desired trajectory. At this stage of the work, the simulation conditions didn't include any contact model, neither between the feet and the ground, nor between the subject's hands and the parallel bars. For now, ground reaction forces have not been computed. The control strategy used by the CMC consists in the combination of a proportional-derivative (PD) controller and a static optimizer.

5   Results

The actuators forces for the four considered models were computed in order to obtain the necessary values to make the system accurately reproduce the desired gait cycle[3]. The graphs in Fig. 5 and Fig. 6 present the simulation results for the real device model and the virtually modified prototype, respectively. For both cases, the exoskeleton only (green line) and the human-constrained system (blue line) are compared to each other. Joint angle positions and actuators forces are expressed in Degree and Newton respectively. Time is measured in second. The reference trajectories (dotted line) correspond to the validated OpenSim walking data.

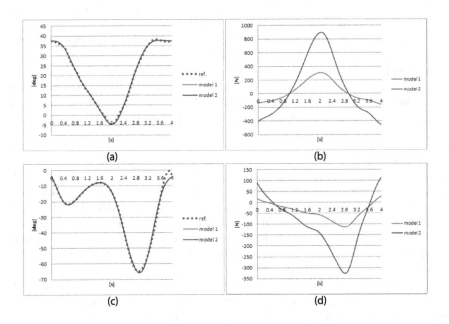

Fig. 5. (a) (c) Angle positions of Hip and Knee joints. (b) (d) Correspondent Upper and Lower actuators forces. The green and blue lines represent the simulation results for the non-constrained and human-constrained models, respectively.

---

[3] A video of the simulation is available at: http://youtu.be/WO9Z--nrwwc

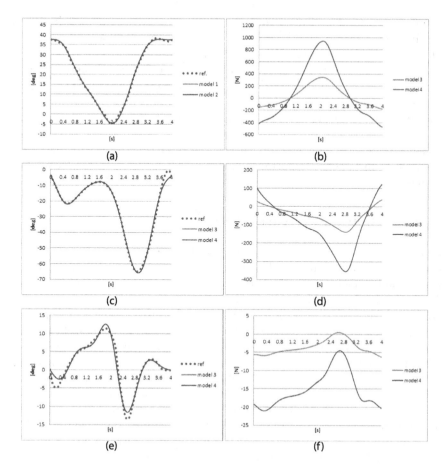

Fig. 6. (a) (c) (e) Angle positions of Hip, Knee, and Ankle joints. (b) (d) (f) Correspondent Upper, Lower, and Ankle actuators forces. The green and blue lines represent the simulation results for the non-constrained and human-constrained models, respectively.

The graphs in Fig. 5 describe the experimental joint angles and the actuators forces obtained by the simulation of the actual device model and its human-constrained version. It can be noticed that the experimental joints positions are the same for both the models and slightly deviate from the reference trajectory. However, the overall trend is followed and the approximation is acceptable for the intended purposes. Considering the forces produced by the upper and lower actuators, the maximum absolute values correspond to the maximum flexion-extension of the hip and knee joints. The obtained values are in line with the results of the dynamic analysis. As expected, the presence of a user in the second model (blue line) increases the actuator forces necessary to move the system. However the trend is quite the same for both the configuration.

After simulating the real system's model, the improved virtual prototype was tested in order to verify the effectiveness of adding an active degree of freedom at the ankle joint. The simulation results are represented in Fig. 6. As in the previous case, the joint trajectories are sufficiently accurate for the intended purpose. The ankle joint's plantar-dorsal flexion is quite different from the reference trajectory in the maximum values but is still acceptable. The upper and lower actuators forces don't significantly change respect to the previous models. As expected, the ankle actuator's force is lower than the other two and the presence of a user significantly increases the values. The simulation results for the modified system confirmed the importance of the ankle joint to make the locomotion more natural.

6   Conclusion and Future Works

A real exoskeleton designed to assist people with lower-extremity paralysis or weakness was analyzed in detail. The aim of the work was to realize a virtual multibody model of the described device in order to study its behaviour and consider interesting design modifications. The model was implemented in a biomechanical simulator called OpenSim and the virtual device was properly constrained to a musculoskeletal model provided by the software itself. Considering a gait analysis data set, a dynamic simulation of the system was performed in order to compute the actuators forces necessary to make the joint angles follow the desired gait pattern. After simulating the actual exoskeleton's model, a new virtual prototype was designed adding an ankle joint. The modified model was simulated using the same approach and the new forces values were compared with the previous one. In conclusion this work has provided an original way to analyze the behaviour of a robotic exoskeleton within a biomechanical simulation platform. A realistic force control was estimated to make the device correctly move. The modelling approach allowed to test potential improvements of the existing system. As a future work, it would be interesting to apply the so found results to the real exoskeleton in order to verify the effectiveness of the present approach. The current results didn't consider the ground reaction forces between the human feet and the ground. The development of a ground contact model for the virtual device would improve the accuracy of the analysis. Finally, to make the gait cycle more natural, it would be interesting to increase the number of DOF of the system.

Acknowledgments. This research has been partially supported by "Consorzio Ethics" through a grant on rehabilitation robotics.

References

1. Kazerooni, H., Steger, R.: The Berkeley Lower Extremity Exoskeleton. Transactions of the ASME, Journal of Dynamic Systems, Measurements, and Control 128, 14–25 (2006)

2. Zoss, A.B., Kazerooni, H., Chu, A.: Biomechanical Design of the Berkeley Lower Extremity Exoskeleton (BLEEX). IEEE/ASME Transactions on Mechatronics 11(2), 128–138 (2006)
3. Steger, R., Kim, S.H., Kazerooni, H.: Control scheme and networked control architecture for the Berkeley lower extremity exoskeleton (BLEEX). In: Proceedings 2006 IEEE International Conference on Robotics and Automation, pp. 3469–3476 (2006)
4. Karlin, S.: Raiding Iron Man's closet. IEEE Spectrum 48, 25 (2011)
5. Mikołajewska, E., Mikołajewski, D.: Exoskeletons in Neurological Diseases-Current and Potential Future Applications. Adv. Clin. Exp. Med. 20(2), 227–233 (2011)
6. Aach, M., Meindl, R., Hayashi, T., Lange, I., Geßmann, J., Sander, A., Nicolas, V., Schwenkreis, P., Tegenthoff, M., Sankai, Y., Schildhauer, T.A.: Exoskeletal Neuro-Rehabilitation in Chronic Paraplegic Patients Initial Results. Converging Clinical and Engineering Research on Neurorehabilitation Biosystems & Biorobotics 1, 233–236 (2013)
7. Pons, J.L., Ceres, R., Calderon, L.: Introduction to wearable robotics. In: Pons, J.L. (ed.) Wearable Robots: Biomechatronic Exoskeletons, pp. 1–2. John Wiley & Sons Ltd., West Sussex (2008)
8. Dollar, A.M., Herr, H.: Lower Extremity Exoskeletons and Active Orthoses: Challenges and State-of-the-Art. IEEE Transactions on Robotics 24(1) (2008)
9. Guizzo, E., Goldstein, H.: The rise of the body bots. IEEE Spectrum 42, 50–56 (2005)
10. Strickland, E.: Good-bye, wheelchair. IEEE Spectrum 49, 30–32 (2012)
11. ReWalk - Walk Again, http://rewalk.com/
12. Quintero, H.A., Farris, R.J., Goldfarb, M.: Control and implementation of a powered lower limb orthosis to aid walking in paraplegic individuals. In: IEEE Int. Conf. on Rehabilitation Robotics, ICORR 2011 (2011)
13. Rex Bionics - Step into the future with Rex, http://www.rexbionics.com/
14. Tsukahara, A., Hasegawa, Y., Sankai, Y.: Standing-up motion support for paraplegic patient with Robot Suit HAL. In: IEEE Int. Conf. on Rehabilitation Robotics, ICORR 2009 (2009)
15. Hayashi, T., Kawamoto, H., Sankai, Y.: Control Method of Robot Suit HAL working as Operator's Muscle using Biological and Dynamical Information. In: 2005 IEEE/RSJ International Conference on Intelligent Robots and Systems (2005)
16. Ferrati Electronic Orthosis, http://www.ferrati.com/
17. Delp, S.L., Anderson, F.C., Arnold, A.S., Loan, P., Habib, A., John, C.T., Guendelman, E., Thelen, D.G.: OpenSim: Open-Source Software to Create and Analyze Dynamic Simulations of Movement. IEEE Trans. on Biomedical Engineering 54(11) (2007)
18. Thelen, D.G., Anderson, F.C.: Using computed muscle control to generate forward dynamic simulations of human walking from experimental data. Journal of Biomechanics (2006); 39(6), 1107–1115 (epub July 14, 2005)

Where Wall-Following Works: Case Study of Simple Heuristics vs. Optimal Exploratory Behaviour

Charles Fox

Sheffield Centre for Robotics, University of Sheffield, UK

Abstract. Behaviours in autonomous agents – animals and robots – can be categorised according to whether they follow predetermined stimulus-response rules or make decisions based on explicit models of the world. Probability theory shows that optimal utility actions can in general be made only by considering all possible future states of world models. What is the relationship between rule-following and optimal utility modelling agents? We consider a particular case of an active mapping agent using two strategies: a simple wall-following rule and full Bayesian utility maximisation via entropy-based exploration. We show that for a class of environments generated by Ising models which include parameters modelling typical robotic mazes, the rule-following strategy tends to approximate optimal action selection but requires far less computation.

1 Introduction

Autonomous behaviours can be categorised according to whether they follow predetermined rules or make decisions. Examples of rule-following robots include mobile robots programmed to follow walls, solve mazes by turning left at every junction, or operate factory equipment using machine vision and if-then rules to generate actions based on the percepts. Decision making agents do not follow such pre-programmed behaviours, but internally simulate the effects of various possible actions in order to choose the best one. For example, classic AI game-playing agents search trees of board games for the best move; Monte Carlo supply simulations search for best actions to operate business supply chains [17] and financial trades [8]; and active SLAM [2] simulates mobile robot explorations to find best movements for building maps. 'Best' is always specified by some utility function, and the agent's sole objective is to maximise this utility function. For example, an active SLAM robot could assign additive utilities to a world state based both on how much map-building is achieved (positive utilities) and how much time and power has been expended on motion (negative utilities).

During our experiments on Active SLAM [3], [6], we noticed a tendency for wall-following behaviour to emerge from the full Bayesian decision computations. This was surprising, because wall following can be obtained so easily from simple rules as well as by these large, expensive computations. This suggested that in at least some environments, full Bayesian computation may be unnecessary and

N.F. Lepora et al. (Eds.): Living Machines 2013, LNAI 8064, pp. 108–118, 2013.

well-approximated by the simple wall-following rule. In the present paper we perform controlled experiments in a simplified class of environments to examine how general this approximation might be.

In purely theoretical decision making, the utility maximisation approach is always the 'correct' one, as it is by definition mathematically optimal. However, practical deployed systems tend to use simple rules instead. Such simple-rule based systems were popular in the Artificial Life research of the 1990s, before the advent of Bayesian machine learning, which has shifted the focus to utility maximisation. In Artificial Life, inspiration was drawn from natural systems such as insects following simple rules such as 'turn left if about to hit something' in order to form apparently complex behaviours [12] which were formalised into robotic systems such as Braitenberg vehicles [1]. Rats are known to perform much of their exploration by wall-following behaviour [10], the implementation of which has been modelled as emerging simply from maximum contact minimum impingement (MCMI) behaviour in conjunction with other biomimetic control systems [14]. State of the art robotic vacuum cleaners still use similar simple rules, such as 'follow walls' or 'if hit something, turn a random angle and move in a straight line' [16].

What is the relationship between these rules and the ideal utility maximisations? From a Bayesian perspective, we consider that an agent's objective when in some state s_0 is to maximise some utility function,

$$U(\{a_t\}_{\forall t}, s_0) = \sum_{t=1}^{\infty} \sum_{s_t} p(s_t|s_0, a_{0:t}) R(s_t), \qquad (1)$$

where s are states of the world (which may include the agent's knowledge state), t is discretised time, a are actions and $R(s_t)$ are (discounted present value) rewards for being in states s_t. In general full planning, the agent must consider and optimise over all of its future action sequences to select a plan or sequence of actions $\{a_t\}_{\forall t}$,

This optimisation is generally computationally infeasible, especially for real-time systems, being exponential compute time in the number of time steps when the infinite sum is truncated. Even if a myopic approach is assumed as an extreme approximation to full planning, and we assume that all utility contribution occurs at the next greedy step,

$$U(a_t, s_t) \approx \sum_{s_{t+1}} (s_{t+1}|s_t, a_t) R(s_{t+1}), \qquad (2)$$

or if a Q-learning temporal difference approach is taken (and ignoring the issue of how to compute Q),

$$U(a_t, s_t) \approx \sum_{s_{t+1}} p(s_{t+1}|s_t, a_t) Q(s_{t+1}), \qquad (3)$$

with

$$Q(s_\tau) \approx R_\tau + \max_{\{a_t\}_{t=\tau:\infty}} \sum_{t=\tau+1}^{\infty} \sum_{s_t} p(s_t|s_\tau, a_{\tau:t}) R(s_t), \qquad (4)$$

we can still be left with an exponentially hard problem to compute this and select an action because the state s_t can itself be a joint variable $s = (s^1, s^2, \cdots s^S)$ having a state space with size exponential in S which must be summed over (e.g. the joint states of many cells in a map). Furthermore, it is sometimes the case that for *each* s, $R(s)$ may be similarly exponentially hard to compute (as will be the case in the present study). Therefore it is imperative that a real-time[1] agent uses some fast approximation to this computation instead of sitting still doing computation. The psychological literature has identified many 'heuristics and biases' [11] and 'simple heuristics' [7] showing cases where humans behave as if following simple rules rather than computing utilities. Daw and Dayan [4] have shown similar heuristic behaviour in rats, and further shown the existence of complexity boundaries in families of tasks which cause the rats to switch from full utility computation to the use of heuristics.

As a case study of this general problem of understanding how simple rules relate to full Bayesian utility computation, we will study in this paper a simple form of greedy active mapping – a component of Active SLAM [2] – problem, and its relationship to the popular wall-following heuristic used in many exploring robots. In what environments does wall-following tend to work so well in practice?

1.1   Active Mapping

Active SLAM is the Bayesian formalisation of the problem of how a robot should act to explore and localise itself to build a map of an environment and its location in it. Active SLAM is defined in terms of maximising the expected reduction in the entropy of an agent's belief about its map of the world including its own location in it, or informally, 'where to look next'. Active *mapping* is a simplified case of this problem which considers only the map of the world, and assumes the agent has a perfect location sensor at all times.

We assume the physical world w is made of discrete cell locations w^i which are either occupied or not, $w^i \in \{0, 1\}$. A map of the world at time t, m_t, is similarly made of discrete cells m_t^i, and an agent's belief about what the true map is, is represented by a distribution $P(m_t|o_{1:t})$ given all available observations up to time t, $o_{1:t}$.

While this belief may be an inseparable joint distribution over all the cells, its entropy can be (crudely but standardly [2]) approximated by assuming independent cells and summing the marginal entropies,

$$H[P(m_t|o_{1:t})] \approx \sum_i H[P(m_t^i|o_{1:t})], \qquad (5)$$

with

$$H[P(m_t^i|o_{1:t})] = P(m_t^i|o_{1:t}) \log P(m_t^i|o_{1:t}) + P(\neg m_t^i|o_{1:t}) \log P(\neg m_t^i|o_{1:t}), \quad (6)$$

---

[1] Where 'real-time' here means having a large negative utility for slow decision making.

where $P(m_t^i|o_{1:t})$ is the occupancy probability for a cell, and $P(\neg m_t^i|o_{1:t}) = 1 - P(m_t^i|o_{1:t})$ is the probability of non-occupancy.

At each time t, greedy active mapping seeks to choose the best action $\hat{a}_t$ which reduces the entropy of the map belief as much as possible,

$$\hat{a}_t|o_{1:t} = \arg_{a_t} \max U(a_t, o_{1:t}) \qquad \cdot (7)$$

$$= \arg_{a_t} \max \left(H[P(m_t|o_{1:t})] - H[P(m_{t+1}|a_t, o_{1:t+1})] \right) \qquad (8)$$

$$= \arg_{a_t} \min H[P(m_{t+1}|a_t, o_{1:t+1})]. \qquad (9)$$

At decision time we do not know what the future observations o_{t+1} will be, so greedy active mapping marginalises the decision over their expectation,

$$\hat{a}_t \approx \arg_{a_t} \min \langle H[P(m_{t+1}|a_t, o_{1:t})] \rangle_{P(o_{t+1}|m_t, a_t)}. \qquad (10)$$

2 Methods

To isolate the phenomenon of interest in our experiments, we made as many simplifying assumptions as possible to produce a micro-world model of more general robotic exploration. We ignore the localisation part of the Active SLAM task and assume perfect (GPS-like) agent knowledge of its location at all times. Rather than model directional distal sensors such as vision or laser range finders, we assume 360 degree proximal sensors (such as touch sensors all around its body). We discretise the world into cells, and use a hexagonal grid as shown in fig. 1. (This has the modelling advantage of each cell having six well-defined, equally important neighbours rather than having to handle diagonal contacts in a square grid).

Fig. 1. A physical map. Black=occupied, white=unoccupied. The world is discretised into hexagonal cell locations. The agent position is shown by the face.

In order to test for advantageous behaviour from wall-following, relative to the Bayesian optimal solution, we wish to compute approximate average entropy changes over all possible physical worlds. To approximate this expectation we draw large numbers of samples of worlds. A simple way to model natural environments with the hex grid is to use the Ising [13] model,

$$P(\{w^i\}_{\forall i}) = \frac{1}{Z} \exp\left(\sum_i \pi(w^i) + \sum_{i,j:neigh(i,j)} \phi(w^i, w^j)\right), \qquad (11)$$

with factors

$$\pi(w^i) = (\pi)^{w^i}(1-\pi)^{1-w^i} \qquad (12)$$

$$\phi(w^i, w^j) = (\phi)^{w^i w^j}, \qquad (13)$$

where $neigh(i,j)$ means that cells i and j are neighbours in the hexagon grid topology and Z is a normaliser. Fig. 2 shows sample worlds drawn from this model with various parameters. The model assigns probabilistic factors π to individual cells and ϕ to neighbouring pairs of cells. The effect of π is to increase the overall number of occupied cells. The effect of ϕ is to increase the contiguity of occupation, i.e. make it more probable that a cell is on if its neighbours are on. Intuitively this models the fact that physical 'stuff' in the world tends to actively clump together into large objects such as walls and furniture, but empty space does not, except as a result of the stuff clustering. By manual parameter search we found that $\pi = 0.05, \phi = 2$ gives similar environmental characteristics to our real robot mazes such as in [3]. Note that we use all positive factors rather than negative energy terms as in statistical physics[13]; the Boolean exponents in the factors act as on-off switches; and the pairwise factors equal one if it is not the case that both neighbours are occupied. To draw samples from eqn. 11 we use standard Gibbs sampling [13], with a burn-in of 100 samples initialised with independent cell draws from the π factors only.

To simplify the simulation further, we sample each decision instance in a new environment, rather than a previous partially mapped one. In a typical exploration of a *completely new* part of an environment, an agent will always have arrived from some direction, so have no (greedy) need to consider moving back in that direction. With proximal sensors, it will sense which cells around it are occupied. Fig 3 shows an example of the agent's mental world corresponding to the physical world of fig. 1, including its previous location that it has moved from. So for each decision, we sample a new world from the Ising occupancy model, place the agent at a random location, and consider the five possible moves it could take, excluding a randomly chosen simulated arrival direction. As we only consider one movement decision per world, we can use very small, 36-cell worlds as in the figures, and allow their edges to wrap around to simplify the mathematics.[2]

As we are investigating only the question of whether wall following is useful, the decisions that interest us are only those that occur when the agent begins next to at least one wall cell. If our sample world and random robot location do not meet this criterion then we discard the simulation and draw a new sample world.

---

[2] This does however have an effect on the characteristics of sampled worlds such as island sizes, however we have chosen parameters to give realistic characteristics after the wrap around was included.

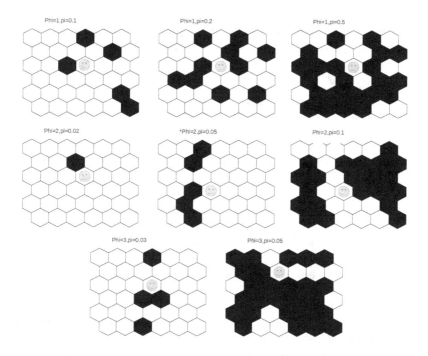

Fig. 2. Physical world samples with various parameters. Parameters around $\phi = 2, \pi = 0.05$ tend to generate worlds most like typical man-made environments, being quite sparse but locally correlated so as to produce structures resembling walls.

At each movement decision, we consider moving to each neighbouring unoccupied cell. We compute (our best approximation to) the Bayesian expected entropy reduction given that move. We also test whether or not the move is a wall-following move (defined as a move in which the pre-move and post-move states share a neighbouring occupied cell). We store records of expected entropy reductions for every wall-following and non-wall-following move considered so that we may compare their averages to determine whether wall-following is significantly more useful than non-wall-following.

For each possible action, the ideal correct sensor values were obtained by consulting the physical world model. This is a further approximation to the expectation in eqn. 10, ideally we would use further samples from the agent's mental model here to integrate over possible physical worlds, but this approximation again speeds up the computation.

2.1   Approximation of Entropies

For both the pre-move and post-move states we must compute eqn. 5. We make use of the standard independence assumption in the entropy sum, but we do

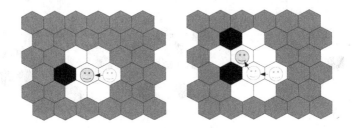

Fig. 3. Mental maps before and after a movement. Grey=uncertain. Outline robots show previously visited locations.

want to retain the joint dependencies,

$$P(m_t|o_{1:t}) \neq \prod_i P(m_t^i|o_{1:t}). \tag{14}$$

This is because we are interested in how the correlations in the world induced by occupancy clumping in natural environments might affect the utility of the wall-following heuristics.

In order to compute eqn. 5 under these assumptions, we must therefore draw samples from the *joint* distribution of map cells, i.e. from the space of possible mental worlds given the observations. This can be done using a 'wake' [9] version of the Gibbs sampler used to generate physical worlds from eqn. 11, i.e. by clamping the observed cells to their observed values and Gibbs sampling from all remaining cells.

The pre-move and post-move mental map joint distributions were each approximated by drawing 100 wake samples from the unobserved cells, using 100 random Gibbs updates between each sample to approximate burn-in. The marginal entropy terms in eqn. 6 were then estimated from Good-Turing ('add-one') adjusted occupancy frequencies [13]. (It has been found empirically that correlated, short approximate burn-ins often work surprisingly well, e.g.. the use of single updates in contrastive divergence learning [9]. A full analysis of this assumption is beyond the scope of the present paper.)

3 Results

Setting $\phi = 2, \pi = 0.05$ which gives typical robot-arena-like environments, we ran the simulation for 100,000 world samples with about 3.5 possible actions per sample. The total computation time using these approximations was around 10 hours on a 3GHz Pentium Duo PC.

We obtained a sample mean expected entropy per cell reduction of 0.001933 for wall-following moves, and -0.001687 for non-wall-following moves.[3] The sample

---

[3] The negative entropy may seem strange but these are *per cell* statistics. The number of uncertain cells changes after a move when new cells are observed which may explain the apparent increase in uncertainty *per cell* rather than total map uncertainty. It is only the difference between the two means that is of interest.

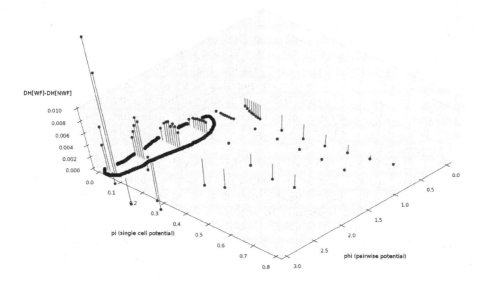

Fig. 4. Results. x and y axes are Ising parameter values. z axis is the sample mean difference between entropy change (DH) given wall following (WF) moves, and entropy change given non-wall-following (NWF) moves. The lasso shows the region of parameter space where wall following gives a significant advantage over non-wall-following.

standard deviations were 0.0127 and 0.0122, and the numbers of observations were 49,241 and 64,039. To avoid any possible bias from simulating multiple actions from the same world sample, we will conservatively use N=10,000 rather than these counts. Assuming both populations standard deviations to be 0.015, a classical t-test gives a significant difference in favour of the wall-following moves. (t=46.7, N=10,000, df=10,000, $p \leq 0.001$).

Having demonstrated that this number of world samples lead to a significant result for our manual-parametrised world, we then ran a further 70 repeats of the experiments (700 hours computation time) for a range of (ϕ, π) parameters, to test how general the wall-following rule's usefulness is. The differences in expected entropy reduction means are plotted in fig. 4.

4   Discussion

Where do animals [10] and practical robots [6] benefit from using simple wall-following behaviour in place of full Bayesian utility maximisation? The results show that for proximal sensing agents exploring previously unvisited regions in a realistic environment class having sparse occupancy but strong local correlation, under simplifying assumptions, that wall-following actions do have significantly better outcomes than non-wall-following actions. This suggests that wall-following provides a fast approximation to optimal entropy exploratory behaviour in this type of environment.

Fig. 4 suggests that there is a 'sweet spot' in the parameter space where wall following works better than non-wall-following. This has been indicated by the hand-drawn lasso. It is interesting to see that the centre of this lasso corresponds roughly to the parameters ($\pi = 0.05, \phi = 2$) that were originally chosen by hand to give real-environment-like characteristics. Fig. 2 shows samples of physical worlds with various parameters, corresponding to parameter coordinates inside and outside the lasso. The ($\pi = 0.05, \phi = 2$) sample in the centre perhaps looks more like a typical room environment than any of the others in terms of density and connectivity. (Note that the hex grid wraps around, so correlations around the edges count like any others.)

Our simulations used the assumption that all exploration was in some 'new' area, i.e. that only one previously-visited cell is nearby, that it is a neighbour, and that we have just moved to the present location from it. This assumption is valid for agents just beginning to explore a new region of a map, but not for agents building up a detailed map over a longer time and revisiting areas. A pure wall-follower starting at the wall would never discover an object in the centre of a room for example. Revisiting also often occurs as an optimal action in full Active SLAM systems, where the agent must re-localise from time to time as well as explore. Our simulations do not include tasks where particular locations must be visited (e.g. vacuuming a room, finding a power charger), rather their utility is only in mapping itself. Further work would be needed to test the utility of wall following in these more complex situations. An intriguing idea would be to test for similar effectiveness of other simple rules in these tasks, against full utilities, for example attempting to compare the Roomba's default combination of wall-following and random-straight-line wall-bouncing against an Active-SLAM approach to room coverage. The greedy active mapping utility used in our Bayesian utility maximisation is itself an approximation to non-greedy full planning, the latter would be expected to behave differently from the former in long-term exploration, but is so computationally hard as to be infeasible to simulate to significance even in our micro-world.

The hex-grid worlds used have the nice features that all actions are cleanly classified into wall-following or non-wall-following, and there are typically similar numbers of each that get sampled by our simulation. (49,241 and 64,039 in the main example above.) It is possible that other simple strategies could exist that are more selective that both wall-following and non-wall-following, and provide an even better approximation to optimal behaviour, though such strategies are beyond the scope of the present study.

In using the same (π, ϕ) parameters in both world generation and mental map sampling, we have implicitly assumed that the agent knows the Ising statistics of the class of environments it will be placed in. Such parameters could be learned from experience, but it would be interesting to investigate the effect of mismatches between the physical and mental parameters.

We have made use of large compute power sampling approximations in our results, and it is possible that for some simple worlds such as the Ising hex worlds, there exist analytic solutions for the expected entropy reduction. Further theoretical work could analyse the models used here to search for closed form solutions which may provide further insights into why the rule works. The 700 hour compute time used was necessary to obtain significance in the results (a 6×6 hex world has 2^{36} physical states, and far more mental states, to sample from), and gives some indication of how computationally hard optimal exploration can be, even for very simple worlds (cf. [15], which uses similarly tiny micro-worlds and approximations to compute entropy based explorations.)

While we have demonstrated a particular simple rule approximating optimal utility-based action selection, this work is also intended as an example of the more general case [5]. Can we analyse other well-known heuristics in this way? It is becoming clear that even with large supercomputers many decisions are intractable, and therefore a relevant research agenda for psychology and robotics is to seek out and try to understand useful rules that approximate such decisions. In what classes of environment are they valid and when do they break down? We could try to invent new rules from intuition or data-mining, or look at rules traditionally used by humans to find test candidates.

References

[1] Braitenberg, V.: Vehicles: Experiments in Synthetic Psychology. Bradford Bks. MIT Press (1986)
[2] Carlone, L., Du, J., Ng, M.K., Bona, B., Indri, M.: An application of Kullback-Leibler divergence to active SLAM and exploration with particle filters. In: 2010 IEEE/RSJ International Conference on Intelligent Robots and Systems (IROS), pp. 287–293 (October 2010)
[3] Fox, C., Evans, M., Pearson, M.J., Prescott, T.J.: Towards hierarchical blackboard mapping on a whiskered robot. Robotics and Autonomous Systems 60, 1356–1366 (2012)
[4] Daw, N.D., Niv, Y., Dayan, P.: Uncertainty-based competition between prefrontal and dorsolateral striatum systems for behavioral control. Nature Neuroscience 8, 1704–1711 (2005)
[5] Fox, C.: Formalising robotic ethical reasoning as decision heuristics. In: Proc. First UK Workshop on Robot Ethics, Sheffield (2013)
[6] Fox, C.W., Evans, M.H., Pearson, M.J., Prescott, T.J.: Tactile SLAM with a biomimetic whiskered robot. In: ICRA (2012)
[7] Gigerenzer, G., Todd, P.M., ABC Group: Simple Heuristics That Make Us Smart. Oxford University Press (2000)
[8] Glasserman, P.: Monte Carlo Methods in Financial Engineering. Springer (2003)
[9] Hinton, G.E., Osinderoi, S., Teh, Y.-W.: A fast learning algorithm for deep belief nets. Neural Computation 18 (2006)
[10] Hosterrer, G., Thomas, G.: Evaluation of enhanced thigmotaxis as a condition of impaired maze learning by rats with hippocampal lesions. Journal of Comparative and Physiological Psychology 63(1), 105–110 (1967)
[11] Kahneman, D.: Thinking Fast and Slow. Macmillan (2011)

[12] Langton, C.G.: Studying artificial life with cellular automata. Physica D: Nonlinear Phenomena 22(1-3), 120–149 (1986)

[13] MacKay, D.J.C.: Information Theory, Inference and Learning Algorithms. Cambridge University Press (2003)

[14] Mitchinson, B., Pearson, M.J., Pipe, A.G., Prescott, T.J.: The emergence of action sequences from spatial attention: Insight from rodent-like robots. In: Prescott, T.J., Lepora, N.F., Mura, A., Verschure, P.F.M.J. (eds.) Living Machines 2012. LNCS, vol. 7375, pp. 168–179. Springer, Heidelberg (2012)

[15] Saigol, Z.: Automated Planning for Hydrothermal Vent Prospeting using AUVS. PhD thesis, University of Birmingham, UK (2010)

[16] Strauss, P.: Roombas make dazzling time-lapse light paintings, http://technabob.com/blog/2009/09/29/roomba-time-lapse-art/

[17] Taylor, M., Fox, C.: Inventory management with dynamic bayesian network software systems. In: Abramowicz, W. (ed.) BIS 2011. LNBIP, vol. 87, pp. 290–300. Springer, Heidelberg (2011)

Miniaturized Electrophysiology Platform for Fly-Robot Interface to Study Multisensory Integration

Jiaqi V. Huang and Holger G. Krapp

Department of Bioengineering, Imperial College London, London SW7 2AZ, UK
j.huang09@imperial.ac.uk

Abstract. To study multisensory integration, we have designed a fly-robot interface that will allow a blowfly to control the movements of a mobile robotic platform. Here we present successfully miniaturized recording equipment which meets the required specifications in terms of size, gain, bandwidth and stability. Open-loop experiments show that despite its small size, stable recordings from the identified motion-sensitive H1-cell are feasible when: (i) the fly is kept stationary and stimulated by external motion of a visual pattern; (ii) the fly and platform are rotating in a stationary visual environment. Comparing the two data sets suggests that rotating the fly or the pattern, although resulting in the same visual motion stimulus, induce slightly different H1-cell response. This may reflect the involvement of mechanosensory systems during rotations of the fly. The next step will be to use H1-cell responses for the control of unrestrained movements of the robot under closed-loop conditions.

Keywords: motion vision, multisensory integration, brain machine interface, sensorimotor control, blowfly.

1 Introduction

The blowfly, *Calliphora vicina*, is a model system for understanding how flying insects use motion vision and information from other sensor systems to control their flight and gaze. Only about 10mm in length, the blowfly can achieve translation velocities of 2.5 m/s and angular rotation rates of 1700 deg/s in free flight [1] easily outperforming any man-made micro air vehicle. The blowfly has a suite of sophisticated sensors that supports high flight performance during a variety of behvioural tasks such as high speed chasing, collision avoidance and landing on the spot. These sensors include: the compound eyes, the ocelli, halteres, and antennae [2]. The ocelli are three small lens eyes on the top of the head that signal pitch and roll attitude changes relative to the horizon based on differential light intensity measurements in the dorsal visual field [3]. The halteres are sensory organs that have evolved from a second pair of wings. They also indicate attitude changes by measuring angular rotation rates around all three body axes [4]. The antennae, which are held at a specific angle relative to the head during locomotion, sense gravity when walking and are thought to be involved in measuring air speed in flight [5].

Amongst all the sensory modalities in blowflies, the visual system plays a cardinal role when information obtained in sensory coordinates is converted into commands

N.F. Lepora et al. (Eds.): Living Machines 2013, LNAI 8064, pp. 119–130, 2013.
© Springer-Verlag Berlin Heidelberg 2013

signals sent to the neck and flight motor [2]. Two thirds of the neurons in the fly nervous system are involved in visual information processing whereby the optic lobes occupy approximately 40% of the volume of the head capsule [6]. In the optic lobes, signals from the retina pass through cells located in parallel pathways which extract information, for instance, about visual motion, position of visual targets, or image expansion [7]. The lobula plate, the highest integration level of visual motion, contains individually identified interneurons with large receptive fields covering more than half the visual field. Most of these lobula plate tangential cells (LPTCs) process widefield optic-flow, as induced during the animal's self-motion [7]. Some LPTCs connect to descending neurons which also receive input from other sensor systems and convey control signals to the various motor systems. Due to electrical synapses [8] the multisensory integration at the level of the descending neurons can be assessed by recording the activity of specific LPTCs [9].

A major limitation in research on LPTCs and on multisensory integration using conventional approaches is that during an experiment the animals are restrained and only a limited number of sensor systems are stimulated. To partially overcome this limitation, an experimental platform is needed where the animal is actually moving in space while the activity of LPTCs is recorded. To this end we have started to develop a fly-robot interface where the activity of an individually identified LPTC will be measured and used to control the steering of a robot under closed-loop conditions. On such platform, several sensory systems would be simultaneously stimulated. The first step however requires the miniaturization of conventional recoding equipment to a size that can be mounted on a small robotic platform and still allows us to perform stable extracellular recordings while the fly is driving the robot.

To test our equipment and demonstrate the feasibility of the approach we recorded the signals of the H1-cell, an LPTC that is arguably one of the best-characterized cells of any motion vision pathway [10-15].

Nearly all the studies of the H1-cell have been performed in stationary flies. The exception is the work of Lewen et al. [16-17]. In these studies, the fly was rotated in the laboratory but also outdoors to stimulate the animal with natural light levels and contrast patterns. What these studies did not report is whether the movement of the animal affected the neural responses of the cell or whether signals from other modalities modified its activity – the latter of which was observed to be the case in later studies on LPTCs [9] and neck motor neurons [18].

Here we present such equipment and show its capability to perform stable extracellular recordings from the H1-cell while either the fly was rotated within a stationary visual pattern or the pattern was rotated around the stationary animal.

2   Method

2.1   System Description

Actuators for Fly and Pattern Rotation
To visually stimulate the H1-cell, we used stepper motors (QSH5718-76-28-189, Trinamic©) to either rotate the fly within a stationary cylinder featuring a vertical grating, or kept the fly stationary while moving the cylinder and thus the grating.

The stepper motors were shielded inside two aluminium diecast boxes (Figure 1B), together with Astrosyn© P808A stepper motor drivers and 48V DC power supplies. The drivers received only input pulses buffered by a customized external 5V Schmitt trigger circuit. Both actuator boxes were identical in terms of their internal structure. To monitor the angular position of the blowfly, a potentiometer was fitted on the shaft of the bottom stepper motor. Input voltage for the potentiometer was supplied by a 9V PP3 battery and the output voltage was sampled by a data acquisition card.

Miniaturized Electrophysiology Platform

The mobile electrophysiology platform was machined from aluminium to keep the weight low. It contains six components: one round chassis plate, a central rod as a fly holder, two hollow micromanipulator holders and a fixing bracket for holding these elements in place. The micromanipulator holders work as faraday cages for shielding a custom-designed extracellular recording amplifier. All metal parts are connected to the amplifier ground for electro-magnetic noise reduction. The length of the central rod is the same as the radius of the round chassis to provide the blowfly with an unobstructed fronto-vental visual field down to -45 degree elevation when mounted on the platform. This includes regions of the H1-cell receptive field where the motion sensitivity extends down below the equator [12].

A small MM-1 XYZ triple stage micromanipulator (Intracel©) is bolted on top of each micromanipulator holder. A custom designed electrode clip is screwed on top of the micromanipulator, electrically isolated to the electrode, connected to the amplifier ground, when mechanically clipping the electrode. The electrode is wired to one of the differential amplifier inputs as signal.

A dedicated fly holder is made from two pieces of header pins, with 2.54 mm industrial standardized gap in between pins. The left pin is soldered to a thin wire, connected to the other differential input of the amplifier as reference. It is in contact with the fly hemolymph, which is bathing the brain.

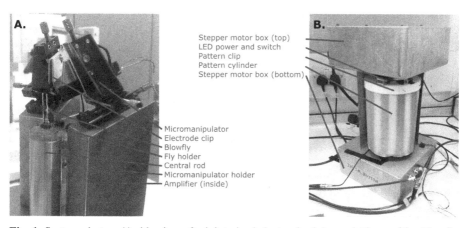

Fig. 1. System photos. A) side view of miniaturized electrophysiology platform with a blowfly held in position. B) the complete setup of the system, the stimulus (pattern cylinder) is placed around the miniaturized electrophysiology platform, both are driven by stepper motors from top and bottom.

Extracellular Recording Amplifier

The amplifier (Fig. 2) is designed for the purposes of compact size, low power consumption and high gain. The size of the amplifier PCB is 18 mm x 30 mm, small enough to be enclosed into the hollowed micromanipulator holder on the platform, for electrical shielding. The circuit is powered by 5V single supply, which is regulated from a 9V PP3 battery. Low power consumption commercial amplifier chips (INA2332 and AD8607) are used, and the gain is configured to 10K overall. The amplifier bandwidth is set by the intrinsic bandwidth of the commercial chips (INA2332: up to 300 KHz, AD8607: up to 400 KHz, for each at gain of 100, from specifications) which is wider than the bandwidth of an H1-cell action potentials.

2.2   Animal Preparation

Female blowflies, *Calliphora vicina*, from 4-11 day old, were used to perform the experiment taken from the laboratory stock (blowflies are kept under a 12 hour light / 12 hour dark cycle). Legs and proboscis were removed to reduce movements of any appendices and wounds were sealed with bee wax to prevent desiccation. Wings were immobilized by placing bee wax at the hinges to reduce vibration due to flight motor action. The back of the fly head was aligned with two pins, and oriented frontally based on the pseudopupil method [19]. The thorax was then pushed down and waxed to the sides of the pins to expose the back of the head capsule. Under 40 x magnification of a stereo microscope (Stemi 2000, Zeiss©), the cuticle is removed from the rear head capsule on both sides. Fat and muscle tissue were removed on the right hand side of the head capsule, so the web from tracheae becomes visible. Ringer solution was used to keep the tissue moist (for recipe see Karmeier et al. [20]).

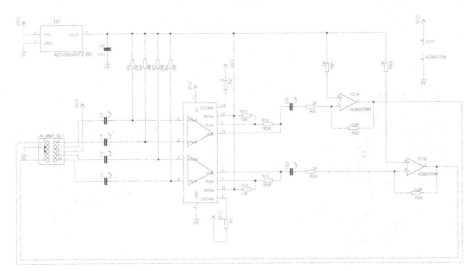

Fig. 2. Schematics of customized extracellular recording amplifier. The amplifier is using 0-5V single supply, for low power. The reference point is set at 2.5V by an accurate regulator AD1582. Differential signals are pre-amplified by INA2332 with a gain of 100 and then amplified at a second stage by AD8607, again with gain of 100.

2.3   Electrophysiology

We used tungsten electrodes for extracellular recording from H1-cell in the right lobula plate (3 MΩ tungsten electrodes, FHC Inc., Bowdoin, ME, USA). The pin used to fix the fly on the left side is connected to the brain tissue across the saline bridge. Both, the recording electrode and the pin for fixing the fly were wired to the differential inputs of the customized amplifier. The signal was amplified 10K times, and then fed into a data acquisition board (NI USB-6215, National Instruments Corporation, Austin, TX, USA), which sampled the data at 20KS/s by a python script (https://github.com/Blueasteroid/Blowfly-robot-interface/ py_H1_calibration_3s_stim.py).

We positioned the signal electrode next to the H1-cell axon on the side of the brain that is contralateral with respect to its dendric input arborizations. The H1-cell increases its spike rate when experiencing contralateral back-to-front motion (preferred direction) and is inhibited during motion in the opposite direction (anti-preferred direction [21]. Recordings were only accepted to be at sufficient quality when the amplitude of the recorded H1-cell action potentials was at least 2 times higher than the largest amplitudes of the background noise, i.e. signal noise ratio (SNR) > 2:1.

2.4   Stimulation

We designed a pattern clip that was manufactured on a 3D printer and mounted on the shaft of the stepper motor that moved the pattern cylinder. The pattern clip is based on a 156 mm diameter disk, where the circumference is 492 mm. Black and white stripes were printed on A4 paper, concatenated and trimmed to 492 mm, and then fitted on the pattern clip to build the cylinder. A spatial wavelength of 30 degree was used for the stimulus pattern that consisted of black and white vertical stripes lined up inside the cylinder.

Three LEDs were installed inside the cylinder, on the top, driven by a 9V PP3 battery. During the experiments the light in the laboratory was switched off and the room was dark. After switching on the LEDs, we tested whether the H1-cell would respond to any movements performed outside the cylinder – which it didn't.

The stimulus was controlled by a DAQ card (NI USB-6211) via the python script. The script controlled pulse sequences as input to stepper motor drivers, generating constant angular velocity and sinusoidal acceleration/deceleration of the pattern cylinder or blowfly recording platform. The sinusoidal acceleration/deceleration is required to reduce the vibration generated by sudden velocity changes.

A single trial of motion stimulation was set to 10 seconds segmented into: 0.5s no motion, 3s null direction motion (ND stimulation => inhibiting spiking in the H1-cell), 3s preferred direction motion (PD stimulation => increasing the spike rate in the H1-cell) and 3.5s no motion. In the case of recording the right contralateral H1-cell, ND motion means count-clockwise rotation of pattern or clockwise rotation of the blowfly, viewed from above, and vice versa for PD motion. ND motion and PD motion were driven by the same constant angular velocity. In order to measure temporal frequency tuning curves, we systematically varied the angular rotation of the stimulus pattern, which – divided by the fixed spatial wavelength of the pattern – resulted in a temporal frequency range from 1 Hz to 10 Hz. Different temporal frequencies were applied in a shuffled order within two blocks where the sequence in the second block was inverted. Each experimental trail was repeated 10 times each for rotations of the fly and the pattern cylinder, respectively.

3   Result

3.1   Recording Stability: Data Analysis in Time Domain

The first test of the system addresses the question whether the custom amplifier circuit is suitable for recordings of extracellular neural signals. Fig. 3A shows a trace of the stimulus pattern velocity, with the response of the H1 cell shown in Fig. 3B. Background noise in the signal can be distinguished from the large action potentials (spikes) which are converted into unit pulses when their amplitude exceeds a set threshold voltage (the red line in Fig. 3B). Action potentials are shown at higher time resolution in Fig. 3C. The signal to noise ratio is high enough to separate H1-cell action potentials without using a spike sorting algorithm. Activity in the cell (in t=0.5-3.5s) is inhibited by front-to-back motion and increased (in t=3.5-6.5s) by motion in the opposite direction. During periods with no pattern motion, the cell is spontaneously active. The response properties shown are in agreement with those previously observed in H1-cell recordings [21].

The second test of the system was concerned with the mechanical stability of the recording equipment which, due to positional changes of the signal electrode relative to the recorded cell, may have an impact on the recording. The quality of an extracellular neural recording depends on the distance between the electrode tip and the neuron and is affected by micrometer movements. Vibrations induced by the motor, for example, could therefore cause the SNR of the recording to degrade, or even damage the neuron.

Fig. 3D shows the instantaneous spike rate during a recording from the H1-cell. The spike rate calculation was based on the inter spike interval (ISI) which corresponds to the inverse of the time interval between consecutive spikes. Pattern motion in the H1 cell's preferred direction initially increased the ISI spike rate to 200 Hz which then adapted to around 100 Hz while the stimulus pattern was still moving. When the pattern motion was stopped the spike rate returned close to the spontaneous activity level. Throughout the recording the rate stays below 400 spikes per second, which indicates that the stimulus caused responses within the physiologically plausible range. The consistent shape of consecutive spikes clearly indicated that we recorded from one individual neuron only, rather than from two or more separate neurons.

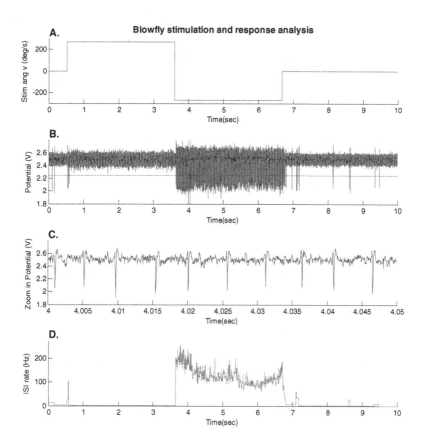

Fig. 3. Stimulus and response over time in an individual experiment. A) Stimulation angular velocities (positive angular velocity corresponds to anti-preferred direction of H1-cell, negative angular velocity corresponds to preferred direction). B) H1-cell action potential signal with threshold (red line) and zoom in area (between black lines). C) Temporal "zoom-in" on a H1-cell action potential sequence. D) Time-continuous spike rate of the H1-cell, computed from the time interval between two consecutive spikes (ISI).

3.2   H1-Cell Status: Spontaneous Spike Rate Monitoring

An additional quality criterion is the ability to maintain a stable recording over an extended period. To test long term stability of our recordings we performed an experiment 200 times longer than the one shown in Figure 3. In Figure 4 we present the peak-rate and spontaneous firing rate of H1-cell obtained in such an experiment. The top panel (Fig. 4A) shows the stimulus trace used. During the first 1000 seconds of the recording the fly was rotated while the background pattern was kept stationary.

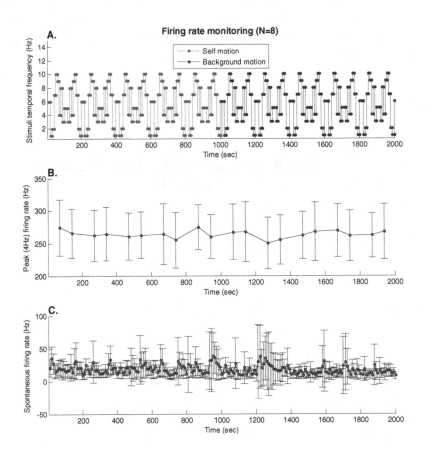

Fig. 4. Spike rate monitoring. A) Temporal frequencies of stimulations of a complete experiment for one animal. B) Mean and standard deviation of peak firing rate at 4Hz temporal frequency stimulation. C) Mean and standard deviation of spontaneous firing rate after PD stimulation of each trial.

For the next 1000 seconds the fly was kept stationary while the pattern was rotated. The peak firing rate in response to a 4Hz temporal frequency stimulus is presented in Fig. 4B. The data show a similar mean and variability of the firing rate for the entire duration of the experiment, indicating sufficient stability of the recording. We also examined the spontaneous firing rate, averaged over a 3.5 seconds time interval after stimulus onset, to see whether or not the H1-cell is damaged by vibrations due to the action of the step motor. In the case of damage by the electrode tip, the cell would be expected to fire constantly at maximum rate as sodium ions are free to enter the cell and cause further opening of voltage-gated sodium channels resulting in a massive depolarization of the cell membrane. Alternatively, zero spike rates may also indicate damage to the cell or excessive movement of the electrode tip away from the cell.

The mean spontaneous activity was observed to remain below 50 Hz throughout the experiment (Fig 4C), which is consistent with previous recordings from healthy H1 neurons [13].

3.3   Differences between Self Motion and Background Motion

Fig.5 presents the response of the H1-cell in 2 different flies (A and B) to visual motion at different temporal frequencies, induced either by rotation of the fly (fly rotation) or of the pattern around the fly (pattern rotation).   The peak response of the cell

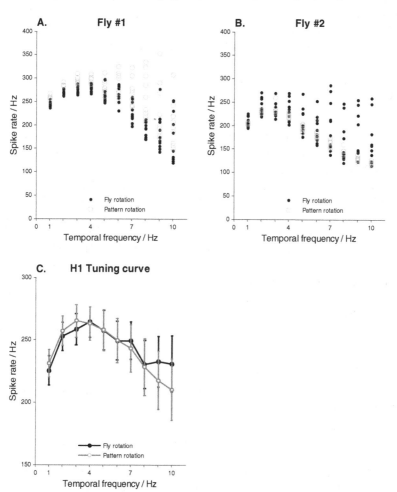

Fig. 5. Self motion and background motion comparison. The mean spike rate plotted against the temporal frequency of the stimulus. A-B) Trials of two different animals. Each dot represents the average spike rate obtained for PD stimulation at a given temporal frequency. C) Mean temporal frequency tuning curve averaged across animals (N=8, n=10) as a result of rotating the fly (filled symbols) and rotating the pattern (open symbols), respectively.

in both flies is found to be between 3-4 Hz, regardless of the method of stimulation, which is consistent with previously published data [13][17]. The response drops with increasing temporal frequency, with around half the peak spike rate observed at 10 Hz. The H1-cell in these two animals were found to behave differently at high frequencies, depending on whether self-motion or background motion was employed; in Fig. 4A, the variability in the response is high when the pattern is rotated. The data presented in Fig. 4B shows the opposite: a higher variability of response to rotations of the fly at high frequencies.

The mean response averaged over 8 flies (and 10 repetitions of stimulation of each fly) is shown in Fig. 4C. No clear differences between the responses to the two types of motion were apparent. Though we found a slight increase in the mean spike rate at 10 Hz when the fly was rotated, this difference was not statistically significant, due to the high variability of the responses under both stimulation regimes.

4   Discussion

In this paper, we achieved two tasks: 1. Validate a miniaturized electrophysiology platform for H1-cell recording while the animal is moving. 2. Systematically study whether H1-cell responses induced by rotating the fly or by rotating the pattern result in different responses.

The result of stimulus-induced and spontaneous spike rate of the H1-cell suggest that the miniaturized electrophysiology platform is performing well and allows for stable recordings over long time intervals. The rigid platform structure guaranteed very little positional slip between cell and electrode as can be inferred from a stable spike amplitude distribution throughout the experiments. The electrode did neither damage the target cell, nor did it slip away which would have meant a dramatically decreased value of the spike amplitudes.

In terms of electrical shielding, the aluminium body of the micromanipulator holders proved to be an efficient way of eliminating most of the electromagnetic noise, from 50 Hz main power noise to high power, high frequency stepper motor noise, which is one of the most essential working conditions of the high amplification application. The customized on-board extracellular recording amplifier has achieved the designed 10K gain with sufficient bandwidth for recording H1-cell signals. The low power feature of the amplifier enabled the whole system to be supplied by a single 9V PP3 battery, which, in the future, will benefit closed-loop studies using an autonomous robot platform [22].

The next step will be to study multisensory integration. We have been recording the H1-cell signal from the fly in two circumstances: first, in rotated flies within a stationary visual pattern and second, in a fixed fly within a rotating visual pattern. The results were nearly identical over most of the dynamic range tested and with respect to both the mean spike rate and the variability of the data. Only at high oscillation frequencies the data obtained in rotated flies assumed slightly higher values compared to the data gathered during pattern rotation. A possible explanation would be that during fly rotation, mechanoreceptive systems may be stimulated and unspecifically modify the activity of the network of horizontal motion sensitive neurons, which ultimately exerts an effect on the spiking in the H1-cell. Along those lines it would make sense that at high oscillation frequencies, where activation of mechanosensory systems would yield the strongest effect, the responses became higher. One potential explanation for an increased variability could be linked to the fly's attempt to change its locomotor state – e.g. trying to escape. In these situations the octopamine level in the hemolymph would increase, causing a boost in the processing of visual motion stimuli [23].

In future studies we will establish a closed-loop fly-robot interface and focus on multisensory integration. Different sensory modalities will be selectively blocked or switched off for comparison of the H1-cell responses in fully intact animals. The halteres, for instance, will be the first choice for blocking. Another identified LPTC to record from would be the spiking V1-cell, which has previously been shown to reflect stimulation of the ocelli [9] and might possibly receive indirect input from the haltere system via an ascending pathway [24]. The fly-robotic interface will be tested at higher angular velocities, to observe H1-cell activity during saccade-like turns [25]. Later, we will perform dual channel recordings on the platform, to see the mutual interaction between both H1-cells [26] or the H1-cell and other spiking neurons, while the system rotates and translates freely. Hopefully, soon, we'll see a fly steering a robot.

Acknowledgments. We'd like to thank Kit Longden, Martina Wicklein, Ben Hardcastle, Kris Peterson, Naveed Ejaz, Daniel Schwyn, Pete Swart for all the helps, experience sharing and discussion on the work presented. This work is partially supported by a research grant US AFOSR to HGK.

References

1. Bomphrey, R.J., Walker, S.M., Taylor, G.K.: The Typical Flight Performance of Blowflies: Measuring the Normal Performance Envelope of Calliphora vicina Using a Novel Corner-Cube Arena. PLoS ONE 4(11), e7852 (2009)
2. Taylor, G.K., Krapp, H.G.: Sensory systems and flight stability: What do insects measure and why? Adv. Ins. Phys. 34, 231–316 (2007)
3. Krapp, H.G.: Ocelli. Current Biology 19(11), R435–R437 (2009)
4. Dickinson, M.H.: Haltere-mediated equilibrium reflexes of the fruit fly, Drosophila melanogaster. Philos. Trans. R. Soc. Lond. B. Biol. Sci. 354(1385), 903–916 (1999)
5. Matsuo, E., Kamikouchi, A.: Neuronal encoding of sound, gravity, and wind in the fruit fly. J. Comp. Physiol. A. Neuroethol. Sens. Neural Behav. Physiol. 199(4), 253–262 (2013)
6. Strausfeld, N.J.: Atlas of an Insect Brain. Springer, New York (1976)
7. Krapp, H.G., Wicklein, M.: Central processing of visual information in insects. In: Masland, R., Albright, T.D. (eds.) The Senses: a Comprehensive Reference, vol. 1, pp. 131–204. Academic Press (2008)

8. Strausfeld, N.J., Seyan, H.S.: Convergence of visual, haltere, and prosternai inputs at neck motor neurons of *Calliphora erythrocephala*. Cell and Tissue Research. 240, 601–615 (1985)

9. Parsons, M.M., Krapp, H.G., Laughlin, S.B.: A motion-sensitive neurone responds to signals from the two visual systems of the blowfly, the compound eyes and ocelli. Journal of Experimental Biology 209, 4464–4474 (2006)

10. Huston, S.J., Krapp, H.G.: Visuomotor transformation in the fly gaze stabilization system. PLoS Biol. 22 6(7), e173 (2008)

11. Hausen, K.: Functional characterization and anatomical identification of motion sensitive neurons in the lobula plate of the blowfly *Calliphora erythrocephala*. Z. Naturforsch 31, 629–633 (1976)

12. Krapp, H.G.: Neuronal matched filters for optic flow processing in flying insects. Int. Rev. Neurob. 44, 93–120 (2000)

13. Maddess, T., Laughlin, S.B.: Adaptation of the motion-sensitive neuron H1 is generated locally and governed by contrast frequency. Proc. R. Soc. Lond. B 225, 251–275 (1985)

14. Borst, A., Haag, J.: Neural networks in the cockpit of the fly. J. Comp. Phys. A 188, 419–437 (2002)

15. Longden, K.D., Krapp, H.G.: Octopaminergic modulation of temporal frequency coding in an identified optic flow-processing interneuron. Front. Syst. Neurosci. 4, 153 (2010)

16. Lewen, G.D., Bialek, W., de Ruyter van Steveninck, R.: Neural coding of naturalistic motion stimuli. Network 12, 317–329 (2001)

17. Nemenman, I., Lewen, G.D., Bialek, W., de Ruyter van Steveninck, R.R.: Neural Coding of Natural Stimuli: Information at Sub-Millisecond Resolution. PLoS. Comput. Biol. 4(3), e1000025 (2008)

18. Huston, S.J., Krapp, H.G.: Nonlinear Integration of Visual and Haltere Inputs in Fly Neck Motor Neurons. J. Neurosci. 29(42), 13097–13105 (2009)

19. Franceschini, N.: Sampling of the visual environment by the compound eye of the fly: fundamentals and applications. In: Snyder, A.W., Menzel, R. (eds.) Photoreceptor Optics, pp. 98–125. Springer (1975)

20. Karmeier, K., Tabor, R., Egelhaff, M., Krapp, G.H.: Early visual experience and the receptive-field organization of optic flow processing interneurons in the fly motion pathway. Vis. Neurosci. 18, 1–8 (2001)

21. Krapp, H.G., Hengstenberg, R., Egelhaaf, M.: Binocular contributions to optic flow processing in the fly visual system. J. Neurophysiol. 85(2), 724–734 (2001)

22. Ejaz, N., Peterson, K., Krapp, H.: An experimental platform to study the closed-loop performance of brain-machine interfaces. J. Vis. Exp. 10(3791) (2011)

23. Longden, K.D., Krapp, H.G.: State-dependent performance of optic-flow processing interneurons. J. Neurophysiol. 102(6), 3606–3618 (2009)

24. Haag, J., Borst, A.: Electrical coupling of lobula plate tangential cells to a heterolateral motion-sensitive neuron in the fly. J. Neurosci. 28(53), 14435–14442 (2008)

25. Kern, R., van Hateren, J.H., Michaelis, C., Lindemann, J.P., Egelhaaf, M.: Function of a fly motion-sensitive neuron matches eye movements during free flight. PLoS Biol. 3(6), e171 (2005)

26. Weber, F., Machens, C.K., Borst, A.: Disentangling the functional consequences of the connectivity between optic-flow processing neurons. Nat. Neurosci. 15(3), 441–448, S1–S2 (2012)

Property Investigation of Chemical Plume Tracing Algorithm in an Insect Using Bio-machine Hybrid System

Daisuke Kurabayashi[1], Yosuke Takahashi[1], Ryo Minegishi[1], Elisa Tosello[2], Enrico Pagello[2], and Ryohei Kanzaki[3]

[1] Tokyo Institute of Technology, Tokyo, 152-8552, Japan
dkura@irs.ctrl.titech.ac.jp
http://www.irs.ctrl.titech.ac.jp
[2] IAS-Lab, Dept. of Information Engieering, University of Padua, 35131 Padua, Italy
[3] Research Center for Advanced Science and Technology, The University of Tokyo,
Tokyo, 153-8904, Japan

Abstract. In this study, we investigated an aspect of the chemical plume tracing behavior of an insect by using a bio-machine hybrid system. We implemented an experimental system by which an insect brain was connected to a robot body. We observed th neural responses to external disturbances and transitions at changes in the motor gain of the robot body. Based on the results of the experiments, we identified a simple control model for the angular velocity of the behavior. We subsequently investigated the effect of the rotational velocity by using information entropy in computer simulations.

Keywords: chemical plume tracing, bio-machine hybrid system, silkworm moth.

1 Introduction

In this study, we investigated an aspect of an insect's behavior locating a chemical source. In air, chemicals form several plumes rather than a smooth gradient. Locating a chemical source by following these plumes is known as the chemical plume tracing (CPT) problem [1][2]. CPT is potentially important for artificial systems because it can be applied to the location of the source of pollution, finding people trapped under debris after large earthquakes, etc. Because the dynamics of chemicals in the atmosphere are quite complex, we found it beneficial to investigate the adaptive behaviors exhibited by an animal to solve the CPT problem.

Animals effectively use CPT for foraging, mating, localizing, etc. In contrast to robots, most animals, including insects, have the ability to solve CPT problems. In this study, we investigated the CPT performance of insects by using an insect brain machine-interface system, called a bio-machine hybrid system. We then considered the effects of motion parameters by using computer simulations and information entropy.

N.F. Lepora et al. (Eds.): Living Machines 2013, LNAI 8064, pp. 131–142, 2013.
© Springer-Verlag Berlin Heidelberg 2013

2   Instruments and Methods

2.1   Silkworm Moth *Bombyx mori*

In this study, we employed an adult male silkworm moth, Bombyx mori (Fig. 1). Silkworm moths have more than 5,000 years of history as domestic insects, and their genome [3], nervous system [4], and behaviors [5] have been thoroughly investigated.

An adult silkworm moth does not drink, eat, or make any voluntary movements. The only thing it does is attempt to locate a female for mating using the pheromones that she releases. Thus, the correspondence between input stimuli and exhibited actions is very clear. This characteristic of the silkworm moth makes it a suitable subject for CPT research.

Fig. 1. Adult male silkworm moth *Bombyx mori*

A silkworm moth has a set of actions, called programmed behavior, which it uses to locate a female [5]. As shown in Fig. 2, it walks straight when it receives a stimulus. This action is called *surge*. It then changes its walking direction from left to right as if it is looking around (*zigzag*). Finally, it continues turning for a period of time (*loop*). If the moth receives another stimulus during this programmed behavior, it restarts the sequence from *surge*.

Kanzaki et al. reported the typical parameters of this programmed behavior [6]. A moth does not move until it receives pheromone stimuli. During *surge*, it walks for 0.5 [s] at 26 [mm/s]. A *zigzag* procedure typiaclly contains three turns, whose durations are 1.2 [s], 1.9 [s], and 2.1 [s]. The angular velocity during *zigzag* or *loop* is 1.0 [rad/s].

The research says that a moth arranges its programmed behavior adaptively based on environmental conditions. The CPT behavior is understood qualitatively based on previous research, but how the moth arranges its actions according to its environment remains unclear. Therefore, we built a system to observe a moth's behaviors and neural activities simultaneously.

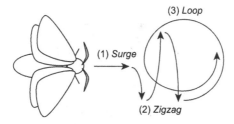

Fig. 2. Programmed behavior of silkworm moth

2.2 Bio-machine Hybrid System

We developed an experimental system called a bio-machine hybrid system [7]-[9], in which a living brain controls a robot body. This system can be categorized as a closed-loop experimental platform (i.e. [10][11]).

Fig. 3 illustrates the implemented hybrid system. At the front of the body, the system is equipped with a chamber where the head of a living moth is set in wax. Through micro glass electrodes, neural signals containing motion commands are sent to a micro controller on the robotic body and are translated into wheel movements. Using this system, we can observe the neural activities corresponding to the behaviors on line.

Although it has wings, a silkworm moth cannot fly. According to the observed trajectories, to simplify the problem, we applied the kinematic model of a two-wheeled mobile robot to that of a moth. In other words, we used a two-wheeled mobile robot as the body of the experimental system. By observing the neural signals of the neck motor neurons at the second cervical nerve b (2nd CNb)[9][12], we counted the firing ratio of neurons and translated them into motion commands for the left and right electric motors. Note that the 2nd CNb consists of 5 motor neurons and innervates the neck muscles, which contract during horizontal side-wise head movement[13][14]. Based on the previous works [9],[12]-[14], we measured all of the activities of these 5 neurons. We had already obtained some results for locating a pheromone source in a simplified environment by using the experimental system [15][16].

2.3 Preparation of Moths

Moths were prepared as follows. After cooling at $4°$ C for 30 [min] to achieve anesthesia, we removed the abdomen, all of the legs, the dorsal part of the thorax, and the wings. Owing to the operation method, the moth was mounted ventral-side-up in a chamber of the robotic body. The ventral part of the neck was opened to expose the cervical nerves and ventral nerve cords. We employed two micro glass electrodes to record and observe the neural activities of the 2nd CNb on both the left and right sides. For this purpose, we filled the chamber with normal saline and covered it with wax.

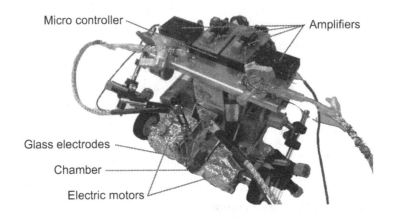

Fig. 3. Proposed bio-machine hybrid system

3   Experiments by Using Bio-machine Hybrid System

We conducted some experiments in order to investigate the properties of the control system in a brain of a moth with the robotic body.

First, we observed the compensations against external disturbances. We introduced unexpected motions to the hybrid system, and then observed the neural responses from the left and right 2nd CNb neurons. We tested the sensitivity to two types of disturbances, rotational and translational (Fig. 4). We gave the robot forced movements, yaw rotations and forward-and-backward translations, and recorded the neural responses. In these experiments, we observed clear responses in the 2nd CNb to the rotational disturbances, but not to the translational ones. We also found that the neural responses were proportional to the angular velocity of the imposed movement. Therefore, in this study, we focused on the rotational disturbances and the identification of a control system for the angular velocity in a moth.

Fig. 5(a) shows a possible block diagram for the hybrid system. In our current system, we could not measure the intention of a moth, which is expressed as r in the figure. Thus, we only focused on the exhibited angular velocity. We considered a transfer function from the external noise to the output. Therefore, we simplified the diagram, as shown in Fig. 5(b), and identified the feedback transfer function, which is expressed as PCS in the figure.

In order to simplify the problem, we employed a linear output error model (1), where q is the shift operator.

$$y(k) = \frac{b_1}{1 + f_1 q^{-1} + f_2 q^{-2}} d(k) + w(k) \tag{1}$$

For identification, we assumed that w is white noise. We identified the parameters for the model in two cases: the robot body (P) with normal motor gain and

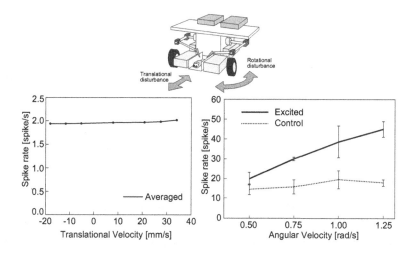

Fig. 4. Neural response to disturbances

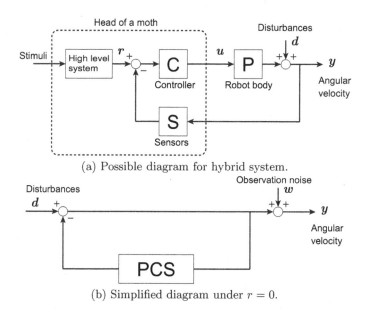

(a) Possible diagram for hybrid system.

(b) Simplified diagram under $r = 0$.

Fig. 5. System expressions for identification

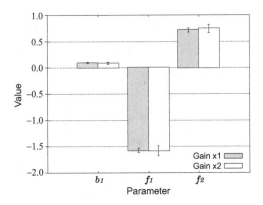

Fig. 6. Results of identification

body with twice the normal gain. The results shown in Fig. 6 indicate that the parameters were the same for the two cases. From these results, we considered that the parameter values in a natural moth were important.

We then conducted additional experiments. During the CPT movements, we changed the gains to rotate the electric motors on the robot body. We prepared a wind tunnel, as illustrated in Fig. 7. The bio-machine hybrid system was set 600 [mm] away from the nozzle of a pheromone injector.

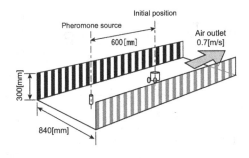

Fig. 7. Wind tunnel for experiments

At the beginning of the experiment, the normal gain was used for the hybrid system. After 10 [s] of CPT movements, we instantly doubled the motor gain for another 10 [s]. We then restored the original motor gain for an additional 10 [s].

Although we changed the motor gain, the hybrid system worked well, as if it adapted to the changes in motor gain, and arrived at its goal position. As shown in Fig. 8, the moth arranged the firing ratio of the neurons. According to the results, we considered that a moth regulates its motor responses. We then attempted to evaluate the effectiveness of the value of rotational velocity in the modeled behavior [6] by using a simplified artificial model.

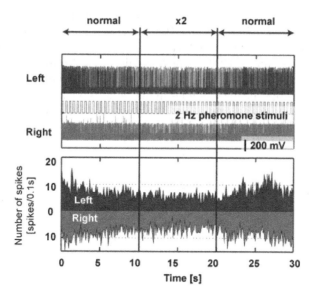

Fig. 8. Adaptation of neural activities to changed motor gain

4   Evaluation of Rotational Velocity

4.1   Information Entropy

According to the experiments, we hypothesized that the values of the rotational velocity in the behavior of a moth were adjusted appropriately. In [6], the rotational velocity of a natural moth was reported to be approximately 1.0 [rad/s]. Thus, we attempted to evaluate the effectiveness of the rotational velocity in CPT.

To validate what was obtained from previous experiments, the entropy of the system was studied. If the behavior of a silkworm moth was correct, the entropy had to reflect the statements made in [17] and [18]: its value had to decrease as the moth searched for the location of the pheromone source. The definition of entropy S for a probability distribution $P(r_j)$ is as follows:

$$S = \sum_j P(r_j) \ln P(r_j) \qquad (2)$$

where r_j is the pheromone source location.

Formula 2 intuitively shows that the information entropy is large when the probability density distribution spreads uniformly: in fact, the maximum value of S is $\log N$, corresponding to the situation in which the same probability characterizes the whole field. This means that the entropy has a large value when little information is known about the goal position, whereas the entropy is small when the approximate goal position is known. Mathematically, $S = 0$ if and only if there is no uncertainty in the field, i.e., the location r_j of the source is known. In this case $P(r_j) = 1$ and $P(r_i \neq r_j) = 0$ for all other events.

In order to study the rate of reduction of the entropy in the estimated probability distribution during CPT, a Java simulator was developed. This system used the same physical distribution model used in [17], because we did not have a concrete model of the chemical plumes in our wind tunnel. It was composed of a moth, initially situated 500 [mm] away the pheromone source, and a pheromone source located at position r_o $(0,0)$ (Fig. 9).

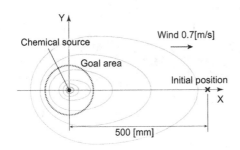

Fig. 9. Overview of the simulation field

The source emits detectable particles at a rate $R = 1.0$ [mm^2/s], and every stimulus has a finite lifetime $\tau = 150$ [ms], is propagated with an isotropic effective diffusivity $D = 1.0$ [mm^2/s], and is advected by a wind current characterized by a speed of approximately 0.7 [mm/s]. The wind has been taken to blow in the negative X-direction. The moth reaches the source if it reaches a goal area defined as a circle with a radius of 100 [mm] from the pheromone source. In this kind of system, the agent was modeled as a spherical object with a small linear size $a = 0.01$ [mm]. It should be rememberd that the system has the purpose of simulating the bio-machine hybrid system, so the agent represents the system and not the real moth. However, moving into such media the agent will experience a series of encounters at rate $R(r|r_0)$:

$$R(r|r_0) = \frac{R}{\ln \frac{\lambda}{a}} e^{\frac{(x_0-x)V}{2D}} K_0(\frac{|r - r_0|}{\lambda}), \tag{3}$$

where

$$\lambda = \sqrt{\frac{D\tau}{1 + \frac{V^2\tau}{4D}}} \tag{4}$$

and K_0 is the modified Bessel function of order zero. In detail, $x_0 - x$ corresponds to the length along which the wind blows, and $r - r_0$ is the distance that separates the current location of the agent from the source.

The agent is asked to perform the motion described at the beginning of this section, and the rate $R(r|r_0)$ allows the simulator to compute the position reached by the agent at each moment. An array of positions input to a

Matlab system makes it possible to evaluate the trajectories and calculate the entropy of the system at every point of the map. At every point, the system will calculate the probability distribution and, as a consequence, the entropy of the system. At the beginning, the estimated probability of the location of a chemical source will be distributed uniformly over the entire space; thus, all of the points will have equal probabilities. Then, this distribution will vary over time with the same trend as a Gaussian function with expected value μ and variance σ, formulated according to the distance from the pheromone source:

$$P(r|r_0) = \frac{\frac{1}{\sqrt{2\pi\sigma^2}}e^{-\frac{(r-\mu)^2}{2\sigma^2}}P(r|r_0)}{\sum_i \frac{1}{\sqrt{2\pi\sigma^2}}e^{-\frac{(r_i-\mu)^2}{2\sigma^2}}P(r_i|r_0)} \tag{5}$$

4.2 CPT Performance

We applied rotational velocities of 0.5 [rad/s], 1.0 [rad/s], 2.0 [rad/s], 3.0 [rad/s], 4.0 [rad/s], and 5.0 [rad/s] to the simulations. We conducted 20 trials for each velocity.

These trials included a search time upper bound equal to 300 [s]. It has been shown experimentally that this leads to an optimal tradeoff between exploration and exploitation. In fact, on average, in 70 % of the cases, the moth reaches the goal area within this time.

Using an upper bound of 600 [s] leads to reaching the source in all 20 samples, but an increase in the bound is not proportional to the improvement obtained: doubling the time results in an improvement of only +30 %. Likewise, decreasing the bound results in drastic reduction in performance: only 40% of the agents reach the source if the upper bound is decreased by 60 [s].

The transition of the information entropy observed is shown in Fig. 10. Figure 11 also shows the ratio to reach the goal area and the average search time.

The values plotted in Fig. 10 are the averages calculated after 20 tests, and the different lines correspond to different angular velocities. The entropy decreases over time as the agent gets closer to the source. It has a maximum value when the agent is located at the farthest position and a minimum value when the moth reaches the goal.

The minimum search time, as shown in Fig. 11, is that corresponding to an angular velocity of 0.5 [rad/s]. However, in that case, the success ratio was quite low. In Fig. 10, we can find that the moth agent did not survey the environment because the information entropy in that case decreased slowly. Thus, if the moth agent reached the goal area by chance, it spent a shorter amount of time. In the case of 1.0, the moth agent performed well, with a high success ratio and short searching time (Fig. 11). In this case, the information entropy shown in Fig. 10 decreased quickly and achieved a low value. We consider that the moth agent in this case did well both in surveying the environment and reaching the pheromone source. If a moth can move at 4.0 [rad/s] or more, it will achieve the highest success ratio, but it needs a longer time than that of other cases.

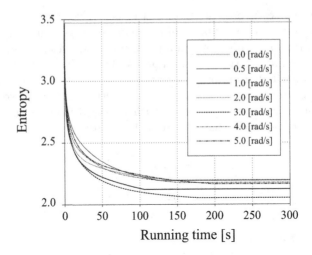

Fig. 10. Information entropy during CPT behavior with different values of rotational velocity

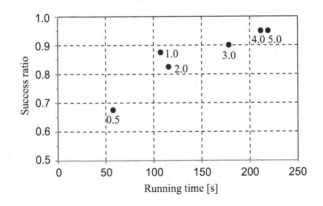

Fig. 11. Success ratio and search time. The values indicate the corresponding rotational velocity.

The best balance among the success ratio and searching time seemed to be 1.0 [rad/s], confirming the statement made by Kanzaki et al [6].

From the perspective of information entropy, the behavior of a silkworm moth, including the parameter settings obtained by our experimental system, was well adapted for achieving CPT.

5   Conclusions

In this study, we investigated the chemical plume tracing (CPT) behavior of an insect by using a bio-machine hybrid system. We implemented an experimental system by which an insect brain was connected to a robotic body. We observed the neural responses to external disturbances and adaptation to changes in the motor gain of the robotic body. Based on the results of experiments, we carried out the simple identification of a control model for the angular velocity of the behavior. We then investigated the effect of the rotational velocity by using the information entropy in computer simulations.

Based on the results of the experiments, we hypothesized that the values of the parameters in the programmed behavior of a moth were regulated as the best values. We then attempted to evaluate the effectiveness of the parameters by using computer simulations with a simple behavior model of the bio-machine hybrid system. From the perspective of information entropy, the behavior of a silkworm moth including the parameter settings obtained by our experimental system was well adapted for achieving CPT.

Acknowledgement. This study was partially supported by Grants-in-Aid for Scientific Research, MEXT Japan 25420212, JSPS Japan 09J01188 and 12J10557. The experiments in this presentation were approved based on Article 22 clause 2 of the safety management rule of Tokyo Institute of Technology.

References

1. Farrell, J.A., Pang, S., Li, W.: Chemical Plume Tracing via an Autonomous Underwater Vehicle. IEEE J. Oceanic Eng. 30(2), 428–442 (2005)
2. Naeem, W., Sutton, R., Chudley, J.: Chemical Plume Tracing and Odour Source Localisation by Autonomous Vehicles. J. Navigation 60(2), 173–190 (2007)
3. The International Silkworm Genome Consortium: The Genome of a Lepidopteran Model Insect, the Silkworm Bombyx mori, Insect Biochem. Mol. Bio. 38(12), 1036–1045 (2008)
4. Wada, S., Kanzaki, R.: Neural Control Mechanisms of the Pheromone-triggered Programmed Behavior in Male Silkworm moths Revealed by Double-labeling of Descending Interneurons and Motor Neurons. J. Comp. Neurol. 484(2), 168–182 (2005)
5. Obara, Y.: Bomby mori Mating Dance: an Essential in Locating the Female. Appl. Entomol. Zool. 14(1), 130–132 (1979)
6. Kanzaki, R., Sugi, N., Shibuya, T.: Self-generated Zigzag Turning of Bombyx mori Males during Pheromone-mediated Upwind Walking (Physology). Zool. Sci. 9(3), 515–527 (1992)
7. Kanzaki, R., Ando, N., Emoto, S., Minegishi, R., Kurabayashi, D., Toriihara, S.: Investigation of Adaptive Behavior of Insects through Constructing Bio-machine Hybrid System. Comp. Biochem. Physiol. B 148(3), 350 (2007)
8. Minegishi, R., Takashima, A., Kurabayashi, D., Kanzaki, R.: Study on Sensory Feedback during Odor Searching Behavior of Silkworm moth using a Brain-machine Hybrid System. In: 9th Int. Congr. Neuroethol., vol. 460 (2010)
9. Minegishi, R., Takashima, A., Kurabayashi, D., Kanzaki, R.: Construction of a Brain-machine Hybrid System to Evaluate Adaptability of an Insect. J. Robot. Auton. Syst. 60(5), 692–699 (2012)

10. Emoto, S., Ando, N., Takahashi, H., Kanzaki, R.: Insect Controlled Robot — Evaluation of Adaptation Ability —. J. Robot. Mechatro. 19(4), 436–443 (2007)
11. Martinez, D., Chaffiol, A., Voges, N., Gu, Y., Anton, S., Rospars, J.P., Lucas, P.: Multiphasic *On/Off* Pheromone Signalling in Moths as Neural Correlates of a Search Strategy. PLoS ONE 8(4), e61220 (2013)
12. Minegishi, R.: Neuroethological Study on Making Adaptive Behavior of an Insect using a Brain-machine Hybrid System, Ph.D. dissertation, Dept. Adv. Interdiscip. Stud., The Univ. of Tokyo (2012)
13. Mishima, T., Kanzaki, R.: Coordination of Flipflopping Neural Signals and Head Turning during Pheromone-mediated Walking in a Male Silkworm Moth *Bombyx mori*. J. Comp. Physiol. A 183(3), 273–282 (1998)
14. Kanzaki, R., Mishima, T.: Pheromone-triggered 'flipflopping' Neural Signals Correlate with Activities of Neck Motor Neurons of a Male Moth. Bombyx Mori, Zool. Sci. 13(1), 79–87 (1996)
15. Takashima, A., Minegishi, R., Kurabayashi, D., Kanzaki, R.: Estimation of Feedback System during Programmed Behavior Exhibited by Silkworm Moth. In: The 2nd World Congr. Nature and Biol. Insp. Comput., pp. 92–97 (2010)
16. Minegishi, R., Takahashi, Y., Takashima, A., Kurabayashi, D., Kanzaki, R.: Study of Adaptive Odor Searching Behavior using a Brain-machine Hybrid System. In: XXIV Int. Congr. Entomol., S1013W01 (2012)
17. Vergassola, M., Villermaux, E., Shraiman, B.I.: 'Infotaxis' as a Strategy for Searching without Gradients. Nature 445(7126), 406–409 (2007)
18. Kim, P., Nakamura, S., Kurabayashi, D.: Hill-climbing for a Noisy Potential Field using Information Entropy. Paladyn. J. Behav. Robot. 2(2), 94–99 (2011)

NeuroCopter: Neuromorphic Computation of 6D Ego-Motion of a Quadcopter

Tim Landgraf[1], Benjamin Wild[1], Tobias Ludwig[1], Philipp Nowak[1],
Lovisa Helgadottir[2], Benjamin Daumenlang[1], Philipp Breinlinger[1],
Martin Nawrot[2], and Raúl Rojas[1]

[1] Freie Universität Berlin, Institut für Informatik, Arnimallee 7, 14195 Berlin,
Germany
[2] Freie Universität Berlin, Neuroinformatik, Königin-Luise-Str. 1, 14195 Berlin,
Germany

Abstract. The navigation capabilities of honeybees are surprisingly complex. Experimental evidence suggests that honeybees rely on a map-like neuronal representation of the environment. Intriguingly, a honeybee brain exhibits approximately one million neurons only. In an interdisciplinary enterprise, we are investigating models of high-level processing in the nervous system of insects such as spatial mapping and decision making. We use a robotic platform termed NeuroCopter that is controlled by a set of functional modules. Each of these modules initially represents a conventional control method and, in an iterative process, will be replaced by a neural control architecture. This paper describes the neuromorphic extraction of the copter's ego motion from sparse optical flow fields. We will first introduce the reader to the system's architecture and then present a detailed description of the structure of the neural model followed by simulated and real-world results.

Keywords: neural networks, neuromorphic computation, biomimetics, self-localization.

1 Introduction

For decades, honeybees have been used as a model organism to study navigation. They display a rich set of complex navigational capabilities [4,9]. Bees integrate the path they travelled over extended periods of time and keep a robust approximation of the direction to their hive [16]. Recent experimental evidence suggests that - throughout their foraging carreers - bees form a complex map-like memory [10,11,12]. Intriguingly, this map might tell the bees where landmarks are (in an allocentric coordinate system) as opposed to the simplistic alternative that only route memories are formed and triggered by landmarks. Bees display remarkably robust and flexible navigation. They can switch from outbound to inbound routes depending on environmental conditions, choose different targets to visit and even fly novel shortcuts between known locations.

It is desireable to understand the neuronal mechanisms behind the cognitive processes like mapping and decision making in the honeybee for two main

N.F. Lepora et al. (Eds.): Living Machines 2013, LNAI 8064, pp. 143–153, 2013.
© Springer-Verlag Berlin Heidelberg 2013

reasons: First, on the level of neuronal computation we might discover basic computational motifs and structures responsible for general navigation tasks such as path integration, sensor fusion, state estimation and alike. Second, we want to build robots with similar cognitive capabilities eventually. Recent robotic systems, whether flying or on the ground, already implement methods for simultaneous localization and mapping (SLAM). However, we are interested in the evolution of a neural programming paradigm, which we believe will emerge as an alternative to conventional programming techniques. The paradigm involves three steps: First a complex "brain" is programmed by defining the principal connection scheme of different neural "black boxes" for certain low-level and cognitive functions of a robotic agent. This raw brain is then trained in a simulation that approximates reality sufficiently for the task to be solved and uses techniques as e.g. structural evolution and reinforcement learning to arrive at a brain structure with satisfying performance (e.g. [1]). Then, the matured brain can be copied to the robot that in turn will learn in the real world to fine-tune its brain.

Today, it is still impractical to use complex (spiking) neural networks on autonomous robots since the processing power necessary can not be carried on-board. In the near future, analog neuromorphic hardware (sub-theshold VLSI circuits) will become a powerful alternative to conventional clocked processing units in mobile robots due to their massive parallel processing capabilities and low power consumption [6,13]. Currently, existing analog chips implement only a low number of neurons. The inter-variability of the transistors is still unfeasibly high due to the production process and lacking temperature robustness [14]. However, a great deal of time and financial investments is dedicated to the development of better hardware and industrial production processes. Eventually, neuromorphic hardware will complement conventional processing units in mobile robots.

In an interdisciplinary project with roboticists and neuroscientists ("Project NeuroCopter") we are investigating neural models of honeybees' navigational capabilities and use 3D world simulations and a flying robotic platform to test them. The neural modules are yet simulated in software using Matlab for ANN-models and the neuron simulator IQR [2] for their spiking counterparts. The copter (or the simulated agent in the 3D simulation) sends sensory data to the brain simulation via an interface module that translates numeric data to spikes. The neuro-modules are continuously executed and the activity of pre-defined output neurons is read out and sent back to the real or simulated agents via the interface module. Our long-term goal is to use analog neuromorphic hardware on the copter to reach full autonomy.

This paper presents our approach starting from the mechanics and electronics of our copter, followed by a detailed description of a neuro-module for the computation of the copter's six-dimensional ego motion vector. This module is the first within a computation scheme that represents an initial model of landmark mapping and localization refinement as depicted in Figure 1. We will conclude by discussing future prospects, new neuro-modules and new models of complex insect navigation.

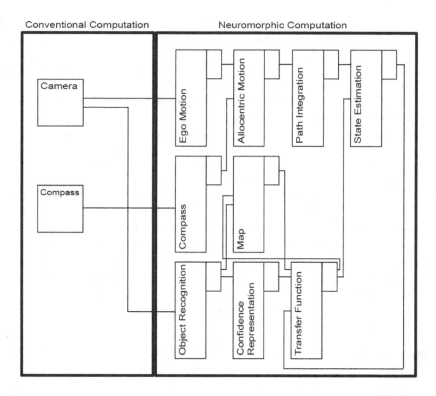

Fig. 1. This figure depicts a connection scheme modeling a central aspect of honeybee navigation: landmark mapping and self-localization based on known landmarks. The connection scheme is composed of neuro-modules, each receiving input at the left side of the respective box and each having a set of output neurons, symbolized by the smaller box to the right. Each module might contain an arbitrary structure and wiring. In the upper left part the neuro-module for ego-motion extraction receives optical flow input from the conventional computer. Its output, a six-dimensional vector coded in an ensemble of neurons, is fed into a module that rotates the ego-motion according to a compass value and translates the motion into an allocentric coordinate system. The output is integrated in the following module which computes the current estimate of the copter's position. This estimate is fed into a neural module responsible for tracking different hypotheses of the own location (e.g. [3]). When landmarks are recognized (either within the neuromodule or outside) their identity is represented by the output of the object recognition module. Its location is fed into the central map, an associative network that receives input from the transfer module which integrates and distributes information among various modules. The confidence module represents the confidence in the position stored in the map. Low confidence drives the transfer network to project the copter's location (as represented by the state estimator) to the map. Is the sensed object known (high confidence) the transfer network updates the state estimation with the output of the map. Medium confidence values result in a mix of both strategies.

2   Mechanics and Electronics of Neurocopter

The copter is a custom built quadrotor. Four carbon fiber tubes are plugged into a central aluminum cross joint milled and turned down to a weight of 30 g. At the end of each tube a brushless DC motor (AC2836-358) is affixed with a custom holder. Each motor is driven by an ESC (electronic speed control, jDrones AC20-1.0) connected to an ArduPilotMega (APM) flight control board which is commercially available and serves as low-level motor controller and flight stabilizer. The board is connected to a GPS unit (MTek GTPA010) and offers common intergrated accelerometer, gyroscope and barometer. The APM connects to a 2.4 GHz 8 Channel radio receiver (Turnigy 9X8CV2) that works with a remote control (Turnigy TGY 9X) for manual / failsafe steering. It also connects through a serial interface with a Linux embedded Cortex A8 based PC (IGEP v2 DM3730). The IGEP runs a message dispatcher that receives UDP packets from the ground station and either collects data from the APM via the serial interface or executes programs on the Linux device itself. The ground station is used for high level control, such as setting semi-automatic behaviors such as starting, landing, position hold, altitude hold or waypoint following. The ground station also collects sensory data for visualization and forwarding to the brain simulation. A common webcam (Logitech Pro 9000) is connected to the IGEP via USB. The total weight of the system is 1340 g and the additional payload possible is approximately 1.5 kg.

Fig. 2. Photograph of the flying NeuroCopter. The quadrotor frame is built from carbon fiber tubes and an aluminum cross joint. The flight control electronics based on the Arduino platform, motors and power supply are standard components. A Linux board is used for wireless communication, camera readout and basic computations. Most processing is done on the ground station's computer.

3   Neuromorphic Extraction of Ego Motion

3.1   General Setup

On the embedded system, we compute a sparse set of image features [15] which are tracked with a pyramidal version of the tracker described in [8]. For each cell of a 5 x 5 grid optical flow vectors are averaged (Figure 5 depicts a sample image taken in flight). The resulting set of 25 motion vectors are sent down to the ground station via Wifi. The ground station translates the new data set and feeds it to a neural network.

3.2   Simulated Optical Flow

Before developing the neural module we asked whether the optical flow output can be unambiguously assigned to the respective ego-motions. For example, a roll motion produces mostly parallel flow on the image plane as would do a translational sidewards motion (which might occur due to gusts). We therefor simulated a three-dimensional terrain and modelled the camera in use. Artificial optical flow was computed by moving or rotating the camera and projecting a number of ground features onto the image plane before and after the camera motion. Figure 3 depicts the resulting flow in the simulated world.

The optical flow on the image plane was then averaged in a 5 x 5 grid of bins. Each action $\vec{a}$ (a six-dimensional ego-motion vector) thus is assigned a 50-dimensional data point $\vec{b}$ (25 bins, each having a two-dimensional vector in the plane). Consequently, we wanted to learn which pairs of actions $(\vec{a}_i, \vec{a}_j)$ have a high dissimilarity but produce a pair of similar flow outputs $(\vec{b}_i, \vec{b}_j)$. To this end, all data points in the input and output space were normalized such that the absolute maximum actions were vectors of unit length. In the next step, two (euclidian) distance matrices for all pairs in each vector space were computed. Each entry in the distance matrices expresses the similarity of actions or results. Low values denote very similar vectors. To make distances comparable between the two vector spaces, again, the distances were normalized to values between 0 (equal) and 1 (maximum dissimilarity). In the next step all inter-space distances were computed resulting in a set of action-pair-to-result-pair distances. The largest values in this set indicate action pairs having high dissimilarity but corresponding to similar result pairs (or vice versa, although this is another problem not adressed here).

By evaluating the largest distance pairs, we found that some complex motions (high rotational and translational components) might counterbalance the optical flow. If, for example, the copter would rotate into a positive pitch (lean forward), the flow on the image plane would point upwards. If the copter moves forward parallel to the ground surface an opposed optical flow is produced. If the altitude of the copter, the rotational and the forward speed fall into a certain range the flow cancels out and the inverse computation would produce a wrong result. However, this movement (within the erroneous range mentioned) might not happen often under normal operating conditions (a forward pitch normally preceeds the translational motion but strong winds might push the copter in the rotational movement). Secondly, roll and sideward motions or pitch and forward motions do produce similar but not equal optical flow fields: a neural network might well be able to separate both actions with feasible precision.

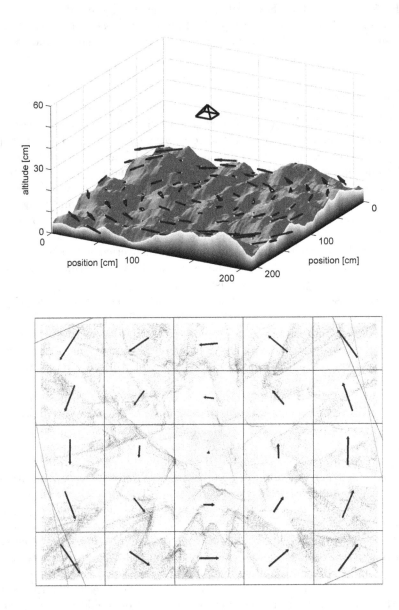

Fig. 3. Rotating a virtual camera (symbolized by the pyramid in the left figure) over a surface introduces image motion on the camera plane. The left figure depicts a three-dimensional surface with a selection of point features (white dots). We simulate the optic flow by projecting those ground features onto the camera plane before and after a certain camera motion. The camera image is divided into 25 bins in which the flow vectors are averaged (blue arrows in the right figure). Those average vectors are the input to the neural network for the ego-motion computation.

We trained a three-layered feed-forward artificial neural network (ANN) using the simulated ego-motion / action pairs. We then tested the reconstruction of a simulated flight by iterating the following commands over a number of time steps:

- move the camera according to a certain (smoothed) six-dimensional ego-motion function
- obtain the flow field by tracking the ground features' projection on the image plane
- average flow vectors in each of the 5 x 5 bins
- feed 50-dimensional input vector into ANN and read result vector

4   Results

The resulting ego-motion could then be integrated over time to reproduce the flight curve. Figure 4 visualizes both the trajectory computed from the input velocities and the network's reproduction. Apparently, the result needs improvement. The reproduction follows the input upwards but substantially deviates from it. The subspace of possible inputs that although being similar are supposed to produce differing outputs could not be learned with optical flow input alone. Feeding the copter's altitude to the network improves the network's performance drastically. This is intuitive: translations in high altitude produce only small flow vectors, whereas rotations are independent of scene depth. It is possible to significantly improve the networks output by updating an initial altitude hypothesis using the ego-motion's upward component and feed this altitude approximation back to the network. This value will accumulate even small errors

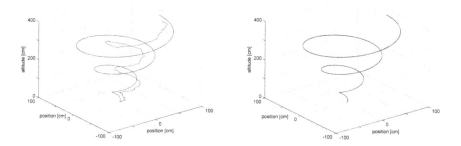

Fig. 4. Integrating the output of the neural network yields a flight trajectory that should preferably equal the trajectory created by integrating the input motions. The left image shows the resulting flight reconstruction of a network that was trained without altitude information. Using the output of the network to update an altitude estimation which is fed back to the network in the following time step yield a precise reproduction of the input.

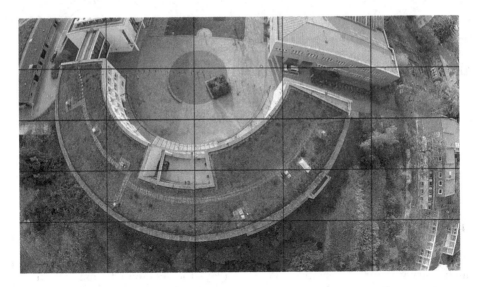

Fig. 5. This figure shows a camera frame captured in-flight over the Institute of Computer Science at Free University Berlin. The frame is overlaid with optical flow features (green points) and average flow vectors (in red) in each of the 25 image bins.

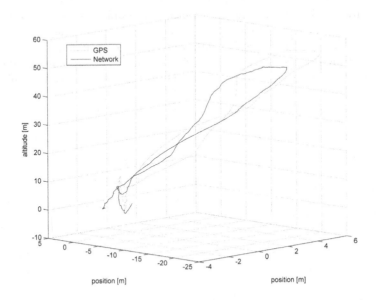

Fig. 6. The figure depicts the flight path reconstruction of a 70 seconds video sequence [7] recorded under strong wind conditions. The path, as computed by the neural module (blue line) resembles closely the trajectory flown (GPS reference trajectory, green dash-dotted line). Fast maneuvers and high flight speeds however lead to significant accumulation of error at the highest point of the path.

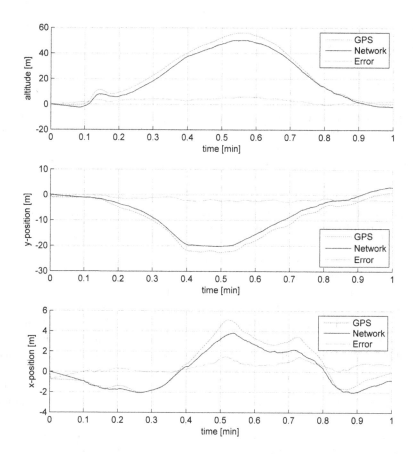

Fig. 7. The reconstruction of the flight path for x, y and z-coordinates. The blue curves depict the network's reconstruction, the green dash-dotted line represents the GPS reference data and the red line describes their difference. Maximum errors for altitude, y-position and x-position are: 5.9 m, 2.9 m, 1.5 m, respectively.

in the computation over time. It is conceivable to either use the copter's barometer or GPS sensor to refine or define the altitude value entering the network. It would also be conceivable to recalibrate the value by static up- or down-movements that produce imploding or exploding flow vectors of a length that is inversely proportional to the altitude.

We tested the ego-motion network module on our copter without XXX feeding back altitude information from the barometer. The test flight was controlled manually by a human pilot. Optical flow was computed on-board and sent down to the ground station. The flow data was fed into the proposed network and the resulting ego-motions were integrated to reproduce the flight trajectory taken. Onboard sensory data was also logged at 50 Hz (gyroscope and accelerometer) and 5 Hz (GPS), respectively. Figure 7 shows the resulting flight path. A video of the flight sequence with an optical flow overlay can be downloaded at [7]. The reproduction of the flight path resembles closely the actual flight path taken with respect to planar coordinates over ground and altitude.

5   Conclusions and Future Work

This paper shows a neural network structure that computes a six-dimensional ego-motion vector from sparse optical flow fields. The network does not require additional sensors other than the camera used. It yields a sufficiently precise estimation of the motion state of the copter that can be processed further in our neural processing scheme.

Since GPS and barometer on the copter are rather imprecise our confidence in a validation using these references is low. We are currently integrating a sonar for higher altitude precisison and are implementing a conventional, non-neural optical flow odometer. Both sensory systems will be used to increase precision of the state estimation. However, the proposed neural system does not necessarily have to be of high precision. One of our aims is to form a memory of the position of landmarks in the neural network. A landmark that is discovered the first time will be assigned the current allocentric self-position. The more often a landmark is revisited from different directions the better the stored position will get and the higher the confidence in this value. If the copter encounters a known landmark the position will tune the self-localization network and improve the copter's own position estimate.

The focus of this work lies in the formulation of our general approach to investigate neural processing units for navigation in bees. The neural module for the extraction of the robot's ego-motion exemplifies the approach on the module level. We are currently experimenting with prototypes of the neuro-modules for state estimation and the transfer function as shown in 1. The scheme however is not complete. The neural network might be able to map landmarks and use this information to refine the copter's postion estimate. Other cognitive processes, such as decision making or motivation for certain behaviors are not yet integrated into the model. In the near future conventionally implemented functionality will be replaced step by step by neural modules. In the upcoming summer period we are planning to conduct behavioral experiments with the copter. We will investigate the question whether the model can reproduce behaviors observed in flights of real bees (e.g. novel shortcuts). Consequently, we will test different models of insect navigation (e.g. [5]) and compare the resulting behavioral repertoire to the one found in nature.

Acknowledgement. Partial funding was received from the German Ministry of Education and Research within the collaborative project Insect Inspired Robots (grant 01GQ0941 to M.N. and R.R.).

References

1. Teller, M.V.A.: Neural Programming and an Internal Reinforcement Policy
2. Bernardet, U., Blanchard, M., Verschure, P.F.M.J.: IQR: a distributed system for real-time real-world neuronal simulation. Neurocomputing 44-46(null), 1043–1048 (2002)
3. Bobrowski, O., Meir, R., Eldar, Y.C.: Bayesian filtering in spiking neural networks: noise, adaptation, and multisensory integration. Neural Computation 21(5), 1277–1320 (2009)
4. Collett, T.S., Collett, M.: Memory use in insect visual navigation. Nature Reviews. Neuroscience 3(7), 542–552 (2002)
5. Cruse, H., Wehner, R.: No need for a cognitive map: decentralized memory for insect navigation. PLoS Computational Biology 7(3), e1002009 (2011)
6. Indiveri, G., Linares-Barranco, B., Hamilton, T.J., Van Schaik, A., Etienne-Cummings, R., Delbruck, T., Liu, S.-C., Dudek, P., Häfliger, P., Renaud, S., et al: Neuromorphic silicon neuron circuits. Frontiers in Neuroscience 5 (2011)
7. Landgraf, T., Wild, B., Ludwig, T., Nowak, P., Helgadottir, B., Daumenlang, L., Breinlinger, P., Nawrot, M., Rojas, R.: Test Flight for Living Machines 2013 (2013)
8. Lucas, B.D., Kanade, T.: An Iterative Image Registration Technique with an Application to Stereo Vision. Imaging 130(x), 674–679 (1981)
9. Menzel, R.: The honeybee as a model for understanding the basis of cognition. Nature Reviews Neuroscience 13(11), 758–768 (2012)
10. Menzel, R., Greggers, U., Smith, A., Berger, S., Brandt, R., Brunke, S., Bundrock, G., Hülse, S., Plümpe, T., Schaupp, F., Schüttler, E., Stach, S., Stindt, J., Stollhoff, N., Watzl, S.: Honey bees navigate according to a map-like spatial memory. Proceedings of the National Academy of Sciences of the United States of America 102(8), 3040–3045 (2005)
11. Menzel, R., Kirbach, A., Haass, W.-D., Fischer, B., Fuchs, J., Koblofsky, M., Lehmann, K., Reiter, L., Meyer, H., Nguyen, H., et al.: A common frame of reference for learned and communicated vectors in honeybee navigation. Current Biology 21(8), 645–650 (2011)
12. Menzel, R., Lehmann, K., Manz, G., Fuchs, J., Koblofsky, M., Greggers, U.: Vector integration and novel shortcutting in honeybee navigation. Apidologie 43(3), 229–243 (2012)
13. Pfeil, T., Grübl, A., Jeltsch, S., Müller, E., Müller, P., Petrovici, M.A., Schmuker, M., Brüderle, D., Schemmel, J., Meier, K.: Six networks on a universal neuromorphic computing substrate. Frontiers in Neuroscience 7 (2013)
14. Poon, C.-S., Zhou, K.: Neuromorphic silicon neurons and large-scale neural networks: challenges and opportunities. Frontiers in Neuroscience 5, 108 (2011)
15. Shi, J., Tomasi, C.: Good features to track. In: Proceedings of IEEE Conference on Computer Vision and Pattern Recognition CVPR 1994, pp. 593–600. IEEE Comput. Soc. Press (1994)
16. Wehner, R., Srinivasan, M.V.: Path integration in insects (2003)

A SOLID Case for Active Bayesian Perception in Robot Touch

Nathan F. Lepora, Uriel Martinez-Hernandez, and Tony J. Prescott

Sheffield Center for Robotics (SCentRo), University of Sheffield, UK
n.lepora@sheffield.ac.uk

Abstract. In a series of papers, we have formalized a *Bayesian perception* approach for robotics based on recent progress in understanding animal perception. The main principle is to accumulate evidence for multiple perceptual alternatives until reaching a preset belief threshold, formally related to sequential analysis methods for optimal decision making. Here, we extend this approach to active perception, by moving the sensor with a control strategy that depends on the posterior beliefs during decision making. This method can be used to solve problems involving Simultaneous Object Localization and IDentification (SOLID), or 'where and what'. Considering an example in robot touch, we find that active perception gives an efficient, accurate solution to the SOLID problem for uncertain object locations; in contrast, passive Bayesian perception, which lacked sensorimotor feedback, then performed poorly. Thus, active perception can enable robust sensing in unstructured environments.

Keywords: Active perception, tactile sensing, localization, robotics.

1 Introduction

Twenty five years after Bajcsy's landmark paper on active perception [1], it remains the case that most machine perception involves static analysis of passively sampled data. Certainly, there has been progress on passive approaches to pattern recognition in relation to machine learning and uncertainty, and there is a diverse body of work on active vision; nevertheless, a search through recent progress in robot vision, audition or touch reveals the majority of papers still rely on wholly forward perceptual processes without any sensorimotor feedback.

Why this slow uptake, when early arguments for active control of perception were compelling [1, 2] and, as Bajcsy said, it should be axiomatic that perception is active? One factor might be the required complexity of the robot hardware, which must involve actuated sensors and sensorimotor control loops. However, this should not be a barrier, because the technology is readily available and many standard robots have these capabilities, *e.g.* the iCub [3]. A more likely explanation is that researchers have focussed on sensing problems, such as identification, that can be solved adequately in many scenarios without introducing active methods for sensorimotor control. That being said, conventional robotics is reaching an impasse with present methods, such as poor performance in unstructured environments, which is preventing wider robot utilization beyond traditional factory settings [4].

N.F. Lepora et al. (Eds.): Living Machines 2013, LNAI 8064, pp. 154–166, 2013.

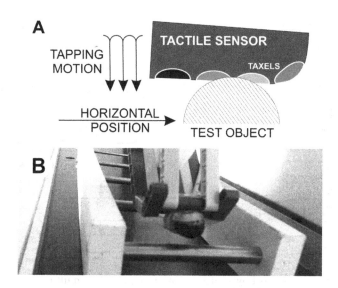

Fig. 1. Experimental setup. (A) Schematic of tactile sensor tapping against a cylindrical test object: the fingertip taps down and then back up again to press its pressure-sensitive taxels (colored) against the test object; each tap is then followed by a small horizontal move to sample object contacts over a range of positions. (B) Forward view of the experiment showing the fingertip mounted on the arm of the Cartesian robot. This experimental setup is ideal for systematic data collection to characterize the properties of the sensor interacting with its environment.

In a series of papers [5–10], we have formalized an approach for robot perception based on recent progress in understanding animal perception [11, 12]. The main principle is to accumulate evidence for multiple perceptual alternatives until reaching a preset belief threshold that triggers a decision, formally related to Bayesian sequential analysis methods for optimal decision making [13]. Here we describe how this perception approach extends naturally from passive to active perception and some implications of this theory of active Bayesian perception.

Our proposal for active Bayesian perception is tested with a simple but illustrative task of perceiving the location (horizontal position) and identity (diameter) of a test rod using tapping movements of a biomimetic fingertip at unknown contact location (Fig. 1; the colored regions are the pressure-sensitive taxels). We demonstrate first that passive perception can solve this task, but the perceptual acuity and reaction time depend strongly on the location of the fingertip relative to the rod. We then show that an active 'fixation point' control strategy can substantially improve the robustness, accuracy and speed of the perception, by moving the fingertip to locations with good perception independent of the starting position. Thus we demonstrate that active perception can enable appropriate perceptual decision making in an unstructured environment.

Related arguments have been presented in two other papers: the active perception method has been applied to texture identification under unknown contact

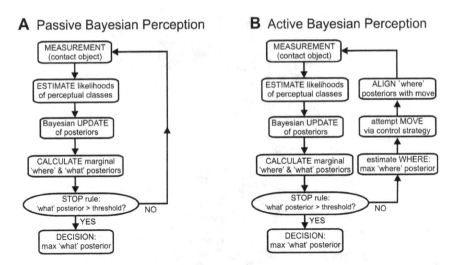

Fig. 2. Passive and active Bayesian perception applied to simultaneous object localization and identification. (A) Passive Bayesian perception has a recursive Bayesian update to give the marginal 'where' and 'what' posterior beliefs, with decision termination at sufficient 'what' belief. (B) Active Bayesian perception has the same recursive belief update, while also actively controlling the sensor location according to a strategy based on those beliefs; furthermore, when the sensor moves, it is necessary to re-align the 'where' component of the beliefs with the new sensor location. The two algorithms differ only in the sensorimotor control loop for active Bayesian perception.

depth [9], and a more detailed, systematic treatment of SOLID is given in [10] for a 2D (horizontal and vertical) 'where' and 'what' scenario.

2   Methods

The main goal of this work is to advance our understanding of the role of active perception for situated agents that seek to determine the 'where' and 'what' properties of objects. We refer to the computational task that must then be solved by Simultaneous Object Localization and IDentification (SOLID), to emphasize a similarity with SLAM of having two interdependent task aims, in that knowledge of location aids the computation of identity (mapping) and similarly that knowledge of object identity (mapping) aids localization.

Passive Bayesian perception accumulates belief for distinct 'where' and 'what' classes by making successive taps against a test object until at least one of the marginal 'what' posterior beliefs crosses a belief threshold, when a 'where' and 'what' decision is made. The passive nature of the perception means that the 'where' position class is constant over this process (Fig. 2A).

Active Bayesian perception also accumulates belief for the 'where' and 'what' perceptual classes by successively tapping until reaching a predefined 'what' belief threshold. In addition, it utilizes a sensorimotor loop to move the sensor according to the online marginal posterior beliefs during the perceptual process (Fig. 2B). For example, the sensor could be controlled with a 'fixation point' strategy, in which the marginal 'where' beliefs are used to infer a best estimate for current location and thus a relative move towards a preset 'good' target position on the object to improve the perceptual decision making.

2.1   Algorithms for Bayesian Perception

Our algorithm for active Bayesian perception is based on including a sensori-motor feedback loop in an existing method for passive Bayesian perception [5]. Both methods assume that the sensor makes a discrete contact measurement (here a tap) onto an object, from which the joint likelihoods of object location and identity are used to update the prior to posterior beliefs for those perceptual classes. In active perception, a control strategy repositions the sensor before each contact, taking input from the updated beliefs and outputting the sensor move.

Because these methods are applicable to any simultaneous object localization and identification task, this section is presented in a general 'where' and 'what' notation. A general SOLID task has N_{loc} distinct 'where' location classes x_l and N_{id} distinct 'what' identity classes w_i, totalling $N = N_{loc}N_{id}$ joint 'where-what' classes $c_n = (x_l, w_i)$. Each contact against a test object gives a multi-dimensional time series of sensor values $z = \{s_k(j) : 1 \leq j \leq N_{samples}, 1 \leq k \leq N_{channels}\}$, with indices j, k labeling the time samples and sensor channels. The tth contact in a sequence is denoted by z_t with $z_{1:t-1} = \{z_1, \cdots, z_{t-1}\}$ its contact history.

Measurement model and likelihood estimation: The likelihoods of all perceptual classes are found using a measurement model of the contact data, which we find by applying a histogram method to training examples for each perceptual class [5, 6]. First, the sensor values s for channel k are binned into $N_{bins} = 100$ intervals, with sampling distribution for each perceptual class c_n given by the normalized histogram over all training data in that class:

$$P(b|c_n, k) = \frac{h(b, k)}{\sum_{b=1}^{N_{bins}} h(b, k)}, \qquad 1 \leq k \leq N_{channels}, \qquad (1)$$

where $h(b, k)$ is the histogram count for bin b ($1 \leq b \leq N_{bins}$) in sensor channel k. Then, given a test tap z, we construct a measurement model from the mean log likelihood over all samples in that tap

$$\log P(z|c_n) = \frac{1}{N_{samples}N_{channels}} \sum_{k=1}^{N_{channels}} \sum_{j=1}^{N_{samples}} \log P(b_k(j)|c_n, k), \qquad (2)$$

where $b_k(j)$ is the bin occupied by sample $s_k(j)$. Technically, this measurement model becomes ill-defined if any histogram bin is empty, which is easily fixed by regularizing the bin counts with a small constant ($\epsilon \ll 1$), giving $h(b, k) + \epsilon$.

Bayesian update: Bayes' rule is used after each successive test contact z_t to recursively update the posterior beliefs $P(c_n|z_{1:t})$ for the perceptual classes with the estimated likelihoods $P(z_t|c_n)$ of that contact data

$$P(c_n|z_{1:t}) = \frac{P(z_t|c_n)P(c_n|z_{1:t-1})}{P(z_t|z_{1:t-1})}, \tag{3}$$

from background information given by the prior beliefs $P(c_n|z_{1:t-1})$. The marginal probabilities are also conditioned on the preceding contacts $z_{1:t-1}$ and given by

$$P(z_t|z_{1:t-1}) = \sum_{n=1}^{N} P(z_t|c_n)P(c_n|z_{1:t-1}). \tag{4}$$

Iterating (3,4), a sequence of contacts $z_1, \cdots, z_t$ results in a sequence of posteriors $P(c_n|z_1), \cdots, P(c_n|z_{1:t})$ initialized from uniform priors $P(c_n|z_0) := P(c_n) = \frac{1}{N}$.

Marginal 'where' and 'what' posteriors: For the following methods, we will need the posterior beliefs for just location or identity, rather than the joint beliefs considered so far. Because each class $c_n = (x_l, w_i)$ has a 'where' location x_l and 'what' identity w_i component, these beliefs can be found by marginalizing

$$P(x_l|z_{1:t}) = \sum_{i=1}^{N_{\rm id}} P(x_l, w_i|z_{1:t}), \tag{5}$$

$$P(w_i|z_{1:t}) = \sum_{l=1}^{N_{\rm loc}} P(x_l, w_i|z_{1:t}), \tag{6}$$

with the 'where' location beliefs given from summing over all 'what' identity classes w_i and the 'what' identity beliefs over all 'where' location classes x_l.

Final decision on the 'what' posteriors: Here we follow sequential analysis methods for optimal decision making that recursively update beliefs up to a threshold that triggers the final decision [13], as used in passive Bayesian perception [5]. The update stops when the marginal 'what' identity belief passes a threshold, giving a final decision from the maximal *a posteriori* (MAP) estimate

$$\text{if any } P(w_i|z_{1:t}) > \theta_{\rm id} \text{ then } w_{\rm id} = \underset{w_i}{\arg\max} P(w_i|z_{1:t}). \tag{7}$$

This belief threshold $\theta_{\rm id}$ is a free parameter that adjusts the balance between decision speed and accuracy. For $N = 2$, this speed-accuracy balance can be proved optimal [13]; optimality is not known for the many perceptual choices considered here, and so we make a reasonable assumption of near optimality [5].

Move decision on the 'where' posteriors: Analogously to the stop decision, a sensor move requires a marginal 'where' location belief to cross its own decision threshold, with the MAP estimate giving the 'where' location decision

$$\text{if any } P(x_l|z_{1:t}) > \theta_{\rm loc} \text{ then } x_{\rm loc} = \underset{x_l}{\arg\max} P(x_l|z_{1:t}). \tag{8}$$

Here we consider two particular cases (Figs 2A,B), termed:

(A) passive perception: $\theta_{\rm loc} = 1$ (never moves)

(B) active perception: $\theta_{\rm loc} = 0$ (always tries to move).

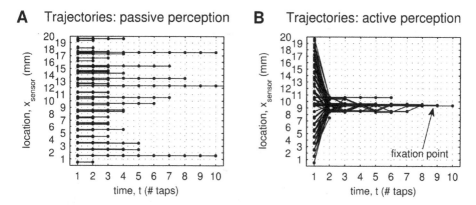

Fig. 3. Example trajectories for passive and active perception. 100 trajectories were selected randomly for each case. (A) Passive perception, with location (x-position) constant over time. (B) Active perception, with trajectories converging rapidly on the central fixation point (10 mm location class) independent of starting position.

For simplicity, we consider a basic movement strategy in which the sensor move Δ depends only on estimated location x_{loc}, although more complex strategies are encompassed by the formalism. Whatever the strategy, the marginal 'where' location belief should be kept aligned with the sensor by shifting the joint 'where-what' posterior beliefs upon each move

$$P(x_l, w_i | z_{1:t}) \leftarrow \begin{cases} P(x_l - \Delta(x_{\mathrm{loc}}), w_i | z_{1:t}), & 1 \leq x_l - \Delta(x_{\mathrm{loc}}) \leq N_{\mathrm{loc}}, \\ p_0, & \text{otherwise,} \end{cases} \quad (9)$$

where we recalculate the beliefs p_0 lying outside the original range by assuming they are uniformly distributed and the shifted beliefs sum to unity. The left arrow denotes that the quantity on the left is replaced with that on the right.

Active control strategy: The final component of the active perception algorithm is to define the control strategy for moving the sensor based on the posterior beliefs. For simplicity, here we consider a 'fixation point' strategy motivated by orienting movements in animals: the sensor attempts to move to a predefined fixation point x_{fix} relative to the object assuming it is at the estimated location x_{loc} on the object, with each move resulting in

$$x_{\mathrm{sensor}} \leftarrow x_{\mathrm{sensor}} + \Delta(x_{\mathrm{loc}}), \quad \Delta(x_{\mathrm{loc}}) = x_{\mathrm{fix}} - x_{\mathrm{loc}}, \quad (10)$$

where x_{sensor} is the actual (unknown) location of the sensor. In practise, only the move Δ need be found, to instruct the sensor how to change location. Example trajectories resulting from this active control strategy are shown in Fig. 3B.

2.2   Data Collection and Analysis

The tactile sensors used in this study have a rounded shape that resembles a human fingertip [14], of dimensions 14.5 mm long by 13 mm wide. They consist

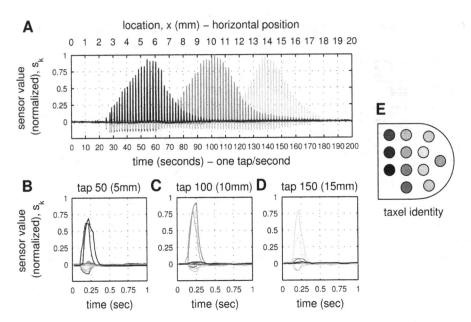

Fig. 4. Tactile dataset (for test rod of diameter 4 mm). (A) Entire dataset, with 200 taps over positions spanning 20 mm. Taps are every 0.1 mm displacement. (B-D) Individual tap data taken from panel A. (E) Taxel layout with color-code for plots A-D.

of an inner support wrapped with a flexible printed circuit board containing $N_{channels} = 12$ conductive patches for the touch sensor 'taxels'. These are coated with non-conductive foam and conductive silicone layers that together comprise a capacitive touch sensor that detects pressure by compression. Data were collected at 8 bit resolution and 50 cycles/sec then normalized and high-pass filtered [14].

The present experiments test the capabilities of the tactile fingertip mounted on a Cartesian robot. This robot moves the sensor in a horizontal/vertical plane in a precise and controlled way onto various test stimuli (~20 μm accuracy), and has been used for testing various tactile sensors [15]. The fingertip was mounted at an orientation appropriate for contacting axially symmetric shapes such as cylinders aligned along an axis perpendicular to the plane of movement (Fig. 1). $N_{id} = 5$ smooth steel rods with diameters 4,6,8,10,12 mm were used as test objects, mounted with their centers offset vertically (by 4,3,2,1,0 mm) to align their closest point of contact with the fingertip in the direction of tapping.

Touch data were collected while the fingertip tapped vertically onto and off each test object, followed by a horizontal move $\Delta x = 0.1$ mm across the closest face of the object (Fig. 1A). The fingertip was oriented so that it initially contacted the rod at its base and finally at its tip. A horizontal x-range of 20 mm was used, giving 200 taps for each of the $N_{id} = 5$ objects, or 1000 taps in total. From each tap of the fingertip against the object, a 1 sec time series of pressure readings ($N_{samples} = 50$) was extracted for all $N_{channels} = 12$ taxels (Fig. 4). All data were collected twice to give distinct training and test sets.

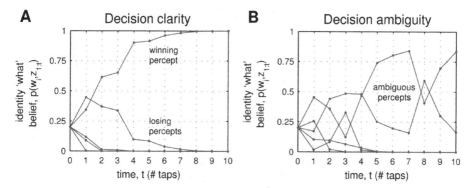

Fig. 5. Example 'what' belief update for perceptual decision making. Evidence from successive taps is integrated to result in accumulating/depreciating marginal beliefs for the $N_{id} = 5$ distinct identity 'what' percepts. The examples show: (A) one clear winning percept and (B) two ambiguous percepts. Using a belief threshold to trigger the decision results in the appropriate number of taps to have a clear winner.

For analysis, the data were separated into $N_{loc} = 20$ distinct location classes, by collecting groups of 10 taps each spanning 1 mm of the 20 mm x-range (tick-marks on Fig. 4A). In total, there were thus $N = N_{loc}N_{id} = 100$ distinct 'where-what' perceptual classes. These were used to set up a 'virtual environment' in which methods for perception could be compared off-line on identical data. A Monte Carlo validation ensured good statistics, by averaging perceptual acuities over many test runs with taps drawn randomly from the perceptual classes (typically 20000 runs per data point in results). Perceptual acuities e_{loc}, e_{id} were quantified using the mean absolute error (MAE) between the actual x_{test}, w_{test} and classified values x_{loc}, w_{id} of object location and identity over the test runs.

3   Results

3.1   Evidence Accumulation for Robot Perception

The 'where' and 'what' perceptual task is to find the location (x-position) of a rod and its identity (diameter) using tactile fingertip data over a sequence of test taps. Example 'what' perceptual beliefs $P(w_i|z_{1:t})$ for tap sequences $z_{1:t}$ of clear and ambiguous data are shown above (Fig. 5). These beliefs begin at equality corresponding to uniform prior beliefs and then evolve smoothly with some rising gradually towards unity and others falling towards zero. In the first example (Fig. 5A), the decision given by the largest perceptual belief remains the same after applying 2 taps or more, while the second example (Fig. 5B) flips between the two leading choices.

There are two common methods for making decisions from sequential data of this type: (i) set in advance the number of taps that will be used, or (ii) set in advance a belief threshold θ that will trigger the decision, so that the decision (reaction) time is a dynamic quantity that depends on the data received.

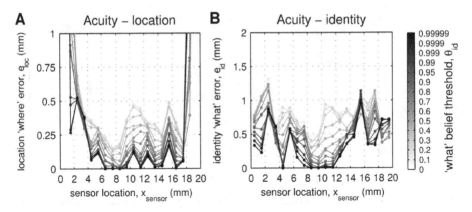

Fig. 6. Dependence of passive perception on sensor location and 'what' belief threshold. The 'where' location errors are shown in (A) and the 'what' identity errors in (B), plotted against sensor location x_{sensor}. The gray-scale denotes the 'what' belief threshold. Each data point corresponds to 1000 decision trials. Perceptual performance improves in the center of the sensor location range and at greater belief thresholds.

Recent progress in perceptual neuroscience strongly supports that animals use a belief threshold to make decisions [11], consistent with the brain implementing sequential analysis for optimal decision making [13]. In accordance, a comparison of these two methods on tactile robot data found that the belief threshold method gave superior performance in perceptual acuity [5]. This can be seen intuitively from Fig. 5: if, for example, a deadline of 10 taps was set in advance, then the decision is unnecessarily slow in situations of clarity (Fig. 5A) and too quick in situations of ambiguity (Fig. 5B). Instead, setting a belief threshold allows the decision time to adjust dynamically to the uncertainty of the evidence.

Both the passive and active methods for perception considered here update beliefs from successive test taps to threshold θ_{id}. They differ, though, in how the sensor responds during the decision process: for passive perception its location is fixed, whereas for active perception it can control changes in location.

3.2 Passive Perception of Location and Identity

This section considers the application of passive Bayesian perception to the 'where' and 'what' perceptual task of identifying rod location (x-position) and identity (diameter). Results are generated with a Monte Carlo procedure using test data as a virtual environment (Sec. 2.2), such that each contact tap passively remains within its initial location class on the object (examples in Fig. 3A).

The 'where' and 'what' decisions for perceiving object location and identity were evaluated over identity belief thresholds θ_{id} from 0.1 to 0.99999 (Fig. 6, colorbar). Initial locations spanned all horizontal position classes x_l. Mean perceptual acuities over all objects improved with identity threshold and towards the center (10 mm) of the location range (Fig. 6), giving minimal errors near zero for identity e_{id} and location e_{loc}. The number of taps to reach a decision also

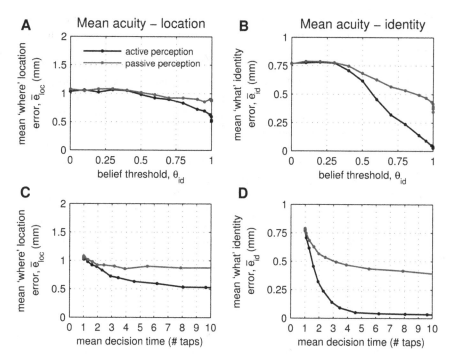

Fig. 7. Comparative perceptual acuity for active and passive Bayesian perception. (A,B) Dependence of the mean absolute errors of location $\bar{e}_{\mathrm{loc}}$ (x-position) and identity $\bar{e}_{\mathrm{id}}$ (rod diameter) upon the identity belief threshold θ_{id}. Passive perception is shown in red and active perception in black. (C,D) Dependence of these perceptual errors upon mean decision time (with threshold an implicit parameter). Active performs better than passive perception, and both improve with increasing 'what' belief threshold.

varied, such that the mean tap number (reaction time) increased with identity threshold and decreased towards the central location range (Figure not shown).

Here we consider situations in which the initial contact location is unknown, with performance measure the mean perceptual errors over all such locations, consistent with location being selected randomly. These mean errors reached their best values $\bar{e}_{\mathrm{loc}} \sim 0.9\,\mathrm{mm}$ and $\bar{e}_{\mathrm{id}} \sim 0.4\,\mathrm{mm}$ at the largest belief thresholds and reaction times (Fig. 7, red plots). Not unexpectedly, the perception is poorer than when choosing just the central contact location (on Fig. 6 at $x_l = 10\,\mathrm{mm}$), emphasizing that passive perception performs poorly because it cannot control contact location.

3.3 Active Perception of Location and Identity

This section considers active Bayesian perception in the same scenario as for passive perception, with a 'where' and 'what' perceptual task of identifying rod location and identity. Results are again generated using a Monte Carlo procedure over a virtual environment (Sec. 2.2), now with an active control strategy that tries to re-locate the sensor to a fixation point in the center of the location range (10 mm) where the passive perception was best (example trajectories in Fig. 3B).

The 'where' and 'what' decisions for perceiving object location and identity were evaluated over the same range of belief thresholds ($\theta_{\rm id}$ from 0.1 to 0.99999) and locations (x_l from 0 to 20 mm) as for passive perception, to permit direct comparison of the active and passive approaches. We again considered the location as uncertain, and thus measured performance by the mean perceptual errors $\bar{e}_{\rm loc}$, $\bar{e}_{\rm id}$ over all initial contact locations. These location-averaged errors reached their best acuities for object location of $\bar{e}_{\rm loc} \sim 0.5$ mm and identity $\bar{e}_{\rm id} \sim 0.1$ mm at the largest belief thresholds and reaction times (Fig. 7, black plots).

Comparing active perception with passive perception, the best mean perceptual errors improve from $\bar{e}_{\rm loc} \sim 0.9$ mm to 0.5 mm for object location (over a 20 mm x-position range) and $\bar{e}_{\rm id} \sim 0.4$ mm to 0.1 mm for object identity (over a 4-12 mm diameter range). Thus, active perception gives far finer perceptual acuity than passive perception when compared under similar conditions of uncertain object location and identity.

4   Discussion

In this paper, we compared active and passive Bayesian perception methods for Simultaneous Object Localization and IDentification (SOLID), or perceiving 'where' and 'what'. We considered a task in which a biomimetic fingertip taps against a smooth steel rod to simultaneously perceive its location (horizontal position) and identity (rod diameter). Active perception can control changes in location of the sensor during the decision making process, whereas for passive perception the location is fixed at where the sensor initially contacted the object. We found that active perception gives far more accurate perception in situations of uncertain object location and identity than passive perception. Thus, active perception is appropriate for sensing in unstructured environments where location is uncertain, and improves performance by compensating the uncertainty in initial sensor placement to, in effect, structure an unstructured environment.

As in other work on active perception [1, 2], the inspiration for our approach was from animal perception. The particular active perception strategy considered here was to fixate the sensor onto an object at a good contact location for perception (here centering the fingertip over the middle of the object); this strategy is analogous to orienting movements found in many perceptual modalities, such as saccadic foveation in vision and head turning for audition. In addition, we used an evidence accumulation method for Bayesian perception [5] that has close relation to leading models of perceptual decision making in neuroscience [11] and also relates to proposals for cortico-basal ganglia function [12]. Bayesian perception, where evidence is integrated to a belief threshold, leads to decision time being a dynamic quantity that depends on the quality of data. Although rare in contemporary approaches to machine learning, having this type of variation in reaction times is ubiquitous in the natural world.

Related methods to those presented here have enabled the first demonstration of hyperacuity in robot touch [8], giving localization acuity finer than the sensor resolution, as is common in animal perception including human touch [16]. Although our previous study also found that active perception helped attain hyperacuity, those methods now seem somewhat *ad hoc* in light of the present study, by not making best use of the 'where' and 'what' aspects of the problem. In the present work, we have developed a principled approach to active Bayesian perception of applicability to simultaneous object localization and identification. Given the taxel spacing is 4 mm, our results verify that passive Bayesian perception is capable of hyperacuity (mean localization error ~ 1 mm) and active Bayesian perception of stronger hyperacuity (mean error ~ 0.5 mm).

In seminal work on active perception, Bajcsy said it is axiomatic that perception (in animals) is active [1]. Robotics is currently in a state of transition from rigidly controlled tasks in predictable structured environments like factory assembly lines, to applications in unpredictable unstructured environments like our homes, hospitals and workplaces. In our opinion, robots will need active perception to accomplish these tasks in unstructured environments, and thus it may also become axiomatic that future robot perception will be active too.

Acknowledgments. We thank Ben Mitchinson and Mat Evans for comments on earlier drafts of this paper. This work was funded by the European Commission under the FP7 project grant EFAA (ICT-270490); UMH was supported by the Mexican National Council of Science and Technology (CONACyT).

References

1. Bajcsy, R.: Active perception. Proceedings of the IEEE 76, 966–1005 (1988)
2. Ballard, D.: Animate vision. Artificial Intelligence 48(1), 57–86 (1991)
3. Metta, G., Sandini, G., Vernon, D., Natale, L., Nori, F.: The icub humanoid robot: an open platform for research in embodied cognition. In: Proceedings of the 8th Workshop on Performance Metrics for Intelligent Systems, pp. 50–56 (2008)
4. Kemp, C.C., Edsinger, A., Torres-Jara, E.: Challenges for robot manipulation in human environments (grand challenges of robotics). IEEE Robotics & Automation Magazine 14(1), 20–29 (2007)
5. Lepora, N.F., Fox, C.W., Evans, M.H., Diamond, M.E., Gurney, K., Prescott, T.J.: Optimal decision-making in mammals: insights from a robot study of rodent texture discrimination. Journal of the Royal Society Interface 9(72), 1517–1528 (2012)
6. Lepora, N.F., Evans, M., Fox, C.W., Diamond, M.E., Gurney, K., Prescott, T.J.: Naive bayes texture classification applied to whisker data from a moving robot. In: The 2010 International Joint Conference on Neural Networks (IJCNN), pp. 1–8 (2010)
7. Lepora, N.F., Sullivan, J.C., Mitchinson, B., Pearson, M., Gurney, K., Prescott, T.J.: Brain-inspired bayesian perception for biomimetic robot touch. In: 2012 IEEE International Conference on Robotics and Automation (ICRA), pp. 5111–5116 (2012)
8. Lepora, N.F., Martinez-Hernandez, U., Barron-Gonzalez, H., Evans, M., Metta, G., Prescott, T.J.: Embodied hyperacuity from bayesian perception: Shape and position discrimination with an icub fingertip sensor. In: 2012 IEEE/RSJ International Conference on Intelligent Robots and Systems (IROS), pp. 4638–4643 (2012)

9. Lepora, N.F., Martinez-Hernandez, U., Prescott, T.J.: Active touch for robust perception under position uncertainty. In: 2013 IEEE International Conference on Robotics and Automation (ICRA), pp. 3005–3010 (2013)

10. Lepora, N.F., Martinez-Hernandez, U., Prescott, T.J.: Active bayesian perception for simultaneous object localization and identification. In: Robotics: Science and Systems (2013)

11. Gold, J.I., Shadlen, M.N.: The neural basis of decision making. Annual Reviews Neuroscience 30, 535–574 (2007)

12. Lepora, N.F., Gurney, K.: The basal ganglia optimize decision making over general perceptual hypotheses. Neural Computation 24(11), 2924–2945 (2012)

13. Wald, A.: Sequential analysis. John Wiley and Sons, NY (1947)

14. Schmitz, A., Maiolino, P., Maggiali, M., Natale, L., Cannata, G., Metta, G.: Methods and technologies for the implementation of large-scale robot tactile sensors. IEEE Transactions on Robotics 27(3), 389–400 (2011)

15. Evans, M., Fox, C., Lepora, N., Pearson, M., Sullivan, J., Prescott, T.: The effect of whisker movement on radial distance estimation: a case study in comparative robotics. Frontiers in Neurorobotics 6 (2013)

16. Loomis, J.M.: An investigation of tactile hyperacuity. Sensory Processes 3, 289–302 (1979)

Modification in Command Neural Signals of an Insect's Odor Source Searching Behavior on the Brain-Machine Hybrid System

Ryo Minegishi[1], Yosuke Takahashi[1], Atsushi Takashima[2], Daisuke Kurabayashi[1], and Ryohei Kanzaki[3]

[1] Tokyo Institute of Technology, Tokyo, 152-8552, Japan
{minegishi,takahashi.y,dkura}@irs.ctrl.titech.ac.jp
[2] Kyushu University, Fukuoka, 819-0395, Japan
takashima@astec.kyushu.ac.jp
[3] The University of Tokyo, Tokyo, 153-8904, Japan
kanzaki@rcast.u-tokyo.ac.jp

Abstract. To investigate the adaptive behavior and the underlying neural mechanisms, we focused on the insect's brain and developed the brain-machine hybrid system. The hybrid system is a mobile robot controlled by recorded neural signals related to steering motor pattern on the robot. We manipulated the motor output of the robot to introduce the rotational disturbances to the hybrid system and acquired the compensatory neural activities. Moreover, we manipulated the motor pattern of the robot during odor source orientation behavior. The moth on the robot maintained the angular velocity and succeeded in odor source localization by modifying the neural activities.

Keywords: Chemical plume tracing, Brain-machine interface, Sensory feedback, Adaptive behavior, Multisensory integration, Insect brain.

1 Introduction

Recent years, along with the improvement of measurement techniques and the processing speed of CPU, the technology to control machines (such as robot arms, mobile robots and computers) using neural activities acquired from a brain (i.e. Brain-machine interface) has been developed [1-4]. Brain-machine interfaces are expected to be used for supporting injured persons and used in rehabilitation for them. In addition to these uses, brain-machine interface can be a studying method to elucidate how the motor pattern is generated in the brain. We can investigate how the brain processes sensory information in response to the dynamic change of the motor outputs of the machine by arbitrarily controlling it working as the artificial body.

In most cases, studies of the brain-machine interface have focused on generation mechanisms of plasticity in mammalian brains (such as rats, monkeys and humans). However, because the networks in the mammalian brains are highly wired, it is not so easy to extract neural signals responsible for the primary commands from mammalian brains and investigate this adaptability during executing behavioral tasks.

N.F. Lepora et al. (Eds.): Living Machines 2013, LNAI 8064, pp. 167–178, 2013.

In this study, we solved this problem by using an insect brain which consists of comparatively small number of neurons (about 10^5-10^6 neurons) [5]. There are some studies that refer to interface between insects and robots [6], [7]. We used "a brain-machine hybrid system" [8] which is a mobile robot controlled by neural activities (Fig. 1) which are sensitive to olfactory stimuli and selective direction of visual motion and investigated compensatory response properties of motor and descending neural signals to the change of motor gain of the robot during odor source searching behavior.

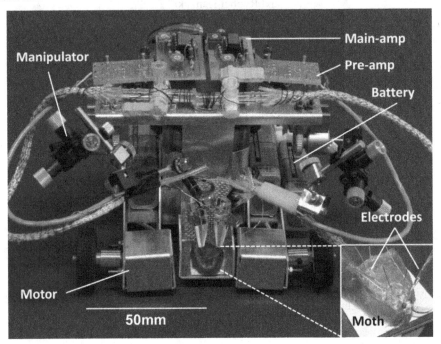

Fig. 1. Brain-Machine Hybrid System

2   Brain-Machine Hybrid System

The brain-machine hybrid system consists of two parts: a part for recording neural activities from the male silkmoth's head and a part for converting recorded signals into behavior of the mobile robot (e-puck, EPFL, Switzerland) [8].

2.1   Experimental Animal

We used a male silkworm moth (Lepidoptera: *Bombyx mori*) as an experimental animal. The reasons to select a male silkmoth are as bellow.

- Behavior of a silkmoth is only driven by female sex pheromones. Unless it doesn't detect pheromones, it doesn't move. We can also excite their olfactory neural pathways by giving pheromone stimuli.
- A silkworm moth doesn't fly. It walks to approach a female moth by repeating a set of programmed behavior (surge walking, zigzagging, loop turn) upon detecting a female sex pheromone (Fig. 2) [9-11]. This programmed behavior coincides with odor source searching behavior of other flying moths [12]. The walking trajectories can be described in two dimensional coordinate, and it is a good model animal to analyze and model behavioral strategies.
- Silkmoths have been used as model animals in biological studies (such as genetics, neuroethoogy and biochemistry) for a long time. Olfactory neural pathways in the silkmoth's brain are well studied, and activities of some neurons correlate steering motor behavior. They are candidates for command signals to control the robot [13-16].

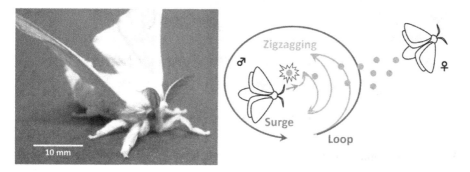

Fig. 2. Silkworm moth and its behavioral pattern

2.2   Recording Neural Activities

To record command signals of the silkmoths during odor source searching behavior, we focused on the activities of the 2nd cervical nerve b (2nd CNb) (Fig. 3). The 2nd CNbs are a pair of motor neurons innervating to the neck muscles swinging the moth's head. They have connections with the descending interneurons (DNs) conveying command signals from the brain to the thorax where actuators such as legs and wings are placed [15]. Because steering information during odor source searching behavior is generated in the brain and is conveyed to the thorax, it is assumed that the 2nd CNbs also convey steering information [13-15]. Based on simultaneous observations of head swing and walking during odor source searching behavior, the angular velocity of the moths' turning walk and angle of head swing were found to be in agreement [13], [17]. Moreover, the 2nd CNb is a nerve bundle consisting of only 5 motor neurons, and it is not so difficult to analyze recorded signals from the 2nd CNb by sorting the units [8].

Neural activities were recorded as previously described [13], [16]. We recorded neural signals by applying the glass micro electrodes (30-40 μm) to suck the cut end of the left and right 2nd CNbs using syringes. In the recording setup, we set the moth ventral-side up condition for suction recording. Moths could track pheromone even if they walked hanging on the ceiling of the wind tunnel (personal observation), so there has no effect of the upside-down condition of this preparation. In this paper, we set the right and left viewed from the ventral side as right and left to the moth.

We acquired neural activities by using amplifiers that we made [8]. We designed our instrumentation amplifiers with an input resistance of 100 MΩ, a gain of 80 dB (variable), a frequency bandwidth of 150 Hz-3.2 kHz.

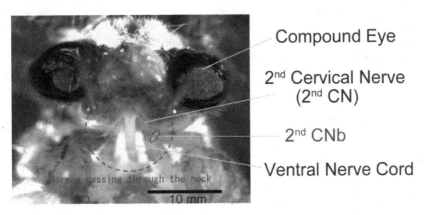

Fig. 3. Anatomy of the 2nd CNb

2.3   Setting a Spike-Behavior Conversion Rule from Visual Response Properties

In general, flying insects follow directional optical cues by rotating their body axis and rotating their heads (optomotor responses) for stabilizing their body position during flight [18-22]. Though silkmoths don't fly, they also follow the optical cues by horizontally rotating their heads. To convert the 2nd CNbs' activities into steering behavior of the robot, we investigated response properties of optomotor responses to suggested visual stimuli. As visual stimuli, we suggested the sinusoidal black and white stripe patterns 50 mm in width on a 24 inch display (GL2450HM, benQ, Japan) in front of the moth's head (distance 35 mm) and moved the pattern sidewise at a certain speed (0.25-1.5 rad/s) [23].

Spiking rate of the 2nd CNbs increased according to the angular velocity of the suggested patterns moving around the head position of the moth (Fig. 4). The 2nd CNbs exhibited excitatory responses to visual stimuli with ipsilateral direction and inhibitory responses to visual stimuli with contralateral direction comparing with the spontaneous spiking rate. These responses mean that the moth follows the direction of the optical flow by contracting its neck and turning its head in the same direction as the visual motion.

Based on the anatomical study that sidewise movement of a head is caused by complementary contraction of left and right neck muscles [13] and the correspondence of the head angle to the angular velocity of the body axis during silkmoth's walking behavior [13], [17], we set a spike-behavior conversion rule that the angular velocity of the robot was proportional to difference of the left and right 2nd CNbs' spiking rate (Fig. 5) [8]. We also assumed the forward velocity was proportional to sum of the left and right 2nd CNbs' spiking rate. The control cycle is 0.1 s. The proportional constants were given by the angular velocities of the visual stimuli and the corresponding spiking rates.

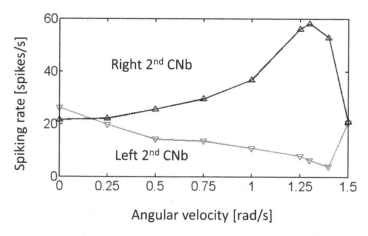

Fig. 4. Responses of the 2nd CNbs to the rightward optical flow stimuli

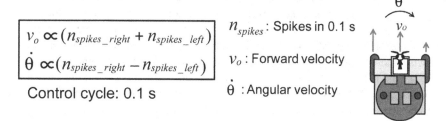

Fig. 5. Spike-behavior conversion rule

3   Disturbance Experiments Using the Hybrid System

On the brain-machine hybrid system, we can manipulate the motor outputs of the mobile robot and investigate the neural responses to the change of behavior. We introduced increments of rotational disturbances (1.0 rad/s) to the motor outputs of the robot while the robot was controlled by the neural activities acquired on the robot (Fig. 6). Disturbances have direction (clockwise and counterclockwise rotations) and different durations with 5 s intervals. We set the hybrid system at the center in a cylindrical arena. The diameter of the arena is 600 mm, while the height 526 mm. The inside wall of the arena is patterned with sinusoidal stripes to simulate a visual condition. The width of a stripe is 47.5 mm.

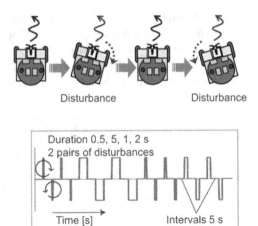

Fig. 6. Protocol of disturbance experiments

Fig. 7 shows typical neural responses to the given increments of disturbances with the direction. We calculated the angular velocity of the robot according to the conversion rule in Fig. 5. Positive and negative values of the angular velocity in Fig.6 indicate counterclockwise and clockwise rotations, respectively. The motor output of the robot was given by the sum of disturbances and neural outputs. When the negative angular velocity (clockwise rotation) disturbances were given, the neural responses exhibited the positive angular velocity to cancel out the disturbances in the motor output. Namely, the disturbances were cancelled out by the neural responses in directions opposite to the disturbances (Fig. 7). Comparing with the spontaneous spiking rate, activities of the left and right 2nd CNbs exhibited excitation during clockwise and counterclockwise rotations, respectively (Wilcoxon signed-rank test, p < 0.05, N = 8) (Fig. 8). Bar graphs in Fig.8 indicate spontaneous spiking rate of the left and right 2nd CNbs and spiking rate during clockwise and counterclockwise rotational disturbances.

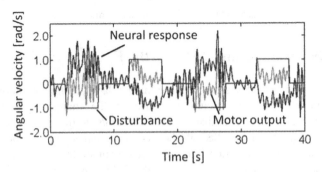

Fig. 7. Typical neural responses to the disturbances

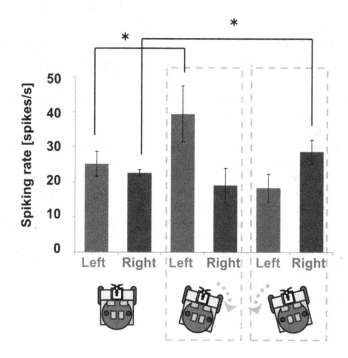

Fig. 8. The left and right 2nd CNbs' response properties to rotational disturbances (*: Wilcoxon signed-rank test, $p < 0.05$)

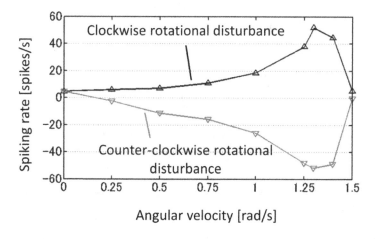

Fig. 9. Difference between the spiking rate of the left and right 2nd CNbs in response to the disturbances

Fig. 9 shows the difference of the left and right 2nd CNbs in response to the rotational disturbances. The left 2nd CNb shows excitatory responses to the clockwise rotational disturbances, and the difference of the spiking rate between the left and right 2nd CNb becomes positive (Fig. 9). The response pattern to the disturbances in Fig. 9 corresponds to the response pattern to the optical flow stimuli shown in Fig.4. In addition, these responses to the disturbances disappeared when the compound eyes of the moth were covered with aluminum foil. These properties mean that the visual feedback works to compensate for the disturbances.

Moreover, we changed the proportional constants of the spike-behavior conversion rule into the angular velocity double and half conditions and investigated the neural responses to the disturbances as in normal condition (Fig. 6). Fig.10 shows spiking rate of the left 2nd CNbs in response to the disturbances under the different angular velocity gain conditions (normal gain: x 1, double gain: x 2, half gain: x 0.5). Comparing with the spontaneous spiking rates, the 2nd CNbs exhibited excitation under every angular velocity gain conditions in response to the disturbances (Fig. 10). However, the amount of spiking rates differed depending on the gains. Comparing with the normal gain condition, spiking rates increased in the half gain condition, and spiking rates decreased in the double gain condition (Fig. 8). Moths on the hybrid system could cancel out the disturbances even though the motor gain was altered, moths on the robot cancelled out the disturbances by increasing or decreasing the spiking rate.

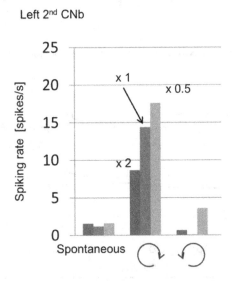

Fig. 10. Spiking rate of the 2nd CNb in response to the clockwise and counterclockwise rotational disturbances

4   Odor Source Orientation Experiments

We tested whether the compensatory responses to the disturbances (Fig. 10) were also observed in odor source searching behavior. We manipulated the proportional constants of the spike-behavior conversion rule and switched them between the normal angular velocity gain and the double angular velocity gain every 10 s during odor source localization experiments. Orientation experiments were held in a wind tunnel with its inside wind speed 0.65 m/s, and the odor source was set at 500 mm upwind from the starting point of the hybrid system (Fig. 11) [8]. We controlled the releasing frequency of the pheromone at 2 Hz by controlling an electric valve. Odor stimuli were delivered in the conventional way used in the previous study [11]. To simulate the visual condition in the wind tunnel, we put a 300 mm square checkerboard pattern on the ceiling and 120 mm width black and white sinusoidal stripes on the inside wall of the wind tunnel. We set a video camera (HDR-XR520V, SONY, Japan) to record movement of the robot to analyze its trajectories.

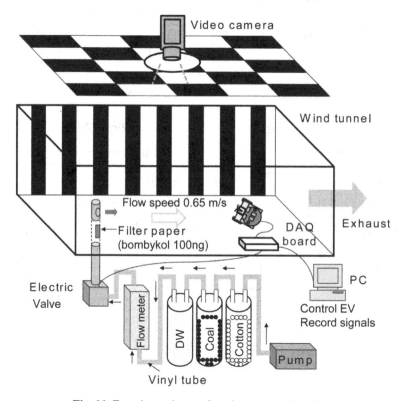

Fig. 11. Experimental setup for odor source orientation

Moths on the hybrid system inhibited neural responses in response to the change of the motor gain and succeeded in odor source orientation (Fig. 12). In Fig. 12, the upper graph indicates neural activities of the left and right 2nd CNbs, and the lower graph indicates the histogram of the left and right 2nd CNbs. The upward histogram indicates spiking rate of the left 2nd CNb, and the downward histogram indicates the right 2nd CNb. Activities in both of the 2nd CNbs were inhibited during the angular velocity double gain orientation (Wilcoxon signed-rank test, $p < 0.05$, $N = 8$)

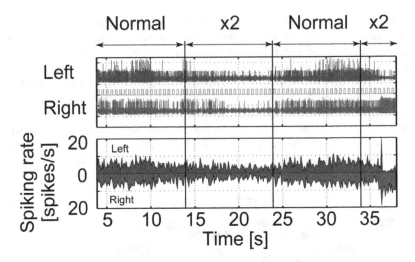

Fig. 12. Neural activities of the left and right 2nd CNbs during odor source searching behavior

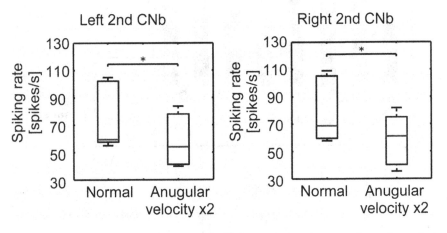

Fig. 13. Spiking rate of the left and right 2nd CNbs during the normal angular velocity conditions and the angular velocity double conditions

(Fig. 13). Average angular velocities acquired on the hybrid system during orientation experiments were 1.1 ± 0.3 rad/s (normal gain) and 1.3 ± 0.4 rad/s (double gain), and there was no significance between the two angular velocities. This means that moths can compensate motor gain of the hybrid system, and the behavior of the hybrid system is not simply decided by the spike-behavior conversion rule. Moths adjust their angular velocity to their appropriate value using the sensory feedbacks caused by their behavior.

5   Conclusion

In this study, we investigated the modification of the insect's command neural activities in response to the manipulation on the sensory feedbacks using the mobile robot controlled by the neural signals recorded from the directional selective neurons in the silkmoth cervical connective. We focused on the neck motor neurons corresponding to the deflection, lateral and sidewise translation movement in the fly's neck motor system [24]. As a result, an insect showed the ability to compensate for the given disturbances of the angular velocity, and performed compensatory responses to the changes of the motor gains even in the odor source searching behavior. This indicates that insects have the appropriate angular velocity during odor source searching behavior and compensate for the disturbances using visual feedback caused by the self-movement. In a previous work, it was also reported that moths exhibited the ability to compensate for the extraordinary behavioral conditions [25]. There are several studies that flying insects control their odor source searching behavior using optomotor responses [26-27]. However, the neural mechanisms underlying the multimodal sensory processing is poorly understood. Our experimental platform can be the effective tool to elucidate the mechanisms. In addition to the investigation of the neural mechanisms, to validate the hypothesis that the moths have appropriate angular velocity, we are now analyzing the contribution of the angular velocity to the effective odor source searching behavior in simulation experiments and experiments using the robot with gaseous sensors available in real environments.

Acknowledgements. This study was supported by Japanese KAKENHI MEXT 17075007 and Japanese KAKENHI JSPS 09J01188 and 12J10557. The experiments in this presentation were approved based on Article 22 clause 2 of the safety management rule of Tokyo Institute of Technology.

References

1. Chapin, J.K., Moxon, K.A., Markowitz, R.S., Nicolelis, M.A.L.: Real-Time Control of a Robot Arm Using Simultaneously Recorded Neurons in the Motor Cortex. Nat. Neurosci. 2(7), 664–670 (1998)
2. Wessberg, J., Stambaugh, C.R., Kralik, J.D., Beck, P.D., Laubech, M., Chapin, J.K., Kim, J., Biggs, S.J., Srinivasan, M.A., Nicolelis, M.A.L.: Real-time Prediction of Hand Trajectory by Ensembles of Cortical Neurons in Primates. Nature 408, 361–365 (2000)
3. Velliste, M., Perel, S., Spalding, M.C., Whitford, A.S., Schwartz, A.B.: Cortical Control of a Prosthetic Arm for Self-Feeding. Nature 453, 1098–1101 (2008)
4. Song, W., Giszter, S.F.: Adaptation to a Cortex-Controlled Robot Attached at the Pelvis and Engaged during Locomotion in Rats. J. Neurosci. 31(8), 3110–3128 (2011)
5. Strausfeld, N.J.: Atlas of an Insect Brain. Springer, Berlin (1999)
6. Ejaz, N., Krapp, H.G., Tanaka, R.J.: Closed-loop Response Properties of a Visual Interneuron Involved in Fly OptomotorControl. Front. Neural. Circuits. 7(50), 1–11 (2013)

7. Halloy, J., Sempo, G., Caprari, G., Asadpour, M., Tache, F., Durier, V., Canonge, S., Ame, J.M., Detrain, C., Correll, N., Martinoli, A., Mondada, F., Siegwart, R., Deneubourg, J.L.: Social Integration of Robots into Groups of Cockroaches to Control Self-Organized Choices. Science 318, 1155–1158 (2007)
8. Minegishi, R., Takashima, A., Kurabayashi, D., Kanzaki, R.: Construction of a Brain-Machine Hybrid System to Evaluate Adaptability of an Insect. Robot. Auton. Syst. 60, 692–699 (2012)
9. Kramer, E.: Orientation of the Male Silkmoth to the Sex Attractant Bombykol (Book style with paper title and editor). In: Denton, D.A., Coghlan, J. (eds.) Mechanisms in Insect Olfaction, pp. 329–335. Academic Press, New York (1975)
10. Obara, Y.: *Bombyxmori* Mating Dance: an Essential in Locating the Female. Appl. Entomol. Zool. 14(1), 130–132 (1979)
11. Kanzaki, R., Sugi, N., Shibuya, T.: Self-Generated Zigzag Turning of *Bombyxmori*Males during Pheromone-Mediated Upwind Walking. Zool. Sci. 9(3), 515–527 (1992)
12. Kanzaki, R.: Coordination of Wing Motion and Walking Suggests Common Control of Zigzag Motor Program in a Male Silkworm Moth. J. Comp. Physiol. A. 182(3), 267–276 (1998)
13. Mishima, T., Kanzaki, R.: Coordination of Flipflopping Neural Signals and Head Turning during Pheromone-Mediated Walking in a Male Silkworm Moth *Bombyxmori*. J. Comp. Physiol. A 183(3), 273–282 (1998)
14. Mishima, T., Kanzaki, R.: Physiological and Morphological Characterization of Olfactory Descending Interneurons of the Male Silkworm Moth, *Bombyxmori*. J. Comp. Physiol. A 184(2), 143–160 (1999)
15. Wada, S., Kanzaki, R.: Neural Control Mechanisms of the Pheromone-Triggered Programmed Behavior in Male Silkmoths Revealed by Double-Labeling of Descending Interneurons and a Motor Neuron. J. Comp. Neurol. 484(2), 168–182 (2005)
16. Iwano, M., Hill, E.S., Mori, A., Mishima, T., Mishima, T., Ito, K., Kanzaki, R.: Neurons Associated With the Flip-Flop Activity in the Lateral Accessory Lobe and Ventral Protocerebrum of the Silkworm Moth Brain. J. Comp. Neurol. 518(3), 366–388 (2010)
17. Kanzaki, R., Mishima, T.: Pheromone-Triggered 'Flipflopping' Neural Signals Correlate with Activities of Neck Motor Neurons of a Male Moth. *Bombyxmori*. Zool. Sci. 13(1), 79–87 (1996)
18. Reichardt, W.: Nervous Integration in the Facet Eye. Biophys. J. 2, 121–143 (1962)
19. David, C.T.: OptomotorControl of Speed and Height by Free-Flying *Drosophila*. J. Exp. Biol. 82, 389–392 (1979)
20. Srinivasan, M.V.: Insect as GibsonianAnimals. Ecol. Psychol. 10(3-4), 251–270 (1998)
21. Srinivasan, M.V., Poteser, M., Kral, K.: Motion Detection in Insect Orientation and Navigation. Vision Res. 39(16), 2749–2766 (1999)
22. Kern, R., Egelhaaf, M.: OptomotorCourse Control in Flies with Largely Asymmetric Visual Input. J. Comp. Physiol. A 186(1), 45–55 (2000)
23. Straw, A.D.: Vision Egg: An Open-Source Libraryfor Realtime Visual Stimulus Generation. Front. Neuroinform. 2(4), 1–10 (2008)
24. Strausfeld, N.J., Seyan, H.S., Milde, J.J.: The neck motor system of the fly *Calliphoraerythrocephala* I. Muscles and motor neurons. J. Comp. Physiol. A 160, 205–224 (1987)
25. Ando, N., Emoto, S., Kanzaki, R.: Odour-Tracking Capability of a SilkmothDriving a Mobile Robot with Turning Bias and Time Delay. Bioinspir. Biomim. 8(1), 1–14 (2013)
26. Baker, T.C., Willis, M.A., Phelan, P.L.: Optomotor Anemotaxis Polarizes Self-Steered Zigzagging in Flying Moths. Physiol. Entomol. 9(4), 365–376 (1984)
27. Vespui, R., Gray, J.R.: Visual Stimuli Induced by Self-Motion and Object Motion Modify Odour-Guided Flight of Male Moths (*Manducasexta* L.). J. Exp. Biol. 212, 3272–3282 (2009)

Perception of Simple Stimuli Using Sparse Data from a Tactile Whisker Array

Ben Mitchinson[1,*], J. Charles Sullivan[2], Martin J. Pearson[2],
Anthony G. Pipe[2], and Tony J. Prescott[1]

[1] The University of Sheffield, Sheffield, UK
b.mitchinson@shef.ac.uk
[2] Bristol Robotics Laboratory, Bristol, UK

Abstract. We introduce a new multi-element sensory array built from tactile whiskers and modelled on the mammalian whisker sensory system. The new array adds, over previous designs, an actuated degree of freedom corresponding approximately to the mobility of the mystacial pad of the animal. We also report on its performance in a preliminary test of simultaneous identification and localisation of simple stimuli (spheres and a plane). The sensory processing system uses prior knowledge of the set of possible stimuli to generate percepts of the form and location of extensive stimuli from sparse and highly localised sensory data. Our results suggest that the additional degree of freedom has the potential to offer a benefit to perception accuracy for this type of sensor.

Keywords: Tactile sensing, Whiskers, Perception, Robotics.

1 Introduction

Many small mammals (including rats, mice, and shrews) sport an array of long, touch-sensitive, facial whiskers which they rely on to locate and identify objects and obstacles in their environment [1]. This system has long been a popular model of biology amongst neuroscientists, but it also has characteristics attractive to biomimetic engineers. Most obviously, whiskers operate without the use of any sort of radiation. This means that they can be operated within opaque fluids (e.g. turbid water or smoke-filled air, which would stymy visible-light sensing) and that they are not affected by ambient radiative noise. Less obviously, whiskers report local geometry very directly, so that the well-known ambiguities familiar from, say, visual sensory processing do not arise, and the recovery of geometry is computationally trivial. Thus, artificial sensory whiskers have more recently been built and studied both as individual sensors and as sensory arrays; see Prescott *et al.* [2] for a recent review.

The design of our own multi-whisker robotic platforms has increased in sophistication—see Figure 1 for illustrative examples—alongside our understanding both of the biological whisker sensory system and of the importance of

---

* Corresponding author.

N.F. Lepora et al. (Eds.): Living Machines 2013, LNAI 8064, pp. 179–190, 2013.

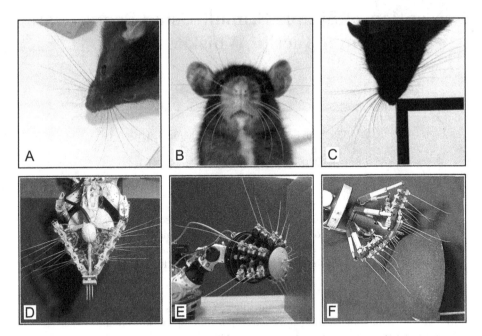

Fig. 1. *Rats and robots.* Features of the mystacial whisker morphology/functionality of rats include bilateral whisker arrays (A), coverage all round the snout (B), and reshaping the mystacial pad to move the bases of the whiskers in response to local geometry (C). These features, respectively, have inspired design features of three generations of whiskered robot: Scratchbot (D), Shrewbot (E) and the BIOTACT G2 Sensor (F).

morphological and functional features to an artificial system. The whiskers of 'Scratchbot' (Figure 1D) were more-or-less sideways-facing, had limited degrees of freedom (DOFs; three per side) [2], and the platform was used to investigate active sensing on small temporal and spatial scales [3]. Scratchbot exhibited 'blind spots', however, owing to the limited coverage of the area around the snout by whiskers, so that unsupervised exploration was impossible. 'Shrewbot'[1] (Figure 1E) added all-around-the-snout whisker sensing, as well as independent actuation of each whisker so that active sensing could be optimised per-whisker [4]. This allowed unsupervised operation on long timescales (for instance, to study tactile SLAM [5]). Using the Shrewbot platform we also demonstrated simultaneous identification and localisation of 3D shapes in a mobile target acquisition task with feedback-controlled interactions [6]. In that study, we used model-based percepts to drive effective animal-like prey capture behaviour from sparse, localised, sensory data. Alongside Shrewbot, we developed a system—the 'BIOTACT G1 Sensor' (not shown in the Figure)—that mounted a similar array of whiskers on a robot arm rather than a mobile platform to facilitate investigations of

---

[1] Though, as the name implies, the design of Shrewbot is actually based on a type of shrew, the functional and morphological features are much the same in rats.

precisely controlled interactions [4]. G1 has been used to demonstrate identification/localisation of both textured surfaces [7] and 2D shapes [8,9].

A shortcoming of these systems that became apparent through experience is the difficulty in bringing many whiskers into contact with convex local geometry (for example, small objects in front of the snout). This is because the bases of the whiskers are arranged in a regular fashion spread along—essentially—the 'length' of the robot, and more rearward whiskers cannot cross over more forward whiskers. Observation of rats exploring such challenging geometries (for example, see Figure 1C) suggests that quite different arrangements of the whisker bases are available to the animal depending on what is being sensed, through control over the shape of the 'mystacial pad' (something like the 'cheek'). A recent, very detailed, study of rat whisker and facial morphology highlighted the centrality of such morphological details to the available sensor-environment interactions with objects of different curvature [10]. In the current study, then, we introduce the next generation BIOTACT Sensor: 'G2' (Figure 1F) adds an additional DOF to each whisker row—control over the angle of the bearer ('cheek') which carries the whisker modules—intended to offer more flexible sensor-environment interactions. We also report the results of a task involving identification/localisation of 3D shapes with four stimulus classes and pre-programmed interactions, confirming and extending our previous studies using Shrewbot and G1, and providing a first assessment of the usefulness of the additional DOF.

2 Methods

2.1 Data Collection

Hardware. G2 is shown in Figure 2 with 18 whisker modules fitted (3 whiskers in each of six rows distributed regularly around the 'snout'). The whisker modules are identical to those used in G1 [4]. Briefly, the module integrates a composite tapered whisker mounted on a rotating shaft, motor and motor controller to cause whisker 'protraction', and Hall effect sensors measuring protraction angle and 2D whisker deflection. Each whisker can be rotated around its base through approximately 100° (Figure 2A/B). Each whisker module is mounted on one of six cheek members, pivoted near the tip of the snout, which can be swung backwards and forwards through approximately 30° (Figure 2B/C) by a linear actuator (Firgelli L12, `firgelli.com`). The whiskers vary in length along each row, from 80mm (at the front) to 165mm (at the back). The G2 sensor is mounted on a 7 DOF robotic arm so that it can be positioned and oriented as required in space. Note that the sensor also sports an array of small whiskers at the front in the centre, visible in the Figures, that are not used in this study.

Experimental Protocol. To perform experimental trials, the sensor is brought by the robot arm into juxtaposition with one of four stimuli, S1-4, so that the sensor 'faces' the stimulus (Figure 2D). S1-4 are, respectively, spheres of diameter 200, 300 and 600mm, and a flat plane. The location of the sensor relative to the stimulus is controlled in one lateral dimension to ten locations, L1-10, 40mm apart.

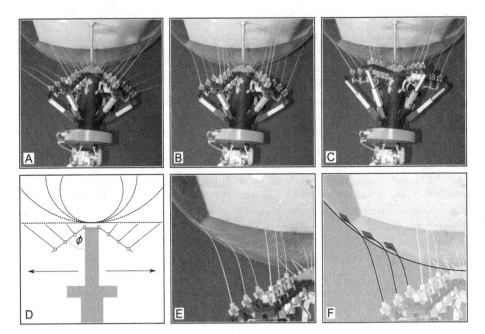

Fig. 2. *G2 Sensor.* Top-down views of G2 contacting stimulus S3 (600mm sphere), with cheeks and whiskers retracted (A), whiskers protracted (B), and cheeks and whiskers protracted (C). *Experimental protocol.* (D) Top-down view of experimental protocol showing key functional geometry of G2 (solid lines; cheeks and whiskers) as well as the four stimuli (dashed lines; three spheres and one plane, S1-4). The variable relative location of sensor and stimulus in one dimension (L1-10) is indicated by arrows; the cheek protraction angle—denoted ϕ—is also indicated. *Sensory perception.* (E) Close-up of whisker-stimulus interaction from panel B. (F) Same image with stimulus and three whiskers highlighted (solid lines) and three 'surfels' indicated (dark grey patches).

At each position, the sensor conducts thirty 'whisks'. Each whisk takes one second, and involves protracting all the whiskers forward until they either reach their maximum protraction angle (after 700ms) or contact a surface; in either event, the individual whisker is retracted. This 'early' cessation of protraction in the case of surface contact is intended to implement a 'minimal impingement' control strategy such as has been proposed as a component of the whisker control strategy used by rats [11]. The cheek angles (ϕ, Figure 2D) are controlled according to one of three strategies, C1-3 (described below), in separate realisations of the protocol. Thus, this study involves a total of $4 \times 10 \times 30 \times 3 = 3600$ whisks, which we hereafter label as trials. We index trials with the variable k, and denote the stimulus for trial k as s_k.

Contact Signals. At each time sample (200Hz), each whisker reports a 2D deflection vector, $\boldsymbol{d}_w(t)$. From this, we derive a 'contact belief' signal, $b_w(t)$, as

$$b_w(t) = \begin{cases} 0, & v_w(t) < 0 \\ 1, & v_w(t) > 1 \\ v_w(t), & \text{otherwise} \end{cases}, \quad v_w(t) = ||d_w(t)||/\eta_w - 1. \tag{1}$$

The parameter η_w is a measure of the sensor noise during non-contacting whisking, and is chosen separately for each length of whisker (of which there are three) based on data from test trials where no stimulus was present: η_w is set by eye to the peak value of $||d_w(t)||$ seen in such test trials plus about 50%. Thus, values of b_w of 0 and 1 are intended to indicate, approximately and respectively, 'certainty of not contact' and 'certainty of contact'.

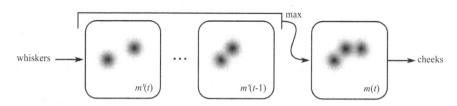

Fig. 3. *Egocentric processing.* 'Whisker-centric' contact signals are mapped into a 3D egocentric map $m'(t)$ through a Gaussian spatial filter. An element-wise maximum over one second's worth of this signal stream forms the (with memory) contact map $m(t)$, which drives the cheeks when the C1 control strategy is in use.

Egocentric Processing. Spatial processing is performed using egocentric maps, each of which is maintained as a 3D array of voxels (cubes of side 10mm). Using the known geometry of the sensor and the instantaneous cheek and whisker angles, the locations of the whisker tips in 3D egocentric space are computed. The contact belief signals for all whiskers, $\{b_w(t)\}$, are then transformed into a map $m'(t)$ at these tip locations through a Gaussian spatial filter of width 25mm, to form an egocentric map of instantaneous contact belief (see Figure 3). An element-wise maximum operation over all samples of $m'(t)$ from the last one second is then applied, to produce a single final map of contact belief, $m(t)$, that has a one second memory (Figure 3). This map $m(t)$ forms the input to a transform, described below, that controls the cheeks for the C1 strategy.

Cheek Control Strategies. The C1 control strategy chooses a protraction angle independently for each cheek such that the whiskers mounted on it are afforded the opportunity, as far as possible, to bring their tips to where contact has recently occurred (i.e. to regions of $m(t)$ that are active). The strategy is written

$$\phi_c = \frac{\sum_e K_c \star U_c \star m(t)}{\sum_e U_c \star m(t)}, \tag{2}$$

where ϕ_c is the angle of the cth cheek, K_c is a 3D array of proposed cheek angles for cth cheek equal in size to $m(t)$, U_c is a weight array of the same size, $\star$

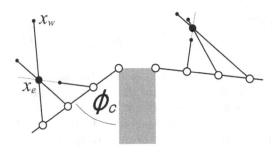

Fig. 4. *Strategy C1.* (Left) Example element of egocentric map at $\boldsymbol{x}_e$ (large black dot). For cheek angle shown, ϕ_c, whiskers are adjusted so tips (small black dots) are as close to $\boldsymbol{x}_e$ as possible. Tip of wth whisker in that configuration is denoted $\boldsymbol{x}_w(\phi_c)$. (Right) A second example, at a different location.

represents element-wise multiplication and $\sum_e$ is the sum over all elements of the 3D array. Thus, ϕ_c is given by an average weighted by $U_c \star m(t)$ over the angles proposed by each element of K_c. The arrays K_c and U_c are generated as follows (see also Figure 4). To compute the eth element of both arrays, we start with the location in space represented by that element, denoted $\boldsymbol{x}_e$. For any value of ϕ_c, the protraction angle of each whisker can be adjusted so as to bring its tip as close as possible to $\boldsymbol{x}_e$. Its tip position in this configuration is denoted $\boldsymbol{x}_w(\phi_c)$. The 'benefit' of a proposed cheek angle for this whisker for this element is then calculated as $u_{e,w}(\phi_c) = \exp(-\|\boldsymbol{x}_e - \boldsymbol{x}_w(\phi_c)\|/r)$ where $r = 75$mm. That is, if the tip of the whisker can reach near to the element's location at some whisker angle, given the proposed cheek angle, a large benefit results; otherwise, a small benefit results. Finally, the preferred cheek angle ϕ_c for this element e of the map is simply that which maximises the average benefit $u_{e,w}(\phi_c)$ across the whiskers w of that cheek. This angle is entered into the corresponding element of K_c, and the average benefit at that value is entered into U_c. To summarize, activity in a region of the map $m(t)$ will tend to drive a cheek so that the tips of the whiskers mounted on it can reach that region (encoded in K_c), but only for regions of the map for which that can actually be achieved (encoded in U_c). Thus, strategy C1 is intended to favour cheek angles that permit whisker-stimulus contact at *some* whisker protraction angle, so as to implement a 'maximal contact' control strategy, proposed by Mitchinson *et al.* [11] as a component of the strategy used by rats. Strategies C2 and C3 set all cheek protraction angles to a fixed value, as if the cheeks were not actuated; the three strategies are summarised below.

- **C1** Cheek angles controlled individually and automatically based on past sensory data to favour contact at the whisker tips (Equation 2).
- **C2** All cheek angles fixed at fully protracted ($\phi_c = 85°\ \forall\ c$).
- **C3** All cheek angles fixed at halfway protracted ($\phi_c = 70°\ \forall\ c$).

2.2   Analysis

Feature Extraction. The data streams from each whisker are initially treated separately. Within each trial, for the wth whisker, we identify the time of maximum protraction angle, and sample $\boldsymbol{d}_w$ and $\boldsymbol{b}_w$ at that time (we expect, then, to have taken these samples when the whisker was undergoing the most strong deflection). For each whisker a 'surfel'—an oriented patch of surface—is computed, as follows (some example surfels are indicated in Figure 2F). First, the amount of physical deflection of the whisker tip is estimated according to $v_w = q_v d_w$ where q_v is an (unknown) gain between deflection sensor data and physical deflection in millimetres. The undeflected location of the whisker tip is calculated according to the known geometry and the current angles of cheek and whisker, and v_w is added to it to give the estimated location of the whisker tip at peak protraction/deflection, x_w. Next, a 'surface parallel' vector, p_w, is estimated by finding a unit vector perpendicular to both v_w and the undeflected whisker shaft (note that the surface normal cannot be uniquely identified from the report of a single whisker at a single time sample). Finally, the surfel $L_w = \{x_w, p_w, b_w\}$ consists of an estimate of a point on the surface, x_w, an estimate of a unit vector parallel to the surface, p_w, and a measure of the belief that contact occurred, b_w. This is, of course, a simplified model of whisker deflection but is expected to be reasonably accurate for the small deflections and contact at or near the tip that are encountered in this experiment.

Candidate Percepts. For the kth trial, the complete set of sensory data extracted can be written as $\mathbf{L}_k = \{L_w\}$, one surfel for each whisker. Four percept models, denoted Ŝ1-4, are fitted to these data. Ŝi corresponds to Si—that is, Ŝ1-3 are models of spheres with radii of 200/300/600mm, whilst Ŝ4 is a model of a plane. The parameters of Ŝ1-3 are the location relative to the tip of the snout of the centre of the sphere; the parameters of Ŝ4 are the distance ahead of the tip of the snout and the angle the plane normal makes with the midline of the snout. The parameters of each model Ŝi are optimised (`fminsearch`, Mathworks MATLAB™) against $\mathbf{L}_k$ for each trial using a cost function

$$J(\hat{S}i, \mathbf{L}_k) = \sum_w \left(J_1(\hat{S}i, L_w) + J_2(\hat{S}i, L_w) + q_p J_3(\hat{S}i, L_w) \right) \tag{3}$$

$$J_1(\hat{S}i, L_w) = b_w |F(\hat{S}i, x_w)| \tag{4}$$

$$J_2(\hat{S}i, L_w) = (1 - b_w) \max(F(\hat{S}i, x_w), 0) \tag{5}$$

$$J_3(\hat{S}i, L_w) = b_w \left| < G(\hat{S}i, x_w), p_w > \right|. \tag{6}$$

In the above, both $F(.)$ and $G(.)$ are functions that involve computing the point on the surface of the model Ŝi nearest the surfel location x_w, denoted x'_w. $F(.)$ returns the signed distance that x_w is inside the model surface (i.e. the distance from x_w to x'_w, when x_w is inside Ŝi, and the negative of this value, otherwise). $G(.)$ returns the normal to the model surface at x'_w. We do not explicitly describe these functions since they are simple geometric computations specific to the form of each percept model (sphere or plane, here). The operator $<, >$ indicates the scalar product. Thus, the three cost function components penalise, respectively, surfels that are distant from the model surface and have high contact belief, surfels that are *inside* the model surface and have low contact belief, and surfels that have high contact belief and are not aligned with the model surface. The overall cost, J, is a weighted sum of the three components, where q_p is an (unknown) weight parameter for the alignment cost component.

Unitary Percept. Finally, a unitary percept for the kth trial, $\hat{s}_k$, is generated simply by selecting the candidate percept $\hat{S}i$ with the lowest weighted cost. That is,

$$\hat{s}_k = \hat{S}i, \ i = \operatorname*{argmin}_i \left(q_i J(\hat{S}i, \mathbf{L}_k) \right), \tag{7}$$

where $\{q_i\}$ are a set of (unknown) weights. These weights are required, for example, because smaller stimuli may elicit lower cost values simply by eliciting a lower number of whisker contacts. Note that we do not need to consider an alternative percept 'no stimulus present' since our experiment does not permit that case (cf. a related investigation where absence of stimulus is a valid case [6]). We can then compute the 'correctness' of the identification,

$$C_k = \begin{cases} 1, \hat{s}_k = s_k \\ 0, \hat{s}_k \neq s_k \end{cases}. \tag{8}$$

3 Results

3.1 Data Collection

In these experiments the whiskers were used to palpate the stimuli, as described in Methods. Therefore, sensory data (non-zero values of contact belief) are available for a short period (some tens of milliseconds) within each whisk/trial (one second), at around the time of maximum whisker protraction. The one second memory in $m(t)$ is necessary so that these transient data can govern the cheek angles during the next whisker-stimulus interaction, bridging the intervening hiatus in data availability (incidentally, the cheek actuators take around the same period of time—one second—to respond to a commanded angle). As a consequence, within each 30 trial set, the cheek angles during trial k are driven by sensory data collected during trial $k-1$. Since within these sets the stimulus and sensor are not moved, this delay—that would also be encountered by a whisking animal—need not be accounted for by the control algorithm. Such a long delay does permit that a large set of controllers would generate oscillation; with the controller described, however, we generally saw instead rapid (one trial or so) convergence to a fixed cheek position, though we did not assess this formally.

3.2 Parameter Optimisation

The analysis described above has six unknown parameters: q_v, q_p, and $\{q_i\}$, $i \in [1, 4]$. In practice, the percept weights can be normalised, so we set $q_1 = 1$, leaving just five unknowns. q_v could in theory be measured, but it is difficult to do so accurately in practice, so we used optimisation to select all five. To make fair comparison across the conditions C1-3, we did this separately for each condition. Thus, we split the 3600 trials collected into training and test sets for each condition. Specifically, of each set of 30 whisks/trials, we discarded the first 2 (control loop settling), used the next 8 for training, and the remaining 20 for testing. We then maximised the total number of 'correct' training trials ($\sum_k C_k$) against the

Table 1. Analysis parameter values chosen by independent optimisation for each condition (C1, C2, C3) and by global optimisation across all conditions (C1-3). $q_1 = 1$ is assumed in all conditions.

Condition	q_v	q_p	$[q_2 \; q_3 \; q_4]$
C1	250	10	[0.75 0.60 0.60]
C2	150	2	[0.70 0.50 0.45]
C3	150	10	[0.85 0.60 0.70]
C1-3	250	10	[0.95 0.90 0.90]

analysis parameters by exhaustive search. This *per-condition* parameter choice included fairly similar values for each condition, as shown in Table 1. We also computed a *global* set of analysis parameters that were optimised across all conditions (C1-3). The remainder of this section reports the results of using these parameter sets to process the 2400 test trials, using the *per-condition* parameter set except where stated.

3.3 Performance

We performed two analyses of the test data. In the first, we provide identification information—that is, we consider only model $\hat{S}i$ for trials with stimulus Si—and review performance in localisation of the stimuli. Results of this analysis are graphed in Figure 5, averaged across 20 test trials. Breaking down

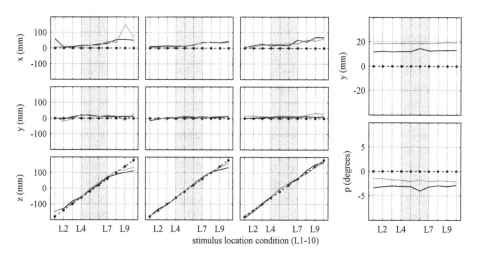

Fig. 5. Columns 1-4 correspond to S1-4 (also, $\hat{S}$1-4, respectively, since correct identification is assumed, here). Columns 1-3 graph normalised location of sphere models $\hat{S}$1-3 (ground truth is dashed line, estimates due to C1/2 are solid black/grey lines; estimate due to C3 was qualitatively similar and is omitted for clarity), averaged across 20 test trials. Column 4 shows location and orientation of plane model $\hat{S}$4 (averaged across whisks, legend as for columns 1-3). Shaded region is L4-7 (see text).

these results by cheek control strategy, mean localisation error averaged over all sphere trials was 37/38/46mm (for C1/2/3, respectively), over all plane trials was 13/19/20mm and 3/2/2°. The graphs show that localisation is poorer at larger distances from smaller spherical stimuli—as expected, since fewer whiskers contact the stimulus under these conditions. In an 'inner region' L4-7, where multiple contacts are most likely, mean sphere localisation error was 29/31/35mm (location did not, as expected, affect performance in locating the plane stimulus). Additionally, the graphs suggest some systematic sources of error—the standard deviation of localisation error across the 20 trials contributing to each point in the plots is not shown, for clarity, but was smaller than the mean error at the majority of parameter points—we return to this in Section 4. Using the *global* parameter set, sphere localisation error was 37/42/43mm and plane localisation error was 13/7/8mm and 3/2/3° (C1/2/3, respectively).

In the second analysis, the percept $\hat{s}_k$ with the lowest weighted cost was chosen for each trial and compared with the true stimulus identity s_k, as described in the equations above. Identification was correct in 92/90/93% of trials (for C1/2/3, respectively). Focusing only on L4-7, these percentages are 99/97/85%. It is interesting that the performance of C3 is actually substantially worse in the inner region than overall—in an 'outer region' (L1-3 and L8-10), identification was correct in 87/85/99% of trials. Examining the confusion matrices in the C3 case for the inner region indicates that 47 of 48 mis-identifications made were mistaking S2 for S1 (i.e. perceiving a 200mm sphere when a 300mm sphere is present), a matter we will return to in Section 4. Using the *global* parameter set, identification was correct in a slightly reduced 88/89/89% of trials.

4   Discussion

We have presented a new multi-element tactile sensory array with an additional DOF (per row of whiskers) over previous sensors, as well as a preliminary quantitative analysis of identification and localisation of four simple stimuli using a model-based technique. Overall, our results confirm and extend those of a previous—smaller and less controlled—study using two stimuli and a mobile platform [6]. That is, this type of sensor and analysis can discriminate amongst such a set of objects with correctness around 90% and, given correct identification, simultaneously locate them with precision on the order of a few tens of millimetres. As illustrated, model-based perception is particularly well suited to whisker sensory systems since it generates complete percepts of extensive stimuli from the highly localised sensory data available from the whisker array.

We also made a first attempt at deriving a control strategy for the additional DOF that favours optimal sensing. Such a strategy could be derived by experiment, but the procedure would be onerous (owing to the large space of possible stimuli). Here, we hand-crafted a controller (C1) intended to favour contact at the tips of the whiskers. Compared with strategy C2 (cheeks fully protracted), C1 seemed to perform marginally better in both localisation and identification analyses. Compared with strategy C3 (cheeks half protracted), the picture was more complex. C1 performed consistently (and more markedly) better at localisation, and also at identification for the 'inner region' (L4-7), where the sensor was most directly opposite the stimulus (99% correct for C1 versus 85% for C3). However, in the 'outer region' (L1-3, L8-10), where fewer whiskers would have been able to contact the stimulus in general, the picture for identification was reversed, with C3 (99% correct) being markedly superior to C1 (87% correct). Interestingly, the great majority of the mistakes made in the C3/inner case (47 of 48) were confusing the second smallest stimulus, S2, for the smallest, S1. Most of the mistakes made in the C1/outer case (55 of 61) were mistaking the smallest sphere, S1, for one of the other stimuli. These results confirm the general point that smaller (or, perhaps, more convex) stimuli are the more challenging to sense for a system with this general morphology.

A confounding systematic source of error is visible in the plots of Figure 5, both in the trend in the x location of spheres and in the fixed offset in the y location of the plane. The value of the fixed offset was affected by the choice of analysis parameters, underlining one probable source of error: imprecision in the whisker deflection model. Another probable source is error in calibration of hardware geometry, of the robot arm, whiskers, or stimuli. The relative error of the localisation measurements made by the system was generally lower in magnitude than the absolute systematic error, so that it is difficult to draw reliable conclusions regarding the relative performance of the different controllers. Overall, then, these results seem too complex to draw strong conclusions about the benefit to sensing offered by the online cheek control strategy tested here (C1). However, they do indicate unambiguously that a *post hoc* control strategy exists that would substantially outperform either of the 'non-actuated' strategies C2 or C3 in identification. That is, to use C1 in the inner region (99% correct) and C3 in the outer region (99% correct). This strategy markedly out-performs all other strategies tested here (99% correct, versus 92/90/93% correct for C1/2/3).

One line of future work, then, will be to seek a controller that can realise this *post hoc* advantage at run-time. In addition, the conditions under which we would expect to see most strongly the advantage of actuated cheeks remain to be tested: these are when stimuli can be presented not only ahead of, but also to one side of, the sensor. An experiment to test under these conditions is more difficult to design, since some sensor/stimulus/cheek combinations will generate collisions between the sensor hardware and the stimulus. Nonetheless, such tests will be necessary if the full gamut of conditions experienced by the exploring animal/mobile-robot is to be addressed. Figure 5 highlights that the location of the sensor relative to the stimulus (L1-10) has an impact on localisation performance. Lepora *et al.* have shown that active control over location using sensory feedback at run-time can improve sensing performance under such conditions [12] (this issue); incorporating such feedback should improve localisation performance of this system, also. Whilst identification performance using each individual strategy did appear to be location-dependent, the very high performance (99% correct) available from the *post hoc* C1/C3 strategy means that it is impossible to gauge whether active sensing of this type would also aid identification (though it seems intuitively likely). Accordingly, any future tests of identification using active sensing will need to employ more finely-graded stimuli. Finally, the use of an improved whisker deflection model and the addition of adaptive calibration may be able to eliminate systematic errors. We expect inclusion of these features to permit substantially more discerning identification and more precise localisation using a whisker array of this type.

Acknowledgements. The authors thank the technical support engineers at Bristol Robotics Laboratory and the anonymous reviewers. This work was supported by the FP7 grants BIOTACT (ICT-215910) and EFAA (ICT-270490).

References

1. Brecht, M., Preilowski, B., Merzenich, M.M.: Functional architecture of the mystacial vibrissae. Behavioural Brain Research 84(1), 81–97 (1997)
2. Prescott, T.J., Pearson, M.J., Mitchinson, B., Sullivan, J.C.W., Pipe, A.G.: Whisking with robots. IEEE Robotics & Automation Magazine 16(3), 42–50 (2009)
3. Pearson, M., Mitchinson, B., Welsby, J., Pipe, T., Prescott, T.: Scratchbot: Active tactile sensing in a whiskered mobile robot. Animals to Animats 11, 93–103 (2010)
4. Pearson, M.J., Mitchinson, B., Sullivan, J.C., Pipe, A.G., Prescott, T.J.: Biomimetic vibrissal sensing for robots. Philosophical Transactions of the Royal Society B: Biological Sciences 366(1581), 3085–3096 (2011)
5. Pearson, M.J., Fox, C., Sullivan, J.C., Prescott, T.J., Pipe, T., Mitchinson, B.: Simultaneous localisation and mapping on a multi-degree of freedom biomimetic whiskered robot. In: ICRA 2013 (to appear, 2013)
6. Mitchinson, B., Pearson, M.J., Pipe, A.G., Prescott, T.J.: Biomimetic tactile target acquisition, tracking and capture. In: Robotics and Autonomous Systems (in press)
7. Sullivan, J., Mitchinson, B., Pearson, M.J., Evans, M., Lepora, N.F., Fox, C.W., Melhuish, C., Prescott, T.J.: Tactile discrimination using active whisker sensors. IEEE Sensors Journal 12(2), 350–362 (2012)
8. Lepora, N., Fox, C., Evans, M., Mitchinson, B., Motiwala, A., Sullivan, J., Pearson, M., Welsby, J., Pipe, T., Gurney, K., et al.: A general classifier of whisker data using stationary naive bayes: application to biotact robots. Towards Autonomous Robotic Systems, 13–23 (2011)
9. Lepora, N.F., Sullivan, J.C., Mitchinson, B., Pearson, M., Gurney, K., Prescott, T.J.: Brain-inspired bayesian perception for biomimetic robot touch. In: IEEE International Conference on Robotics and Automation (ICRA), pp. 5111–5116 (2012)
10. Towal, R.B., Quist, B.W., Gopal, V., Solomon, J.H., Hartmann, M.J.: The morphology of the rat vibrissal array: a model for quantifying spatiotemporal patterns of whisker-object contact. PLoS Computational Biology 7(4), e1001120 (2011)
11. Mitchinson, B., Martin, C.J., Grant, R.A., Prescott, T.J.: Feedback control in active sensing: rat exploratory whisking is modulated by environmental contact. Proceedings Royal Society B: Biological Sciences 274(1613), 1035–1041 (2007)
12. Lepora, N.F., Martinez-Hernandez, U., Prescott, T.J.: A SOLID case for active bayesian perception in robot touch. In: Lepora, N.F., Mura, A., Krapp, H.G., Verschure, P.F.M.J., Prescott, T.J. (eds.) Living Machines 2013. LNCS (LNAI), vol. 8064, pp. 154–166. Springer, Heidelberg (2013)

Learning Epistemic Actions in Model-Free Memory-Free Reinforcement Learning: Experiments with a Neuro-robotic Model

Dimitri Ognibene[1], Nicola Catenacci Volpi[4],
Giovanni Pezzulo[2,3], and Gianluca Baldassare[2,*]

[1] Personal Robotics Laboratory, Imperial College London, UK
[2] Istituto di Scienze e Tecnologie della Cognizione, CNR, Italy
[3] Istituto di Linguistica Computazionale "Antonio Zampolli", CNR, Italy
[4] IMT Institute for Advanced Studies, Lucca, Italy

Abstract. Passive sensory processing is often insufficient to guide biological organisms in complex environments. Rather, behaviourally relevant information can be accessed by performing so-called *epistemic actions* that explicitly aim at unveiling hidden information. However, it is still unclear how an autonomous agent can learn epistemic actions and how it can use them adaptively. In this work, we propose a definition of epistemic actions for POMDPs that derive from their characterizations in cognitive science and classical planning literature. We give theoretical insights about how partial observability and epistemic actions can affect the learning process and performance in the extreme conditions of model-free and memory-free reinforcement learning where hidden information cannot be represented. We finally investigate these concepts using an integrated eye-arm neural architecture for robot control, which can use its effectors to execute epistemic actions and can exploit the actively gathered information to efficiently accomplish a seek-and-reach task.

1 Introduction

When an agent is executing a task in a non-deterministic and partially observable environment its behavior is affected by its limited knowledge. Recent evidence in neuroscience [1–3] indicates that living organisms can take into consideration the confidence in their knowledge and execute actions that allow the decrease of uncertainty if they satisfy a value/cost trade-off. These actions are named *epistemic actions* in cognitive science and in the planning literature, and *information-gathering actions* in operation research.

In robotics, epistemic actions have been applied in several tasks such as navigation (e.g., moving to positions where sensors can perceive to landmarks [4, 5]), active vision (e.g. moving the camera to acquire information given the limited

---

* This research was funded by the European Projects EFAA (G.A. FP7-ICT-270490), Goal-Leaders (G.A. FP7-ICT-270108), and IM-CLeVeR (G.A. FP7-ICT-IP-231722).

N.F. Lepora et al. (Eds.): Living Machines 2013, LNAI 8064, pp. 191–203, 2013.

field of view, the occlusions, and the changes in the environment [6]), tactile exploration [7, 8], and the active use of bio-inspired sensors such as rat whiskers [9]. The intrinsic complexity of the real world, however, may require even more versatile strategies like the execution of actions that change the environment in order to facilitate perception. Some examples are: opening a box or a drawer to see its content; rotating a picture to inspect its back; digging the ground to find root crops or moving the foliage to find fruits. Most of these actions cannot be predefined in the same way sensor controls are, because they are not purely epistemic. Indeed, in addition to changing the agent knowledge they also change the state of the environment and as a consequence they might result to be maladaptive for the agent.

A typical approach to solve the lack of information is using memory of previous perceptions. However, in some situations acting without appropriate knowledge is not useful (e.g., trying to open a safe with a limited number of attempts) so the first actions to execute should be directed to gather information (e.g., asking the opening number). Acting ignoring ignorance is seldom a good strategy. However, it is also not easy to devise at design time which actions an agent should execute to decrease uncertainty or which hidden structure of the environment it will encounter.

2   Epistemic Actions and POMDP Approximations

In the classical AI planning literature the problem of limited knowledge has been faced by adding knowledge preconditions to classical action definitions and through the definition of *epistemic actions* [10]. Knowledge preconditions define "what" information must be acquired, and the epistemic actions define "how". A common characteristic is that epistemic actions change only the agent knowledge of the world, and, differently from *pragmatic* or *ontic* actions, they do not change the world state [10, 11]. However, as we discussed earlier some actions affect both the perception (and knowledge) of the agent and the state of the environment. Given this ambiguity, a common choice is to model an action as a combination of an ontic action and an epistemic action [11].

In cognitive science epistemic actions are actions executed by a bounded agent as ways to overcome its intrinsic limits [12, 13]. When the limits being tackled are of perceptual nature we can use the expression *external epistemic actions* because they can be easily defined, once known the perceptual apparatus of the agent and the structure of the environment, without using information of the internal structure of the agent.

POMDPs [5] formalise the problem of optimising sequential decisions in partially observable stochastic environments. Agents have a complete probabilistic model of the environment, composed by a set of states S (and the related transition probabilities $\tau(s^{t+1}, a_t, s^t)$), and a set of observations (and again the related transition probabilities). At each step an agent executes an action a and receives an observation o and a reward r, finally it updates a probabilistic distribution of the state s, named *belief state* $b(s)$, (for which transition probabilities

$\tau(b_{t+1}, a_t, b_t)$ can be computed). An optimal behaviour is associated with an optimal value function $V(b)^* = E[\sum_{t=1}^{\infty} \gamma^t R_t | B = b]$. When the full state is observable the optimal behaviour and related value function can simply use the state $V(s)^* = E[\sum_{t=1}^{\infty} \gamma^t R_t | S_0 = s]$.

Classical POMDP theory does not make any explicit difference between epistemic and ontic actions. We propose a definition of *external epistemic actions* for the POMDP framework that takes inspiration from its definition in cognitive science and the classical planning literature. A starting point for the definition are the two main characteristics that Herzig and colleagues [11] associate to epistemic actions: *informativeness* and *non-intrusiveness*.

An action is *informative* when its execution reduces the uncertainty of the belief state $b(s)$. Formal definitions can be found in [5, 14]. More recent approaches, using non myopic value of information, can be found in [4, 15]. We define an action e to be *non-intrusive* in the belief state b if the expected value of the belief state reached after executing e, $E[Q(b, e)] = \sum_{b'} \tau(b, e, b') \sum_s b'(s) V^*(s)$, is equal to the value of the current belief state b computed using the Q-MDP approximation $V^*(b) = \sum_s b(s) V^*(s)$ and if the immediate expected reward of executing the action is 0. $(\sum_s |b(s) r(s, e)| = 0)$[1]. The use of $V^*(s)$ in place of $V^*(b)$ is intuitively explainable by the fact that the latter comprises the modification to the internal state of the agent, thus any epistemic action will affect it. This formalises the concept that the execution of a non-intrusive action does not change the reward that the agent can receive.

We define an action e to be *strictly external epistemic* in the belief state b if it is informative and non-intrusive in the belief state b. An action e is *strictly epistemic over observation* o if the action is epistemic for every belief state b for which $P(o|b) \neq 0$. A POMDP is an *MDP-reducible-POMDP* when it admits a policy π_e that for any belief state reduces state entropy to zero in a finite number of steps using only epistemic actions. A MDP-reducible-POMDP can be solved combining the policy π_e with the optimal policy π_{MDP}. The obtained solution can be sub-optimal because of the time spent executing π_e. This condition can be found in real-world tasks. For example, in some active vision problems [16] the algorithms used are based on a phase in which only the point of view is changed till the agent has enough confidence on the observed state [17]. This is a subclass of the MDP-reducible-POMDPs because only camera control actions are executed for information retrieval and the state of the task is unchanged.

3   Epistemic Actions and Reinforcement Learning

In the hypothesis that an agent is working in POMDPs and is endowed with epistemic actions, or even in the more strict condition of a MDP-reducible-POMDP, the problem remains of acquiring a complete stochastic model of the environment before solving it, which can still be complex. A different approach is using a reinforcement learning (RL) based agent which directly learns the

---

[1] Note that this is different from the optimal transition which states that $V^*(s) = max_a(r(s, a) + \gamma \sum_{s'} \tau(s', a, s) V^*(s'))$.

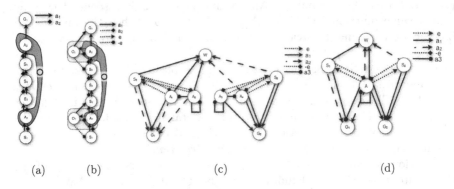

Fig. 1. (a,b) Examples of POMDPs with ambiguity and optimal memory-free policy not learnable through RL. Circles that fall in the same grey shape will give the same ambiguous observation, while circles inside the same rectangle are observations of the same state. The other circles are unambiguous states. (c,d) A POMDP and its underlying MDP problem. Its peculiarity is the presence of a_3 that always brings the agent to the goal from the aliased states A_{1-4}.

action policy by interacting with the environment. Studying how RL, especially model-free and memory-free RL [18, 19], performs when epistemic actions are available in POMDP is interesting for two main reasons:

1. Reinforcement learning has been shown to be able to learn in MDP, however it has also shown to have strong limitations when facing POMDPs [17].
2. Learning a model can allow the execution of epistemic actions related to the hidden states that are correctly represented, but it is a complex problem by itself and cannot help for not represented states.
3. Many greedy POMDP algorithms have been shown to fail in POMDP because they "ignore ignorance" even if they use a complete belief state, so how a model-free [2] reinforcement learning agent can take into consideration its own ignorance and select epistemic actions is an open issue.

Moreover, while it has been shown that in some conditions memory-free agents can effectively behave in partially observable environments [20], limited work has been done to allow the *autonomous development* of such behaviours with similar constraints (but see [18, 21]).

3.1   Learning in POMDP with and without Epistemic Actions

A typical POMDP, originally reported in [22], is shown in Figure 1.a. The states A_1 and A_2 are ambiguous because they result in the same observation O.

---

[2] Note that a model-free, memory-free agent is also belief-free and unaware of its uncertainty. To define epistemic actions for such an agent, an external probabilistic observer has to be used, which receives as input the agent actions and observations. The changes in the uncertainty in the belief state of the observer after each action-observation pair is used to measure the action informativeness.

The optimal policy would be to select the action a_1 in every state. The use of a value iteration algorithm with the assumption that observations are states will result in the agent oscillating around state A_1[3]. Online learning using the exploration approach proposed in [23] obtains the same results without ever learning a path toward the goal. See [24] for more examples of the effect of perceptual aliasing on RL. A similar situation can arise in the POMDP shown in Figure 1.c,d.

Does the addition of epistemic actions allow a model-free memory-free RL agent to learn in these environments? Looking the environment of Figure 1.b, where epistemic actions and the related observations O_1 and O_2 are added, we can see that the agent will still move from state S_3 toward state A_1. So in this case adding epistemic actions does not solve the problem.

Epistemic actions affect learning by increasing the distance between the starting state and the goal, and by increasing the fan out of the resulting graph in comparison to the underlying MDP (each observation is connected to all the states that make it visible). The value of executing an epistemic action depends on the value of all the possible states reached, in other words on the estimated value of the information acquired. This results in an increase of the exploration time during learning. To see this, consider an environment like the one in Figure 1.c,d, and an agent that is endowed with epistemic action e and n_A ontic actions available in all states which, if executed in the right MDP state, result in the reward R (in the figure a_1, a_2 and a_3). Similarly to the previous situation, the agent will not be able to distinguish between observations and states, so the observation A in the POMDP is seen as a single state and is shared by the MDP states A_1, A_2, A_3, A_4. Note that action a_3 if executed several times can take the agent to the solution from every state. During the first trials of learning the epistemic action e will have very low probability of being executed. Consequently the (unambiguous) states, e.g. S_1, will not be evaluated correctly. In the unambiguous state the agent will have also very little probability of learning the right action to do because of the a high number (n_A) of ontic actions. Thus, after the first trials the epistemic actions will not probably increase their values. At the same time several ontic actions in the ambiguous state A will be executed, and so there will be a high chance that one of them leads to a reward. This will result in increasing the probability of reselecting the same action while the probability of selecting an epistemic actions will decrease. Moreover, if there is an action like a_3 available in the ambiguous state which cycles through hidden states and also brings the agent to the goal, it will be easily found and its value will not decrease due to punishments. Even if the policy resulting from a_3 is sub-optimal it requires the exploration of small set of states and will slow down the exploration of other actions.

Given this kind of dynamics, before an epistemic action can acquire a high value the agent should learn how to behave in most of the non-ambiguos states

---

[3] We consider O as a state with transition probabilities resulting from a different mixture of those of state A_1 and of state A_2. The value function obtained in observation O has higher value than in state S_4, so in state S_3 the agent will choose action a_2 and not action a_1.

S_i. It is interesting to note that using a greedy method for POMDPs which ignores uncertainty reduction [5], the action chosen will be one of the ontic actions. With enough learning experience a RL agent might be able to learn a policy comprising epistemic actions because the learnt value of action-state pairs can be comprehensive of the information value.

4 Experimental Results

We used a neuro-robotic system of arm-eye coordination to study experimentally the concepts presented in the previous sections. The architecture of the model (Figure 2.a) integrates two components: (a) an attention control component formed by a bottom-up and a top-down attention sub-component; (b) an arm control component. Only an overview of these components is presented here. For a complete description of the system refer to [25][4].

The setup used to test the model is a simulated version of a real system presented in [24] (see Figure 2.a), formed by a down-looking camera and a 2-DOFs robotic arm. The arm horizontal working plane is formed by a horizontal computer screen where the task stimuli appear. The camera image activates a *periphery map* that implements bottom-up attention. The central part of the image (*fovea*) feeds a *reinforcement-learning actor-critic* [19] component (implemented by two feed-forward neural networks) that learns to predict the positions of relevant visual elements based on the currently foveated cues (top-down attention). A *saliency map* sums up the information from the periphery map with the output of the actor network and selects the next eye movement corresponding to the most active neurons (through neural competition). Each eye fixation point, encoded in a *eye posture map*, suggests a potential arm target to an *arm posture map* which (a) performs the "eye posture → arm posture" inverse kinematic and (b) implements a second neural competition which triggers reaching movements when the eye fixates the same location for about three consecutive time steps. If the reached target is the correct one (red object), the actor-critic component is rewarded. By closely coupling reaching to gaze control, the proposed model embodies the "attention-for-action" principle [26]. This principle states that in organisms attention has the function of extracting the information necessary to control action. This principle might be incorporated in the system as a hard-wired link between an epistemic (visual) action and an ontic (reaching) action. The following experiments will test if the model is able to learn to execute epistemic actions which do not have such a pre-wired link in the architecture.

---

[4] For this work three minor change where made to the architecture: (a) the foveal input was pre-processed in order to separate the different objects on different input units both for the actor and the critic neural networks: this informs the agent on the identity of objects; (b) the action of reaching is not directly punished to satisfy the non-intrusiveness constraint, but for every saccade the agent gets a punishment of -0.0025; (c) Finally PAM, an action-based memory system, was removed to have a model/memory/belief-free agent, so the agent could not rely on the inhibition of return mechanism and on previous estimation of target position, some key properties of PAM.

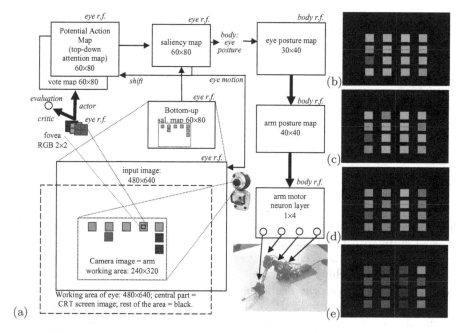

Fig. 2. (a) Eye-arm control architecture.(b-e) Scenes used in the experiment. (b) initial configuration with 0.9 chances of cue covered. (c) some covers removed. (d) map without the covers. (e) another example of map without the covers.

The rationale of this is that the informative role of an action should depend on the task and the environment. In the experiments, reaching actions can change the environment and uncover useful information to accomplish the task.

4.1 Experimental Setup

We used a task inspired from a card game for children. Two different families of 4x4 grid environments were used to train the agent: (a) in the first family the target is randomly positioned and all the other edges of the grid are occupied by cues, each of which has a precise spatial relationship with the target which is constant for all the environments in the family (see figure 2.d,e); (b) the second family is like the previous one but the cues are randomly hidden by grey *cover* (with each cue having 0.1 chance to be free since the beginning, see figure 2.b,c). In both families the target is never covered. When the agent touches a cover with the arm, the underlying cue is revealed. In both families of environments the agent can obtain reward only by touching the target. This will also start a new trial.

To use the nomenclature developed in section 2 is necessary to transform the second family of environments to a POMDP with one state for any possible configuration of the covers and for each position of the gaze and of the target. In this POMDP every state where the agent is not gazing to the target will have

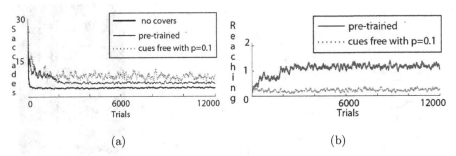

Fig. 3. (a) Average number of saccades per trial (b)Average number of reachings to covers per trial

the same value in the underlying MDP because it is always possible to reach the goal state with the same number of eye and arm actions. So any gazing action from a state where cues or covers are gazed to a non targeted position is non-intrusive. Touching a cue or a cover is always non-intrusive because the target is uncovered from the beginning, thus in no condition removing a cover is necessary to obtain reward. Regarding informativeness, removing a cover is always informative since it gives access to a state that is unambiguous for the agent. Consequently, reaching a cue *is* an epistemic action in this context.

Three runs of 100,000 steps were executed in the two different environments. For each policy learnt with the clean map another run was executed with the corresponding environment with random covers. The data obtained in three runs for each condition were quite similar so only one run for condition is analysed in the following section.

4.2   Results

Figure 3.a shows the evolution of the average number of saccades per trial in the map task with agents fulfilling three different conditions: (a) learning with all the cues uncovered; (b) learning with most of the cues covered; (c) adapting from condition a to condition b. The final average number of saccades per trial for condition a is 3.2, for condition b is 8.0, and for condition c is 5.2. Thus, the agent in c, re-trained after having discovered the value of the cues, faces a simpler task then than the agent in b and so develops a better performance in the environment with the covered cues .

Figure 3.b shows the evolution of the average number of reaching actions on covers per trial (thus only conditions b and c). The final average number of reaching actions on covers for condition b is 0.3 and for condition c is 1.2. The agent in c uncovers a cue or more in most trials. It thus learns to execute strictly epistemic actions. Considering that a reaching action requires about 2-3 saccades to the same spot to be triggered, the exploration policy of the agent in c compared to that of the agent in b is more efficient than what might be supposed on the basis of the simple ratio between the number of saccades. While the agent

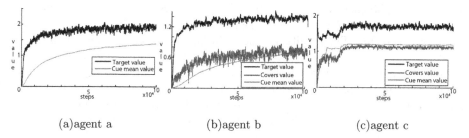

(a)agent a   (b)agent b   (c)agent c

Fig. 4. Values evolution for the various condition of training. In figure a,b and c the average value of the cue was plotted with a grey line.

in b and c is architecturally identical and operates in the same environment, it acquires very different behaviours due to the different training history.

Figures 4.a,b,c show the evolution of the value of the objects in the three conditions. The cues are represented by the evolution of the average value because the epistemic reaching action value depends on the average value of the cue that can be uncovered. The comparison of the three graphs shows that in condition a the values of the cues are learnt faster than in condition b. This means that initially in condition b the agent cannot learn to uncover the cues because it still does not know how to use the information it gets access to. Instead, looking away from the covers (which prevents reaching and uncovering them) can be useful because it can randomly lead to the target. On the other hand, when the agent knows the use of the cues, uncovering them is easily learnt if the agent has not inhibited this behaviour, as shown by Figure 4.c. This is a quite general main result of this research: for a model-free agent, the possibility of learning to use epistemic actions is strongly dependent on knowing how to use the information they deliver. Otherwise the epistemic actions can be inhibited and not explored/exploited anymore even when the agent later acquires the capacity to use the information they deliver. Probably, this might be ameliorated by using a mechanisms like internal simulation (like DYNA [27]). Even if in the condition b

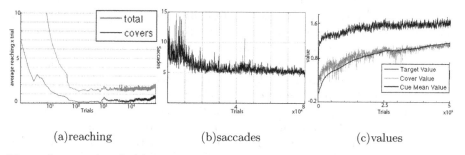

(a)reaching   (b)saccades   (c)values

Fig. 5. Long run trial. (a) Evolution of average reaching per trial. (b) Evolution of average saccades. (c) Evolution of values of objects.

reported above the agent seemed to have converged to a stable performance level, a longer run (500,000 steps) with 10% of cues uncovered (same as condition b) was executed to test if it was able to learn to execute the epistemic actions with additional learning. The results are shown in figure 5. The final average number of saccades per trial is 5.1, similar to that of the agent in c. The final number of reaching actions on covers is 0.7, which means that in most of the trial the agent executes an epistemic action. However the agent takes about 50,000 trials to learn this behavior while the agent pre-trained on the clean map (c) develops the behavior in less than 2,000 trials (to which we must add the trials spent in the pre-training, that are about 12,000).

An experiment with an even longer training was executed with all the cues covered. In this condition even after 750,000 steps the agent was not able to develop the epistemic reaching action. In this condition a simple scanning procedure is really easy to learn for the agent, e.g., moving the gaze to the adjacent left element each time a cover is foveated. Instead, discovering the real underlying structure of the environment results in a complex policy involving a different action for each possible cue. This policy is particularly difficult to learn also because initially all the cues are covered, so moving from an uncovered cue to another position usually brings the agent to another cover.

5   Conclusions

Cognitive science research describes epistemic actions as aiming to change the internal state of the agent to (i) acquire new information from the environment, (ii) facilitate information processing, and (iii) acquire knowledge for better future processing and execution. The distinctive characteristic of epistemic actions is that they are executed by a bounded agent as a means to overcome its intrinsic perceptual, computational, and expertise limits [12, 13].

The work presented in this paper is focused on epistemic actions used to acquire information, named *external epistemic actions*. In psychology these actions have been named *specific exploration actions* [28]. The epistemic actions used to facilitate information processing can instead be named *internal epistemic actions*. A typical example from literature is rotating Tetris game blocks to facilitate visual matching instead of internally simulating their rotation [12]. Another is the use of sensorimotor strategies for discrimination instead of complex internal processing, e.g. scale and rotation invariance [20]. The third and last kind of epistemic actions are named *curiosity epistemic actions*: these are executed by the agent to increase its knowledge of the environment [28, 29].

We provided a formal definition of external epistemic actions for the POMDP framework, together with the concept of MPD-reducible-POMDP. This definition is dependent only on agent perceptual system and environment structure, so it can be applied without knowing the internal structure of the agent. Then we discussed several theoretical issues affecting a simple model-free agent using reinforcement learning in POMDP with epistemic actions, showing that even having MDP-reducible-POMDP is not a sufficient condition to permit learning of the optimal policy. We also showed the issues coming from (a) the initial ignorance of the use of the information that is accessed through an epistemic action and (b) the presence of suboptimal policies which are more information-ally parsimonious and easy to learn. These policies tend to visit a limited set of perceptual states and show that the agent representation of the task, ignoring some of the hidden states, does not match with the actual environment. This is an important issue to consider when modelling organisms' behaviour using rein-forcement learning. In this regards, it would be interesting to further study the preference for informationally parsimonious policies using a principled formal approach based on information theory like the one proposed in [30].

These issues were finally illustrated through a robotic experiment. Using dif-ferent training procedures we showed the importance of knowing how to use the information acquired through an epistemic actions to learn the latter ones. Using an higher degree of partial observability resulted in the use of suboptimal strategies. In this respect, the results showed that it is possible to facilitate the acquisition of epistemic actions using shaping techniques [31]. The experiments also showed that the architecture proposed here is able to merge learnt epistemic actions with ontic actions in a smooth way in several conditions. On the con-trary, most robotic architectures have a separate information acquisition phase followed by an action execution phase, thus limiting their adaptation capabilities.

Previous studies [20, 32] showed that reactive agents whose structure is developed with evolutionary algorithms can produce efficient behaviours even in partially observable environments. The results presented here extend these findings by showing that reactive agents can, in some conditions, learn through direct interaction with the environment how to incorporate epistemic actions in their policies, and this gives substantial advantages when adapting to complex environments with partial observability and a structure unknown at design time.

Another interesting further study can be to consider principles like "free-energy" minimization [33, 34]. When applied to the reduction of the uncertainty on the quantities related to the agent structure, e.g. the perceptual state of the agent, these might also result in the reduction of uncertainty on the environment hidden variables, similarly to epistemic actions.

References

1. Behrens, T.E.J., Woolrich, M.W., Walton, M.E., Rushworth, M.F.S.: Learning the value of information in an uncertain world. Nat. Neurosci. 10(9), 1214–1221 (2007)
2. Kepecs, A., Uchida, N., Zariwala, H.A., Mainen, Z.F.: Neural correlates, compu-tation and behavioural impact of decision confidence. Nature 455(7210), 227–231 (2008)
3. Pezzulo, G., Rigoli, F., Chersi, F.: The mixed instrumental controller: using value of information to combine habitual choice and mental simulation. Front Psychol. 4, 92 (2013)
4. Roy, N., Thrun, S.: Coastal navigation with mobile robots. In: Advances in Neural Information Processing Systems, vol. 12 (2000)
5. Cassandra, A., Kaelbling, L., Kurien, J.: Acting under uncertainty: discrete bayesian models for mobile-robotnavigation. In: Proc. of IROS 1996 (1996)

6. Kwok, C., Fox, D.: Reinforcement learning for sensing strategies. In: Proc. of IROS 2004 (2004)
7. Hsiao, K., Kaelbling, L., Lozano-Perez, T.: Task-driven tactile exploration. In: Proc. of Robotics: Science and Systems (RSS) (2010)
8. Lepora, N., Martinez, U., Prescott, T.: Active touch for robust perception under position uncertainty. In: IEEE Proceedings of ICRA (2013)
9. Sullivan, J., Mitchinson, B., Pearson, M.J., Evans, M., Lepora, N.F., Fox, C.W., Melhuish, C., Prescott, T.J.: Tactile discrimination using active whisker sensors. IEEE Sensors Journal 12(2), 350–362 (2012)
10. Moore, R.: 9 a formal theory of knowledge and action. In: Hobbs, J., Moore, R. (eds.) Formal Theories of the Commonsense World. Intellect Books (1985)
11. Herzig, A., Lang, J., Marquis, P.: Action representation and partially observable planning in epistemic logic. In: Proc. of IJCAI 2003 (2003)
12. Kirsh, D., Maglio, P.: On distinguishing epistemic from pragmatic action. Cognitive Science 18(4), 513–549 (1994)
13. Kirsh, D.: Thinking with external representations. AI & Society (February 2010)
14. Cassandra, A.R.: Exact and Approximate Algorithms for Partially Observable Markov Decision Processes. PhD thesis, Brown University (1998)
15. Melo, F.S., Ribeiro, I.M.: Transition entropy in partially observable markov decision processes. In: Proc. of the 9th IAS, pp. 282–289 (2006)
16. Denzler, J., Brown, C.: Information theoretic sensor data selection for active object recognition and state estimation. IEEE Trans. on Pattern Analysis and Machine Intelligence 24(2), 145–157 (2002)
17. Whitehead, S., Lin, L.: Reinforcement learning of non-markov decision processes. Artificial Intelligence 73(1-2), 271–306 (1995)
18. Vlassis, N., Toussaint, M.: Model-free reinforcement learning as mixture learning. In: Proc. of the 26th Ann. Int. Conf. on Machine Learning, pp. 1081–1088. ACM (2009)
19. Sutton, R., Barto, A.: Reinforcement Learning: An Introduction. MIT Press, Cambridge (1998)
20. Nolfi, S.: Power and the limits of reactive agents. Neurocomputing 42(1-4), 119–145 (2002)
21. Aberdeen, D., Baxter, J.: Scalable internal-state policy-gradient methods for pomdps. In: Proc. of Int. Conf. Machine Learning, pp. 3–10 (2002)
22. Whitehead, S.D., Ballard, D.H.: Learning to perceive and act by trial and error. Machine Learning 7(1), 45–83 (1991)
23. Koenig, S., Simmons, R.G.: The effect of representation and knowledge on goal-directed exploration with reinforcement-learning algorithms. Mach. Learn. (1996)
24. Ognibene, D.: Ecological Adaptive Perception from a Neuro-Robotic perspective: theory, architecture and experiments. PhD thesis, University of Genoa (May 2009)
25. Ognibene, D., Pezzulo, G., Baldassarre, G.: Learning to look in different environments: An active-vision model which learns and readapts visual routines. In: Proc. of the 11th Conf. on Simulation of Adaptive Behaviour (2010)
26. Balkenius, C.: Attention, habituation and conditioning: Toward a computational model. Cognitive Science Quarterly 1(2), 171–204 (2000)
27. Sutton, R.S.: Integrated architectures for learning, planning, and reacting based on approximating dynamic programming. In: Proc. ICML, pp. 216–224 (1990)
28. Berlyne: Curiosity and exploration. Science 153(3731), 9–96 (1966)
29. Baldassarre, G., Mirolli, M.: Intrinsically Motivated Learning in Natural and Artificial Systems. Springer, Berlin (2013)

30. Tishby, N., Polani, D.: Information theory of decisions and actions. In: Perception-Action Cycle, pp. 601–636. Springer (2011)
31. Ng, A.Y., Harada, D., Russell, S.: Policy invariance under reward transformations: Theory and application to reward shaping. In: Proc.of the ICML, pp. 278–287 (1999)
32. Beer, R.D.: The dynamics of active categorical perception in an evolved model agent. Adapt. Behav. 11, 209–243 (2003)
33. Friston, K., Adams, R.A., Perrinet, L., Breakspear, M.: Perceptions as hypotheses: saccades as experiments. Frontiers in Psychology 3 (2012)
34. Ortega, P.A., Braun, D.A.: Thermodynamics as a theory of decision-making with information-processing costs. Proceedings of the Royal Society A: Mathematical, Physical and Engineering Science 469(2153) (2013)

Robust Ratiometric Infochemical Communication in a Neuromorphic "Synthetic Moth"

Timothy C. Pearce[1], Salah Karout[1], Alberto Capurro[1], Zoltán Rácz[2],
Marina Cole[2], and Julian W. Gardner[2]

[1] Centre for Bioengineering, Department of Engineering, University of Leicester, UK
t.c.pearce@le.ac.uk
[2] Microsensors and Bioelectronics Laboratory, School of Engineering,
University of Warwick, UK
marina.cole@warwick.ac.uk

Abstract. An often cited advantage of neuromorphic systems is their robust behavior in the face of operating variability, such as sensor noise and non-stationary stimuli statistics inherent to naturally variable environmental processes. One of the most challenging examples of this is extracting information from advected chemical plumes, which are governed by naturally turbulent unsteady flow, one of the very few remaining macroscopic physical phenomena that cannot be described using deterministic Newtonian mechanics. We describe a "synthetic moth" robotic platform that incorporates a real-time neuromorphic model of early olfactory processing in the moth brain (the macro-glomerular complex) for extracting ratiometric information from chemical plumes. Separate analysis has shown that our neuromorphic model achieves rapid and efficient classification of ratiometrically encoded chemical blends by exploiting early phase chemosensor array transients, with execution times well beyond biological timescales. Here, we test our neuromorphic synthetic moth in a naturally turbulent chemical plume and demonstrate robust ratiometric communication of infochemical information.

Keywords: machine olfaction, ratiometric processing, neuromorphic model, pheromone processing, chemical plumes, infochemical communication, transient processing.

1   Introduction

In insects, olfactory cues in the form of pheromones have long been established to play a crucial role in conspecific and interspecific signalling [1]. In a remarkable example of an olfactory guided behaviour, male moths navigate pheromone-laden plumes comprising fine filamentous structures over large distances in order to locate calling females, necessitating pheromone signals to be detected both rapidly and robustly [2,3] The signal produced by the female is made yet more specific and informative as phero-mones are generally produced in blends, involving more than one molecular cue. The chemical composition, number of components and ratio of concentrations of these

N.F. Lepora et al. (Eds.): Living Machines 2013, LNAI 8064, pp. 204–215, 2013.

blends appears to be important to the pheromone detection task since different species often use similar chemicals in a different context and interfering with the blends can block appropriate behavioural responses [4]. Using ratios in this way confers the important advantage of concentration invariance when communicating through highly intermittent plume structures. Ratiometric processing offers a robust mechanism for concentration invariant communication that may provide the basis for new neuromorphic technologies for infochemical communication [5-8].

In this paper we build a real-time neuromorphic model of early olfactory processing in the moth brain and integrate it together with chemosensor arrays on a robotic platform to test its capability for extracting ratios within an infochemical blend transported on a naturally turbulent plume. Existing neuromorphic olfactory system implementations have typically been tested in laboratory conditions where the stimulus delivery is highly controlled. Thus, we were interested to know whether neuromorphic strategies can provide an effective ratiometric processing solution in the case where the stimulus is instead driven by natural processes and also whether the extracted information can be used to generate robust behaviour in a "synthetic moth".

This builds upon our previous insect olfactory modeling system modeling in two respects – firstly, we develop a spiking neuromorphic hardware implementation of the early insect pheromone processing pathway for real-time processing of chemosensory input, and secondly, we integrate this neuromorphic hardware together with a mobile robot platform and test its capability for extracting blend information from highly intermittent chemical plume structures.

2   Materials and Methods

2.1   Macroglomerular Complex Antennal Lobe Computational Model

To perform ratiometric processing of real-time chemosensor input we first developed a biologically-constrained computational model of the first stage of ratiometric processing in insects, the macro-glomerular complex (MGC) of the antennal lobe (AL), first in Matlab (Mathworks, USA). Probabilistic connectivity patterns were based on morphological studies of the moth brain [9-14] shown in schematic form in Figure 1A and described in detail in [15].

Briefly, two types of ORN (ORN 1 and ORN 2 in Figure 1A) drive excitation within the model, being connected to LNs and PNs through excitatory synapses. The model consists of two glomeruli, where every LN receives input from both ORNs, but each PN of a single glomerulus is activated only by a specific ORN type. LNs have a probability to form inhibitory interconnections between themselves (shown as dashed red connections with inhibitory synapses as circles) and PNs belonging to the same glomerulus have excitatory interconnections (dashed blue and green connections) and provide the main output of the MGC to higher centres of the insect brain. LNs can also make inhibitory connections with PNs of any glomerulus (red connections), again according to a probabilistic connectivity rule. The neuromorphic MGC

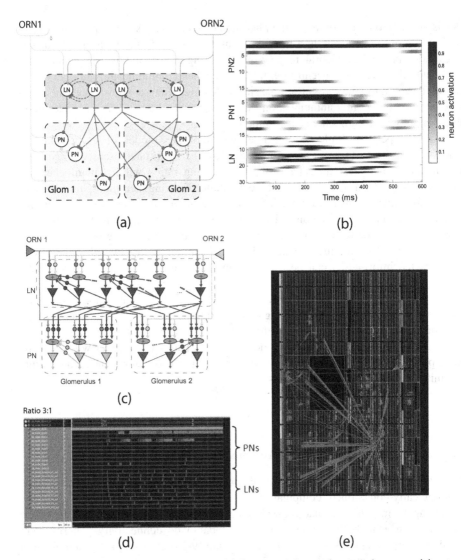

Fig. 1. (a) Schematic of the insect antennal lobe macro-glomerular complex comprising two types of olfactory receptor neurons (ORN) innervating the projection neurons (PNs) of their respective glomerulus (GLOM1 and GLOM2), and the population of local interneurons (LN) - arrows and circles represent excitatory and inhibitory synapses respectively. (b) Example neuronal activation dynamics for the LN and PN population in each glomerulus, showing LCA dynamics (PN1 and PN2 subpopulations correspond to those PNs innervating GLOM1 and GLOM2 glomeruli respectively), (c) Schematic of the neuromorphic MGC model showing the programmable logic neuronal model components and their interconnectivity, (d) Programmable logic MGC AL model core firing events for PN and LN neuron populations during stimulation with excitatory current input in the ratio 3:1 ORN1:ORN2. (e) Field programmable gate array (FPGA) footprint for the MGC model optimized down to 3,095 of the 11,200 available logic slices of a single Xilinx Virtex 5 device.

implementation consisted of 2 glomeruli (each containing 15 uniglomerular PNs) and 30 multiglomerular LNs. Depending upon the connectivity probabilities in the LN population, the model displays two categories of behavior - fixed points attractor (FPA) and limit cycle attractor (LCA) dynamics (see Results). Figure 1B shows an example of LN and PN population activity exhibiting LCA dynamics.

2.2 Spiking Neuromorphic Implementation of MGC AL Model in Programmable Logic

Our MGC AL model was then implemented in programmable logic (FPGA) using an existing construction kit of programmable logic neuronal modelling components, comprising leaky integrate-and-fire neurons and dynamical synapses adapted by spike timing-dependent plasticity [16]. The neuronal modelling construction kit exploits a fine-grained parallelism architecture, achieving iteration speeds well beyond biological timescales. The relationship of the modelling kit components and MGC AL model is shown in Figure 1C (LIF soma depicted as triangles and inhibitory/excitatory synapses as circles). Recurrent feedback in the model that underlies the complex spatiotemporal dynamics shown in Figure 1B occurs is due to both LN and PNs outputs being reciprocally connected to their equivalent cell type inputs. Current summation nodes receiving input from multiple converging synapses (shown as "+" symbols) simply sum multiple EPSC and IPSCs generated at each time step.

Neuronal modelling kit component RTL designs were coded in VHDL and logic synthesis was completed using ISE Design Suite (Xilinx Inc., USA) and Precision Physical (Mentor Graphics, USA) for the target FPGA device (Xilinx Virtex®-5 FX70T). The resulting design is massively parallel in that model kit components share no common computational resources and each computation within each component completes after a single clock cycle, resulting in one model iteration step per clock cycle overall. Thus, the ca. 2 ns clock cycle period of the Xilinx device (550 MHz clock frequency) renders hyper real-time simulation of the neuronal model at least 3 orders of magnitude faster than biological timescales, that could potentially be traded for model scale through time multiplexing should scalability in the model become important. The resulting floorplan of the MGC AL design is shown in Figure 1E and occupies 3,095 of the available 11,200 logic slices available on the device.

2.3 Experimental Procedure and Neuromorphic Programmable Logic Processor Integration

The real-time neuromorphic MGC AL model input was driven by surface acoustic wave resonator (SAWR) chemical microsensors [5,17]. A programmable logic development board based around Xilinx Virtex 5 FPGA device was used as the target platform for the MGC neuromorphic processor (Virtex®-5 FX70T development kit, Xilinx, Figure 2B) for interfacing with the SAWR devices. The Virtex 5 device contained three cores developed for the real-time processing of chemosensor data (Figure 2A). The first core subtracts the common mode signal between a pair of identical SAW sensors, one of which was coated with polymer material and one not, so as to reduce systematic drift in the sensor signals. Thus, the differential signal for each sensor response was derived from both sensors (coated and uncoated generated signals) by subtracting the integrated oscillation signals obtained from the coated SAWR device as a frequency, $f_{coated}(t)$, from that of the uncoated device, $f_{uncoated}(t)$, to generate the synaptic current driving the ORN leaky LIF soma,

$$I(t) = \beta (f_{coated}(t) - f_{uncoated}(t)). \tag{1}$$

The resultant two sensor pair signals were provided as an input to the neuromorphic model core. The output of the AL model (30 PNs) was used as an input to the third and final stage of the chemoreceiver, the ratiometric detection core (see below). The core compared the MGC PN model output trajectories for an unknown ratio input with those stored internally corresponding to known classes using a k-nearest neighbour classifier.

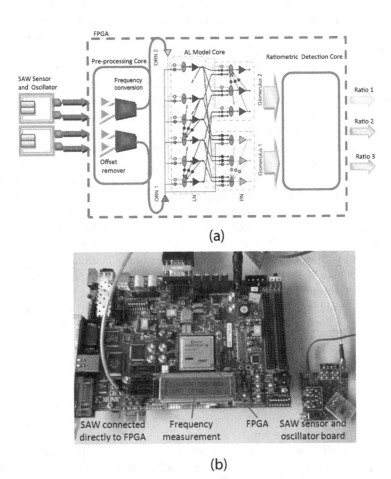

(a)

(b)

Fig. 2. (a) Neuromorphic MGC processor and its chemosensor array integration. (b) Xilinx Virtex®-5 FX70T development board integrated with SAW resonator and associated electronics displaying the current SAW input frequency.

2.4   Ratiometric Detection Core

We applied a simple k-nearest neighbor algorithm using Euclidean distance to classify the blend ratio from the MGC model PN output trajectories as a separate FPGA core. First we calculated an average firing rate trajectory across the entire PN population, as a function of time for each ratio class, $\vec{r}^x(t)$, (ratio class x) from both the PN population activity for each repeat trial y, such that $\vec{r}^{x,y}(t) = (r_1(t), r_2(t), \ldots r_n(t))$ for all PNs in the MGC model

$$\vec{r}^x(t) = \frac{1}{n} \sum_{y \in x}^{n} \vec{r}^{x,y}(t),\tag{2}$$

using repeat trials for each class $y \in x$, where n is the total number of trials in each class. k-nearest neighbour then proceeds by computing the nearest mean class trajectory, $\vec{r}^x(t)$, to a new unseen trajectory $\vec{r}^{unseen}(t)$ by averaging the Euclidean distance between the two trajectories over their length, up to the decision time, T

$$D_x = \frac{1}{T} \int_0^T dt \, \|r^{unseen}(t) - \vec{r}^x(t)\|.\tag{3}$$

We then choose class k, such that D_x is minimized (k-nearest neighbor).

2.5   Robotic Implementation

Wheeled robots (Khepera, K-team Corporation, Switzerland) were used as mobile chemoemitter and chemoreceiver platforms for infochemical communication in realistic, naturally turbulent chemical plumes (Figure 3A). The chemoemitter robot platform encodes ratiometric messages by blending different pre-synthesized compounds with different concentrations in time. In this experiment a dual programmable syringe pump (CMA 402, CMA Microdialysis AB) was used to create blend component sequences delivered to the mixer to release a blend of specific ratios over time into an unsteady turbulent flow. While the syringe pump independently controls the delivery of each chemical to create a specific ratio in time, a portable nebuliser was used to atomize the resultant blend into micron scale droplets for delivery into a moving flow (CompAir Elite Ne-C30 Portable Nebuliser, Omron, UK).

Four dual SAW devices were installed on the chemoreceiver robot for the detection of the volatile mixture ratios (Figure 3A, bottom) - each dual device having a polymer-coated and uncoated (reference) chemosensor driven by high frequency oscillator mixer circuit. The polymers used in this experiment and their corresponding frequency changes due to polymer deposition are as follows SRL330A_B2: PCL (-149 kHz); SRL330B_B2: PVK (-111 kHz); SRL330C_B2: PSB (-101 kHz); SRL330D_B2: PEVA (-105 kHz) [17]. These frequency shift signals in response to chemical input were fed directly to the FPGA development board to (Figure 2) containing the SAW sensor interface cores, the AL neuromorphic model core and AL model output decoding algorithm core. The robot has a controller unit to execute a programmed behavioural response to the specific ratio detected (Table 1).

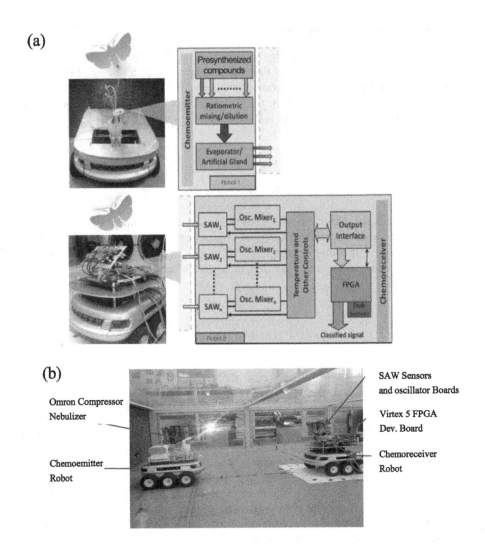

Fig. 3. (a) System architecture for female (top) and male (bottom) synthetic moth. (b) Mobile ratiometric communication scenario. Photograph showing female chemoemitter (left) and male chemoreceiver (right) robots with associated peripherals performing infochemical communication within a wind tunnel.

2.6   Synthetic Moth Infochemical Communication

We selected two pre-synthesized plant volatile compounds (3-methylbutan-1-ol (3M), Hexyl Butanoate (HB)) for delivery into the wind tunnel in specific ratios as the basis for testing infochemical communication. Four different ratios were prepared to demonstrate the capability of the communication system together with a specific robot behaviour to be executed (Table 1). The ratios created were 1:3 (10ml of 3M and 30ml of HB), 2:2 (20ml of 3m and 20ml of HB) and 3:1 (30ml of 3m and 10ml of HB).

Table 1. Corresponding programmed chemoreceiver robot action for every ratio

Ratio	Concentration of chemical 1 (3M)	Concentration of chemical 2 (HB)	Robot Action
0:0	0%	0%	Stop
1:3	25%	75%	Forward
3:1	75%	25%	Reverse
2:2	50%	50%	Move Left then Right

A specialized wind tunnel (approximately 3m (width) x 3m (length) x 0.8m (height)) constructed from transparent polyethylene sheeting. Hexcel material (aluminium honeycomb, with 1 cm diameter) at the inlet of the wind tunnel is used in order to remove the large-scale eddies from the air flow. This hosted both chemotransmitter (female) and chemoreceiver (male) robot and was operated in physiologically relevant flow conditions (0.75 m s^{-1} flow velocity - see [18] for full details – Figure 3B). In the stationary tests both chemoemitter and chemoreceiver robots were separated by a distance of 1.5 m.

3 Results

Without adaptations in synaptic strength and depending upon the probability of connection between LNs, the PN population displayed both FPA and LCA dynamics when stimulated with a synthetic square ORN input pulse of different ratios (Figure 4A and B, respectively). Trajectories of the PN responses to specific ratiometric inputs were represented using PCA applied to the firing rate outputs of the model. FPA dynamics were characterized by a rapid change in the PN population output to a fixed-point attractor in the phase space that was stable for as long as the stimulus was maintained. Since no spontaneous firing was used in the model, the PN population quickly returned to zero activity after stimulus offset (dashed lines in the trajectory figures). LCA dynamics, in contrast, caused PN population trajectories to oscillate about a limit cycle attractor for as long as the stimulus was present. Importantly, for both LCA and FPA modes of behavior, the MGC network produced population dynamics that were found to be specific to the ratio of activity in the ORN inputs in a binary blend. Moreover, the attractor location was found to vary continuously within this phase space depending upon ratio. For FPA dynamics, the network encoded ratios through the spatial identity of neurons, whereas LCA networks produced a PN output continuously varying in time (bottom panels in Figure 4). While both operating regimes were able to encode the pheromone ratio reliably, we observed in previous work [15] that the ratio encoding was faster and more accurate using LCA dynamics, but FPA dynamics were found to be more robust to interference from complex stimulus dynamics. For this reason we selected LCA dynamics to be used in our spiking version of the MGC model integrated on the "synthetic moth".

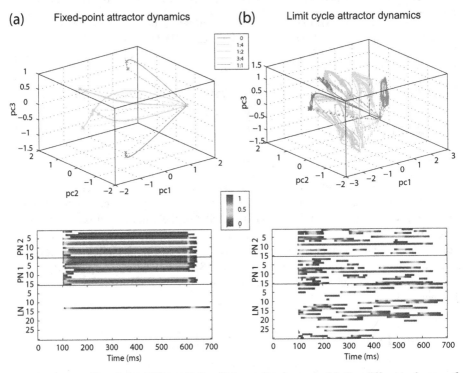

Fig. 4. Trajectories of the PN population firing rates (top panels) for different classes of dynamics observed in the MGC AL model: (a) fixed-point attractor (FPA), and (b) limit cycle attractor dynamics (LCA). For both cases the chemical blend stimulus was a square pulse of fixed ratio and duration 500 ms with onset at 100 ms and offset at 600 ms.

Although our neuromorphic MGC model was able to accurately identify ratios of infochemical blends during controlled square-pulse stimulus delivery, we wanted to test its behaviour when driven by naturalistic stimuli. The statistics of locally sampled concentration in a naturally turbulence chemical plume are highly non-stationary due to the essentially chaotic nature of the underlying physical process of advection. In this physiologically relevant flow regime it poses a far more challenging problem to accurately and robustly extract blend ratios advected within the chemical plume since the internal dynamics of the neuromorphic model can be potentially confounded with those due to the stimulus variation. This potential conflict in dynamics is evidently solved in the brain of the moth and so we were curious to understand if this property is inherited by our neuromorphic model.

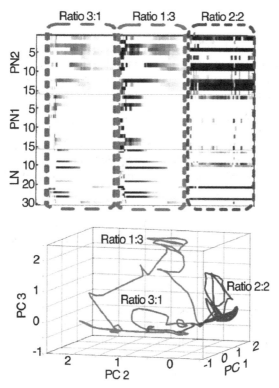

Fig. 5. MGC PN/LN population dynamics (top) generated over time by the male "synthetic moth" in response to three distinct chemical blend ratios released by the female during wind tunnel experiments. PCA plot (bottom) showing the PN trajectories in response to the same ratios. The coloured segments show the times during which the k-NN classifier of PN activity correctly identifies the infochemical blend.

To assess this, specific sequences of ratios were delivered by the female chemoemitter robot inside the wind tunnel, detected and processed in the chemoreceiver neuromorphic model remotely to control behaviour in the male chemoreceiver (Table 1). The resulting continuous trajectory of the output PN population of the neuromorphic model to the ratio sequence 3:1, 2:2 and 1:3 in the male is shown in Figure 5 for three different ratios emitted by the female. The dynamics for the distinct ratios are clearly separable by the ratiometric detection core and could be reliably decoded in the chemoreciever to effect the appropriate behavioural response in the male, successfully demonstrating infochemical communication between the two (http://bit.ly/12bUPAy provides a video of the executed behavioural sequencing in the male). We found no misclassifications of chemical ratios in this experiment using just three components over 1 hr of behavioural testing, but the classification task obviously becomes more challenging with increasing numbers of distinct ratios.

We conclude that the robust processing properties of ratiometric extraction evident in male moths is inherited in our synthetic moth enabling it to reliably extract ratiometric information in order to drive an appropriate behavioural response.

4   Discussion

A number of computational olfactory models have been applied to offline processing of real-world chemosensor input, as reviewed in [19,20]. Some of these bio-inspired and biological constrained model demonstrate clear advantages for chemosensor array processing to overcome the manifold challenges inherent to machine olfaction, beyond existing engineered machine learning solutions. Since these studies focus on offline processing of chemosensor input, they cannot be considered neuromorphic solutions in terms of real-time neural simulation hardware. Separately however, hardware neuromorphic systems have been reported in both programmable logic and custom VLSI, mostly based on mammalian olfactory processing, but at this stage only a small number have been tested with real-world chemosensor input. All of these artificial olfaction neuromorphic systems have been tested with offline data-sets collected in controlled laboratory environments, whereas the current study aims to take such a system out of the laboratory and demonstrate robust performance under realistic uncontrolled conditions, such as within odour plumes and advected environments.

Acknowledgements. Funding for this research was provided under the EU Future and Emerging Technologies Framework Programme (to TCP, MC and JWG) – Biosynthetic Infochemical Communication (iChem). We are grateful to Shannon Olsson for on-going discussions regarding the biological details of the MGC model.

References

1. Wyatt, T.D.: Pheromones and Animal Behaviour, pp. 1–408. Cambridge University Press (2003)
2. Vickers, N.J., Baker, T.C.: Reiterative responses to single strands of odor promote sustained upwind flight and odor source location by moths. Proc. Natl. Acad. Sci. USA 91, 5756–5760 (1994)
3. Vickers, N.J., Christensen, T.A.: Functional divergence of spatially conserved olfactory glomeruli in two related moth species. Chem. Senses 28, 325–338 (2003)
4. Mustaparta, H.: Central mechanisms of pheromone information processing. Chem. Senses 21, 269–275 (1996)
5. Rácz, Z., Cole, M., Gardner, J.W., Chowdhury, M.F., Bula, W.P., Gardeniers, J.G.E., Karout, S., Capurro, A., Pearce, T.C.: Design and Implementation of a Modular Biomimetic Infochemical Communication System. Int. J. Circ. Theor. Appl. (2012), doi:10.1002/cta.1829
6. Rácz, Z., Olsson, S., Gardner, J., Pearce, T.C., Hansson, B., Cole, M.: Challenges of Biomimetic Infochemical Communication. Proc. Comp. Sci. 7, 106–109 (2011)
7. Cole, M., Gardner, J.W., Racz, Z., Pathak, S., Guerrero, A., Munoz, L., Carot, G., Pearce, T.C., Challiss, J., Markovic, D., Hansson, B.S., Olsson, S., Kubler, L., Gardeniers, J.G.E., Dimov, N., Bula, W.: Biomimetic Insect Infochemical Communication System. In: Proceedings 8th IEEE Conference on Sensors, pp. 1–5 (2009)
8. Cole, M., Gardner, J.W., Pathak, S., Pearce, T.C., Racz, Z.: Towards a biosynthetic infochemical communication system. In: Proceedings of the Eurosensors XXIII Conference, pp. 1–4 (2009)

9. Hansson, B.S., Ljungberg, H., Hallberg, E., Löfstedt, C.: Functional specialization of olfactory glomeruli in a moth. Science 256, 1313–1315 (1992)
10. Carlsson, M.A., Hansson, B.S.: Responses in highly selective sensory neurons to blends of pheromone components in the moth Agrotis segetum. J. Insect Physiol. 48, 443–451
11. Hansson, B.S., Anton, S.: Function and morphology of the antennal lobe: New developments. Annu. Rev. Entomol. 45, 203–231 (2000), (2002)
12. Akers, R.P., O'Connell, R.J.: Response specificity of male olfactory receptor neurons for the major and minor components of a female pheromone blend. Physiol. Entomol. 16, 1–17 (1991)
13. Sun, X.J., Tolbert, L.P., Hildebrand, J.G.: Synaptic organization of the uniglomerular projection neurons of the antennal lobe of the moth Manduca sexta: A laser scanning confocal and electron microscopic study. J. Comp. Neurol. 379, 2–20 (1997)
14. Carlsson, M.A., Galizia, C.G., Hansson, B.S.: Spatial representation of odours in the antennal lobe of the moth Spodoptera littoralis (Lepidoptera: Noctuidae). Chem. Senses 27, 231–244 (2002)
15. Chong, K.Y., Capurro, A., Karout, S., Pearce, T.C.: Stimulus and network dynamics collide in a ratiometric model of the antennal lobe macroglomerular complex. PLoS One 7, e29602–117, doi:10.1371/journal.pone.0029602 (2012)
16. Guerrero-Rivera, R., Morrison, A., Diesmann, M., Pearce, T.C.: Programmable logic construction kits for hyper-real-time neuronal modeling. Neural Comput. 18, 2651–2679 (2006)
17. Yang, J., Rácz, Z., Gardner, J.W., Cole, M., Chen, H.: Ratiometric info-chemical communication system system based on polymer-coated surface acoustic wave microsensors. Sensors Actuators B 173, 547–554 (2012)
18. Pearce, T.C., Gu, J., Chanie, E.: Chemical Source Classification in Naturally Turbulent Plumes. Analytical Chemistry 79(22), 8511–8519 (2007)
19. Marco, S., Gutierrez-Gálvez, A.: Recent developments in the application of biologically inspired computation to chemical sensing. In: AIP Conference Proceedings, vol. 1137, p. 151 (2009)
20. Huerta, R., Nowotny, T.: Bio-inspired solutions to the challenges of chemical sensing. Front. Neuroeng. 5, 24 (2012), doi:10.3389/fneng.2012.00024

Bacteria-Inspired Magnetic Polymer Composite Microrobots

Kathrin E. Peyer[1], Erdem C. Siringil[1], Li Zhang[2],
Marcel Suter[3], and Bradley J. Nelson[1]

[1]Institute of Robotics and Intelligent Systems,
ETH Zurich, Switzerland
bnelson@ethz.ch
[2]Department of Mechanical and Automation Engineering,
The Chinese University of Hong Kong
[3]Micro and Nanosystems, ETH Zurich, Switzerland

Abstract. Remote-controlled swimming microrobots are promising tools for future biomedical applications. Magnetically actuated helical microrobots that mimic the propulsion mechanism of *E. coli* bacteria are one example, and presented here is a novel method to fabricate such microrobots. They consist of a polymer-nanoparticle composite, which is patterned using a direct laser writing tool. The iron-oxide nanoparticles respond to the externally applied low-strength rotating magnetic field, which is used for the actuation of the microrobots. It is shown that a helical filament can be rotated around its axis without the addition of a body part and without structuring the magnetization direction of the composite. The influence of the helicity angle on the swim behavior of the microrobots is examined and experimental results show that a small helicity angle of 20 degrees is preferred for weakly magnetized microstructures.

Keywords: Bio-inspired microrobots, swimming microrobots, magnetic actuation, magnetic polymer composite.

1 Introduction

Remote controlled swimming microrobots are an emerging field of research as they are promising tools in medical applications, such as targeted drug delivery [1], or for *in vitro* experimentation, such as single cell manipulation and characterization [2]. The design challenges revolve around the fact that these robots are only a few micrometers in size. Firstly, they navigate at a low Reynolds (Re) number regime, which dictates the type of suitable locomotion method that can be employed. Secondly, the question of power supply has to be solved. Thirdly, the microrobots should be non-toxic and be able to interact safely with living cells. And finally, fabrication methods have to be found to make devices that fulfill the previous three criteria.

There are a number of recently published micro-devices that have successfully addressed some of these design challenges, although not necessarily solving all

N.F. Lepora et al. (Eds.): Living Machines 2013, LNAI 8064, pp. 216–227, 2013.

of them at once. For example, possible means of power supply include the use of temporal and spatially varying magnetic fields or chemical gradients [3],[4]. Chemically fueled devices have been used for *in vitro* experiments and can reach high velocities of several body lengths per second. The difficulty is to establish the chemical gradients in the experimental environment, which is why they cannot easily be implemented *in vivo* or when handling sensitive cells. Low-strength and low-frequency magnetic fields, on the other hand, can be employed *in vitro* as well as *in vivo*. The magnetic field strengths necessary for actuating microrobots lie in order of a few milli Tesla, which is much lower than the field strength of a magnetic resonance imaging (MRI) system (reaching up to 3 Tesla).

Due to their size, microrobots swim at a low Reynolds (Re) number, which is a measure for inertial versus viscous forces in a liquid. Baceria, such as *E. coli*, move in this viscous environment by rotating one or several helical filaments with a molecular rotary motor [10]. The helical filament transforms rotational into translatory motion, similar to a corkscrew (see Fig. 1). Microrobots employing this bio-mimicking helical locomotion method have been published previously, such as the first artificial bacterial flagella (ABFs) by Bell [5], the propellers by Ghosh [6], and more recently the polymer ABFs by Tottori [7]. All these devices are actuated by uniform rotating magnetic fields.

The first two reported prototypes were manufactured by batch fabrication methods that limit the type of shapes that can be achieved [5], [6]. Tottori employed a direct laser writing (DLW) method, which allows the fabrication of almost arbitrarily-shaped structures from a variety of photosensitive polymers [7]. The magnetic material necessary for the robot actuation has to be deposited in a second fabrication step. Recently, Suter et al. developed a magnetic polymer composite (MPC) with superparamagnetic properties [8]. This MPC can be patterned by the same DLW tool. The advantage of using MPC is that the magnetic material is already incorporated into the polymer. Furthermore, the material has been shown to be non-cytotoxic [9].

This paper presents the characterization and modeling of the swim behavior of MPC helical microrobots fabricated using the DLW method and actuated by

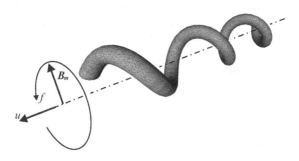

Fig. 1. Schematic of a magnetic helical microrobot. A magnetic field B_m is rotated around the helical axis at a frequency f. The magnetic torque rotates the robot and its helical shape transforms the rotation into a linear velocity u.

uniform rotating magnetic fields. Several prototypes were fabricated and their swim performance tested. It is shown that the helicity angle plays an important role in how the structures magnetize, and this determines the success or failure at achieving corkscrew-type swimming motion.

2 Magnetic Helical Microrobots

2.1 Helical Propulsion

Helical propulsion at low Re numbers has been studied for several decades by biologists interested in the locomotion of microorganisms [11],[12], and recently with respect to microrobots [13]. The linear relationship between the drag forces F_d and torques T_d and the object's velocity U and rotational speed Ω can be represented by a 6 × 6 resistance matrix.

$$\begin{pmatrix} F_d \\ T_d \end{pmatrix} = - \begin{pmatrix} A & B \\ B^T & C \end{pmatrix} \begin{pmatrix} U \\ \Omega \end{pmatrix} \tag{1}$$

A, B, and C are 3x3 matrices and are functions of the object's geometry and fluid viscosity only. A number of methods have been employed to model low Re flows and resistance matrices, such as the method of fundamental solutions, the boundary element method, or the method of regularized stokeslets, which is the method chosen in this paper (see Section 3.3). The resistance matrix of a helix contains non-zero elements in B, which model the coupling between the rotational and translatory motion.

2.2 Magnetic Actuation

Using helical filaments brings advantages when considering the magnetic actuation of microrobots. Abbott *et al.* investigated the scaling of magnetically actuated swimming microrobots and compared magnetic force-driven with magnetic torque-driven devices [14]. It was shown that torque-driven devices scale favorably over force-driven devices as the size of the device decreases to the microscale. This supports the use of helical propulsion for magnetic microrobots as it only relies on the application of a torque to rotate the device rather than on a force to pull it.

A magnetized body aligns itself with the direction of an external magnetic field, in the same manner as a compass needle aligns with the earth's magnetic field lines. This magnetic torque can be used to rotate microstructures in a controlled manner. The magnetic torque T_m acting on a body with volume V is

$$T_m = V \cdot M \times B_m \tag{2}$$

where M is the magnetization and B_m the external magnetic field. The magnetic field vector B_m is rotated at a frequency f. The magnetization M is constant for a permanently magnetized body or it is a function of the applied field and the body geometry for soft-magnetic materials. For simple shapes, such as bodies of revolution, the magnetization can be modeled analytically, but no such solutions exist for a helical body.

In order to actuate the ABF a torque has to be applied in the direction of the helical axis. This can only be achieved if the magnetization is not parallel to the helical axis and is optimal for a magnetization perpendicular to the helical axis. Many previous publications used special means to influence the magnetization of their helical microrobots, such as adding a magnetic head plate [5], permanently magnetizing the structure [6], or aligning the nanoparticles [16]. It is, however, possible to generate an actuation torque even for a helix without aligned magnetization [7]. The structures presented here do not have a pre-aligned magnetization and their response to the magnetic actuation will be discussed in Section 5.

2.3   Microrobot Propulsion Model

To model the behavior of the microrobot, the magnetic and fluid mechanical model of Eq. (1) and (2), respectively, are combined. The equation can be simplified by considering only the rotation around and the translation along the helical axis. This results in a simple 2×2 resistant matrix with scalar entries a, b and c. Furthermore, the external force is assumed to be zero because the helix moves in a horizontal, unobstructed plane, and without the application of magnetic gradient forces.

$$\begin{pmatrix} 0 \\ T_m \end{pmatrix} = \begin{pmatrix} a & b \\ b & c \end{pmatrix} \begin{pmatrix} u \\ 2\pi f \end{pmatrix} \tag{3}$$

At any given low rotational frequency $f < f_{stepout}$ of the magnetic field $\boldsymbol{B_m}$ (see Fig. 1), the magnetic torque T_m is such that it counterbalances the fluidic drag $T_d = -(bu + c2\pi f)$, and the velocity is linearly related to f:

$$u = -\frac{b}{a} 2\pi f = p_{eff} \cdot f \tag{4}$$

The 'effective pitch' $p_{eff} = -2\pi b/a$ describes how far the helix advances forward per rotation. Unlike a corkscrew in solid material, the helix slips when moving in the fluid and p_{eff} is, therefore, only a small fraction of the actual, i.e. geometric, pitch length of the helix.

If the frequency is increased above the step-out frequency $f_{stepout}$, the fluidic drag exceeds the maximal available magnetic torque $T_{m,max}$. The robot falls out of sync with the magnetic field and starts to oscillate backwards and forwards around its axis, which leads to a decrease in velocity. It is therefore desirable to have microrobots with large step-out frequencies.

3   Experimental Methods

3.1   Fabrication

The magnetic polymer composite (MPC) was developed by Suter and consists of
SU-8 50 and a magnetite nanoparticle (Fe_3O_4) suspension. Details on the process
for creating a uniform suspension of the nanoparticles in the viscous SU-8 can
be found in this publication [8]. In order to create 3D helical structures, a DLW
tool by Nanoscribe Inc. was used. The fabrication process is based on the two-
photon polymerization of the photosensitive SU-8 at the focal point of the laser.
A piezoelectric stage moves the substrate along the predefined helical trajectory.
The nanoparticles introduced in the polymer scatter and absorb the laser beam,
and inhibit the polymerization reaction if the concentration is too large [9]. The
line resolution is on the order of a few hundred nanometers. Composite structures
were written with 4 vol.% and 2 vol.% nanopatricle fill factors, but reproducible
results were only accomplished with the 2 vol.% composite. Figure 2 shows SEM
images of a 2 vol.% and 4 vol.% 3-turn helix. In order to increase stability, several
lines were written to achieve a filament thickness of 1.8μm. Helices with helicity
angles of $\Psi = (20, 30, 40, 50)$ degrees were fabricated from the 2 vol.% MPC,
while the helix radius was kept constant at $R = 2.25$μm.

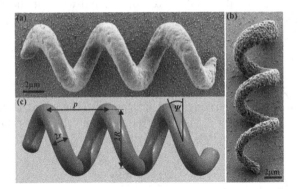

Fig. 2. Helical microrobot prototypes. (a) SEM image of an MPC microrobot with
a nanoparticle fill factor of 2 vol.%. (b) SEM image of an MPC microrobot with a
nanoparticle fill factor of 4 vol.%. Due to the increased quantity of nanoparticles the
4 vol.% MPC does not get polymerized uniformly. The polymerization depends on the
polymer layer depth, which results in a different filament thickness along the structure.
(c) Corresponding CAD model. The main design parameters are the helix radius R,
the filament radius r, the helicity angle Ψ, and the pitch p.

3.2   Experimental Setup

The swim tests were conducted in deionized water in a small tank at the center
of three orthogonally placed Helmholtz coil pairs. Helmholtz coils achieve an
almost uniform magnetic field in their center. Hence, the magnetic forces acting

on the microrobot are negligible. The setup can generate rotating magnetic fields with strengths up to $|B_m| = 10\text{mT}$. All the experiments are performed under a microscope and the images recorded at 30 fps by a camera with a resolution of 640×480 pixels. More details of the experimental setup can be found in previous publications [17].

3.3 Numerical Model

Resistance Matrix. In order to solve Eq. 3, the entries of the resistance matrix have to be known. As mentioned previously, they are only a function of the geometrical parameters and the viscosity of the fluid. There are analytical solutions for a slender helix [13]. The MPC microrobots, however, do not fulfill the slenderness criteria and numerical methods have to be employed. It is possible to superimpose singularity solutions due to linearity of the Stokes equation that governs low Re fluid mechanics. The problem with the superposition of fundamental solutions, of which the most famous is called the stokeslet, is that the solutions are singular at their source point. Cortez adapted this method by introducing a spreading function to regularize the stokeslet, and other fundamental solutions, and thereby removing the singularity at the source point [18]. The resistance of an object can be approximated by a distribution of regularized stokeslets over the surface of the body. The surface of the microrobot was discretized (see Fig. 1) and Cortez's algorithm implemented in C++ to calculate the entries of the resistance matrix.

Magnetization. In order to capture the basic actuation behavior for an optimal microrobot, soft-magnetic properties were assigned to the robot with the easy magnetization axis being perpendicular to the helical axis. This does not represent the non-ideal magnetization direction in the different robot designs, but is sufficient to model the step-out behavior as a function of different externally applied magnetic field strengths. The magnetization model as well as the motion model Eq. 3 were implemented in MATLAB®.

4 Experimental Results

4.1 Frequency-Dependent Swim Behavior

Microrobots with helicity angles of $\Psi = (20, 30, 40, 50)$ degrees were tested and their swim behavior recorded. It has previously been reported that magnetically actuated microswimmers show a dynamic behavior where they exhibit a wobbling around the helical axis with a frequency-dependent precession angle [15]. It was shown that motion changes from tumbling, i.e. a precession angle of 90 degrees, to corkscrew-type motion around the helical axis when the actuation frequency is increased. Fig. 3 shows video excerpts from the swim tests of the different prototypes. Fig. 3 (a) and (b) show the same microrobot ($\Psi = 20$ degree) at different frequencies and the change in wobbling angle β is apparent.

The other microrobots ($\Psi = 30, 40, 50$ degree) all rotate with a precession angle of 90 degrees. Increasing the actuation frequency did not result in a decreasing wobbling angle. These designs were abandoned and the following experiments were conducted only with 20-degree prototypes.

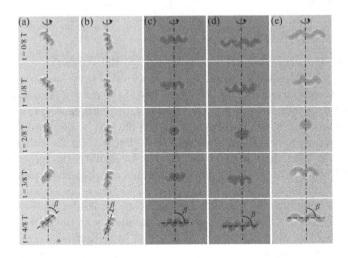

Fig. 3. Tumbling motion of robot prototypes with helicity angles of 20 (a) and (b), 30 (c), 40 (d), and 50 (e) degrees. Each image sequence (top to bottom) shows a half turn; T denotes the period time $T = 1/f$. The dash-dotted line indicates the axis of rotation. (a) and (b) show the same 20-degree helicity angle microrobot at two different frequencies. This prototype exhibits typical frequency-dependent wobbling, which decreases with increasing frequency; i.e., the precession angle decreases from (a) approximately 45 degrees at a $f = 0.5$Hz to (b) approximately 20 degrees at $f = 1$Hz. (c) - (e) shows microrobots with a helicity angles of 30, 40 and 50 degrees, respectively. They tumble at all input frequencies without reorienting to a screw-type motion.

4.2   Influence of Magnetic Actuation

The wobbling angles for several 20-degree microrobot prototype were measured for different frequencies and different magnetic field strengths. As expected, the wobbling decreased with frequency but increased with magnetic field strength (see Fig. 4). With small wobbling angles, the microrobots achieved a corkscrew-type motion, where they were propelled forward. Fig. 5 shows the successful propulsion of two prototypes. They moved simultaneously because they received the same magnetic input. They were steered by changing the direction of the rotating magnetic field.

Fig. 6 shows how the magnetic field strength influences the velocity of the swimmer. The velocity increased with frequency until the step-out frequency $f_{stepout}$ was reached, where the velocity suddenly dropped. The step-out frequency increased with the magnetic field strength. Additionally, a shift of the linear frequency-velocity region to the right was observed when the field strength

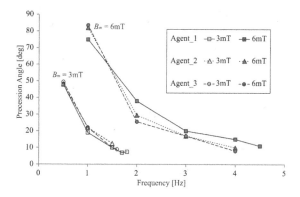

Fig. 4. The precession angle at different actuation frequencies and magnetic field strengths. The precession angle is measured between the axis of the helix and the axis of rotation (see Fig. 3).

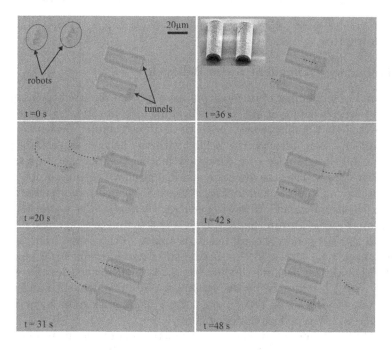

Fig. 5. Video sequence showing the simultaneous steering of two microrobots ($\Psi = 20$ degree). They are steered through two adjacent tunnels on the substrate. The inset shows an SEM image of the two tunnels.

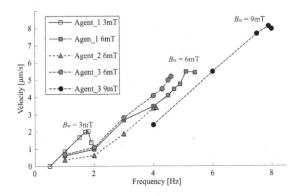

Fig. 6. Frequency-velocity plot at magnetic field strengths $|\boldsymbol{B_m}| = 3$, 6, and 9 mT.

was increased. Finally, Fig. 7 contains the result from a 4vol.% MPC microrobot in comparison to a 2vol.% prototype. The simulation results in Fig. 7 and all the experimental results will be discussed in detail in the next section.

5   Discussion

5.1   Frequency-Dependent Swim Behavior

There have been a number of previous publications that reported the frequency-dependent swim behavior of magnetic helical microrobots [15],[7]. It was, therefore, expected that similar results with the MPC structures presented in this paper would be seen. Instead, the helices with the helicity angles larger than 20 degrees did not stabilize to a corkscrew motion even as the frequency was increased. It appears that the direction and magnitude of the microrobot's magnetization play an important role. Fig. 8 shows a schematic representation of a typical experimental result. At low frequencies, there is a large precession angle, which minimizes the forward motion. Only when the precession angle decreases does the linear frequency-velocity behavior appear. The change from tumbling to a corkscrew-type motion occurs at the stabilization frequency f_{stable}, indicated by the vertical dash-dot line in Fig. 8. At high frequencies the step-out frequency $f_{stepout}$, indicated by the vertical dotted line, is reached and the microrobot loses speed again. For $f_{stable} < f_{stepout}$, there exists a linear velocity region where screw-type locomotion occurs.

The wobbling is increased when the magnetization of the microrobot is not perfectly perpendicular to the helical axis. The results published by Tottori indicate that magnetization of helical structures is strongly influenced by the helicity angle [7]. The further away the magnetization vector is from the desired 90 degrees to the helical axis, the higher the stabilization frequency f_{stable}. If f_{stable} becomes too large, i.e. $f_{stable} > f_{stepout}$, the corkscrew motion disappears and the microrobot goes from tumbling directly into step-out behavior, which is

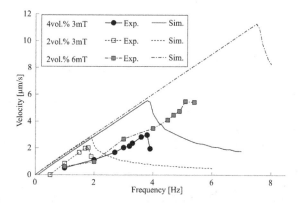

Fig. 7. Comparison of 2vol.% and 4vol.% MPC structures with experimental and simulation results. At $|\boldsymbol{B_m}| = 3\text{mT}$ the step-out frequency of the 4vol.% MPC is approximately double to the 2vol.% MPC. Increasing the external magnetic field has a larger impact than increasing the amount of magnetic material.

the behavior observed with the 30-, 40-, and 50-degree helicity angle structures in Fig. 3. The helicity angle is not the only factor that influences f_{stable}. For example, an overall slender microrobot design can decrease f_{stable} as well.

5.2   Influence of Magnetic Actuation

The magnetic field strength influences the tumbling behavior by increasing the stabilization frequency f_{stable} (from approximately 2Hz to around 4Hz in Fig. 4). A strong magnetic field also increases $f_{stepout}$ and therefore still allows for a screw-type swimming region. Fig. 6 shows the resulting forward velocity gained by increasing the magnetic field strength. The linear corkscrew region is shifted to higher frequencies. It is important to note that the magnetic field does not influence the frequency-velocity slope, i.e. the 'effective pitch' p_{eff}, in the linear corkscrew region. The same holds true for increasing the magnetization of the microrobot. Fig. 7 shows the 2vol.% MPC microrobot in comparison to the 4vol.% MPC. The figure also contains the simulation data of the velocity and step-out frequency from Eq. 3. The simulation assumes a magnetization perpendicular to the helical axis, which is why it predicts a $f_{stable} = 0$ unlike the experimental data. The magnetic material properties of the simulation were tuned to achieve a step-out frequency of approximately 2Hz for the 2vol.% MPC at 3mT.

The simulation predicts that doubling the amount of magnetic nano-particles in the polymer results in approximately double the step-out frequency, which corresponds well to the experimental results. Doubling the applied magnetic field to 6mT should quadruple the step-out frequency. This can also be seen from examining Eq. 2. The maximum magnetic torque is a function of the applied field and the magnetization of the microrobot. For soft-magnetic or super-paramagnetic

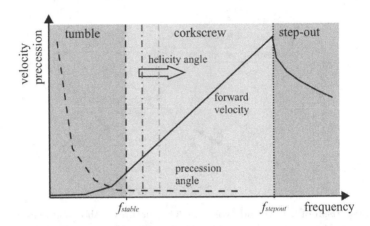

Fig. 8. Schematic representation of magnetic helical swimmers. At low frequency tumbling occurs. Increasing the frequency beyond f_{stable} stabilizes the motion into a screw-type motion. At the limit of the available magnetic torque, the step-out frequency $f_{stepout}$ is reached. The helicity angle Ψ increases f_{stable}, which eventually leads to $f_{stable} > f_{stepout}$, where no corkscrew motion appears as observed in Fig. 3 (c)-(e).

materials, the magnetization itself is a function of the applied field. For small magnetic field strengths this relationship is approximately linear, which results in a quadrupling of $T_{m,max}$, and therefore $f_{stepout}$, when $|B_m|$ is doubled. In the experiments, the step-out frequency for $|B_m| = 6\text{mT}$ is, however, smaller than that. This may be explained by the different magnetic saturation behavior of nanoparticle composited compared to bulk material.

6   Conclusion

Bacteria-inspired swimming microrobots can be actuated and guided remotely via low-strength rotating magnetic fields. Helical swimming microbots made of a nanoparticle-polymer composite have been fabricated and investigated. One of their main advantages is the straightforward fabrication method by direct laser writing and their non-cytotoxicity. It was shown that the helicity angle plays an important role for the swimming behavior. Designs with helicity angles of 30 degrees or larger were not able to achieve corkscrew motion. The 20-degree prototypes, on the other hand, stabilized from tumbling to a screw-type motion, and their maneuverability was demonstrated Fig. 5.

Acknowledgments. The authors thank the financial support from the Swiss National Science Foundation, contract No. 200021_130069.

References

1. Nelson, B.J., Kaliakatsos, I.K., Abbott, J.J.: Microrobots for Minimally Invasive Medicine. Ann. Rev. Biomed. Eng. 28, 55–85 (2010)
2. Zhang, L., Peyer, K.E., Nelson, B.J.: Artificial Bacterial Flagella for Micromanipulation. Lab Chip 10, 2203–2215 (2010)
3. Peyer, K.E., Zhang, L., Nelson, B.J.: Bio-Inspired Magnetic Swimming Microrobots for Biomedical Applications. Nanoscale 5, 1259–1272 (2013)
4. Sanchez, S., Solovev, A.A., Schulze, S., Schmidt, O.G.: Controlled Manipulation of Multiple Cells Using Catalytic Microbots. Chem. Commun. 47, 698–700 (2011)
5. Bell, D.J., Leutenegger, S., Hammer, K.M., Dong, L.X., Nelson, B.J.: Flagella-like Propulsion for Microrobots Using a Magnetic Nanocoil and Rotating Electromagnetic Field. In: Proc. IEEE Int. Conf. Robotics and Automation, pp. 1128–1133 (2007)
6. Ghosh, A., Fischer, P.: Controlled Propulsion of Artificial Magnetic Nanostructured Propellers. Nano Lett. 9, 2243–2245 (2009)
7. Tottori, S., Zhang, L., Qiu, F., Krawczyk, K., Franco-Obregón, A., Nelson, B.J.: Magnetic Helical Micromachines: Fabrication, Controlled Swimming, and Cargo Transport. Adv. Mater. 24, 811–816 (2012)
8. Suter, M., Ergeneman, O., Zürcher, J., Moitzi, C., Pané, S., Rudin, T., Pratsinis, S.E., Nelson, B.J., Hierold, C.: A Photocurable Superparamagnetic Nanocomposite: Material Characterization and Fabrication of Microstructures. Sens. Act. 156, 433–443 (2011)
9. Suter M., Zhang, L., Siringil, E. C., Peters, C., Luehmann, T., Ergeneman, O., Peyer, K. E., Nelson, B. J., Hierold, C.: Superparamagnetic Microrobots: Fabrication by Two-Photon Polymerization and Biocompatibility (submitted for publication)
10. Berg, H.C., Anderson, R.A.: Bacteria Swim by Rotating Their Flagellar Filaments. Nature 245, 380–382 (1973)
11. Taylor, G.: Analysis of the Swimming of Microscopic Organisms. Proc. Nat. Acad. Sci. 209, 447–461 (1951)
12. Lighthill, J.: Flagellar Hydrodynamics. SIAM Rev. 18, 161–230 (1976)
13. Mahoney, A.W., Sarrazin, J.C., Bamberg, E., Abbott, J.J.: Velocity Control with Gravity Compensation for Magnetic Helical Microswimmers. Adv. Robot. 25, 1007–1028 (2011)
14. Abbott, J.J., Peyer, K.E., Lagomarsino, M.C., Zhang, L., Dong, L., Kaliakatsos, I.K., Nelson, B.J.: How Should Microrobots Swim? Int. J. Robot. Res. 28, 1434–1447 (2009)
15. Peyer, K.E., Zhang, L., Kratochvil, B.E., Nelson, B.J.: Non-ideal Swimming of Artificial Bacterial Flagella Near a Surface. In: Proc. IEEE Int. Conf. Robotics and Automation, pp. 96–101 (2010)
16. Peters, C., Ergeneman, O., Nelson, B.J., Hierold, C.: Superparamagnetic Swimming Microrobots with Adjusted Magnetic Anisotropy. In: Proc. IEEE Int. Conf. Micro Electro Mechanical Systems, pp. 564–567 (2013)
17. Peyer, K.E., Mahoney, A.W., Zhang, L., Abbott, J.J., Nelson, B.J.: Bacteria-Inspired Microrobots. In: Kim, M., Steager, E., Julius, A. (eds.) Microbiorobotics, pp. 165–199. Elsevier (2012)
18. Cortez, R., Fauci, L., Medovikov, A.: The Method of Regularized Stokeslets in Three Dimensions: Analysis, Validation, and Application to Helical Swimming. Phys. Fluid. 17, 031504 (2005)

Generic Bio-inspired Chip Model-Based on Spatio-temporal Histogram Computation: Application to Car Driving by Gaze-Like Control

Patrick Pirim

Brain Vision Systems, Paris, France
patrick.pirim@bvs-tech.com

Abstract. A neuromorphic generic chip has been developed for human-like perception, from 1986 onwards. Similarly to the brain, the chip intertwines three aspects of visual perception, respectively related to color vision, to movement detection and to border identification. These so-called *Global*, *Dynamic* and *Structural* perceptions are processed on-line by a family of spatio-temporal histogram computations. The interconnected histograms mimic the brain's "What and Where" mode of visual processing. The chip's capabilities are demonstrated here with an automatic car driving simulation that mimics the human gaze control on the steering wheel.

Keywords: generic chip, spatio-temporal histogram, model-based processing, bio-inspired, neuromimetic, perception, vision, gaze control, car driving, GVPP, BIPS.

1 Introduction

Long ago, we observed that histogram computation is an efficient solution to conventional imaging processing, provided that a correct choice of input signals is made. In such case, it provides good results with little software development.

The goal is to emulate the human perception by a silicon chip. Opposing characteristics between human and silicon computations are the number of connections and speed of computation. With a product of the two latter values, i.e. the number of connections by the speed, is an index of the computational complexity of a perceptual task. By contrast with the brain, the silicon chip can easily sample in time at high speed the sensory signals in accordance with the Shannon's law.

The design of our chip is similar to the brain in that the sensor input is filtered in modalities, themselves divided in criteria. Modalities are defined as *Global* with the three criteria of luminance, hue and saturation, *Dynamic* with the three criteria of temporal variation, direction and velocity of motion, and *Structural* with the three criteria of oriented edges, curvature and texture.

The brain's cerebral cortex simultaneously processes temporal and spatial information at multiple levels of scale. At the higher level, a parietal magnocellular

N.F. Lepora et al. (Eds.): Living Machines 2013, LNAI 8064, pp. 228–239, 2013.

route computes the object's location ("Where") and a temporal parvocellular route computes its characteristics ("What"), both routes interacting [6]. At the lower level, perpendicular to the cortical surface, small columnar-like populations of neurons simultaneously deal with temporal and spatial information [4] [5]. Here, spatio-temporal histograms of selected criteria perform a similar process, each module delivering an elementary "What and Where" information. By an appropriate sequence of criteria associations, a real-life application is emulated.

2   Histogram Computation of a Dynamic Criterion

A first-level of computation is based on histogram computation of a dynamic criterion (Fig. 1). It was implemented in 1986 [14] within the industrial chip HISTO (4 Kgates and 5 Kbits of memory) and initially tested on the luminance criterion of a video (Fig. 1-3).

A criterion is sequenced in time, each frame being itself sequenced as rows Y and columns X. The elementary element is a pixel P. The histogram computation is done frame by frame and, at the beginning of each frame, the histogram memory is reset.

During the whole frame, the histogram computation is divided into four blocks.

1. An histogram block increments the number of times the input criterion In agrees with the validation command Val.

2. A classification block simultaneously sends a binary signal that is validated if the present criterion value is inside two values A and B (see Fig.1).

3. An association block, sets by the user (Req), associates the classified values in order to give the binary validation command Val.

4. A register block saves, at the end of the frame, the histogram's main results :

- $NbPix$, the global number of criterion values validated on this frame (that is the energy of the function),
- $Rmax$ and $PRmx$, the histogram's maximum and its abscissa, and
- $Med(t)$ and $Med(t-1)$, the histogram's medians of the current and previous frames.

At the end of the frame, two automats actuate the classification and the anticipation values :

- for classification, $Rmax$ and $PRmx$ that are used to obtain A and B values,
- for anticipation, the variation of the median ($Med(t)$-$Med(t-1)$) that is added to A and B values.

The register block stores the result of this histogram computation that will provide the status for the next frame computation.

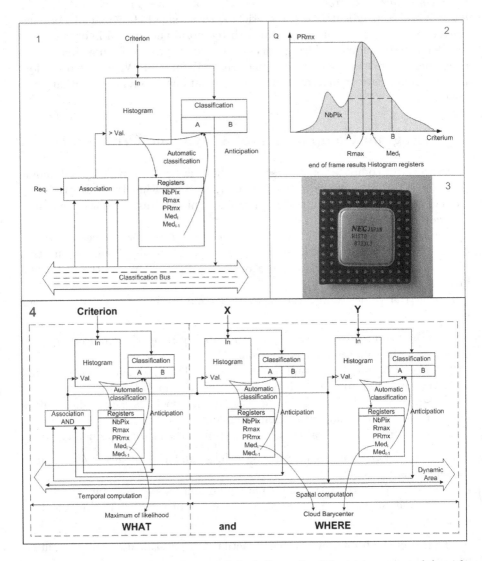

Fig. 1. Histogram computation module 1- Synoptic of the histogram module with four blocks and a general classification bus. 2- Automatic classification by a ratio of the max *PRmx* (*horizontal dotted line*) determining A and B values. Value of the median *Med* is memorized. 3- First microelectronic chip implementation. 4- Spatio-temporal histogram computation. Three histogram modules are linked via a bus, one for object detection ("What" on *left*) and two others for object localization ("Where" on *right*). Note that the block *Association* is shared by all three modules.

3   Spatio-temporal Histogram Computation

The previous organization is efficient for object detection but does not give its position. In order to solve this point, two additional histogram modules were added, one for the X positions (columns of image pixels), the second for Y positions (raws of image pixels). The results of the X and Y classifications are gathered with the criterion classification values into the association block of the histogram criterion module in order to validate its computation.

This looped organization automatically converges, frame by frame, to a maximum of energy ($NbPix$), without any external software intervention, even with moving sensory input.

Then the medians $Med(t)$ give :

- for the criterion's histogram, the maximum of vraisemblance,
- for X and Y histograms, the barycenter of the cloud of criterion's pixels.

This result is an elementary "What and Where" information on this criterion (Fig. 1.4).

The module is drived by a microcontroller μP whose software program performs the three following steps.

1. Initialization :

1.1 choice of input criteria, of X and Y connections and of criterion classification values to the association block,

1.2 in classification blocks, A set to zero and B set to the maximum allowed to the criterion for each histogram module,

1.3 reset of a register counter.

2. Interrupt at end of frame starts a program for:

2.1 reading the $NbPix$ value and comparing it with the minimum defined value as noise threshold ($NThr$) : if $NbPix$ is lower than $NThr$, restart the initialization classification blocks, otherwise continue,

2.2 reading the median of each histogram module in order to get the result ("What and Where" information).

3. In order to improve the robustness, the register counter is incremented during the consecutive frames during which $NbPix$ is higher than $NThr$ and a test validates the result when this counter reaches a given value (for example three successive frames with positive perception).

Then, four states of functioning are observed : 0) a sleeping state (the module is unused), 1) an attentive state (the module is initialized), 2) a detection state (the module starts perceiving) and 3) a pursuit state (the module robustly perceives and follows).

4   Bio-inspired Generic Computation

In a complex environment, it is necessary to perceive enough elementary "What and Where" information at different scales. As in a Lego game, these elementary pieces allow to create an internal representation of the criteria.

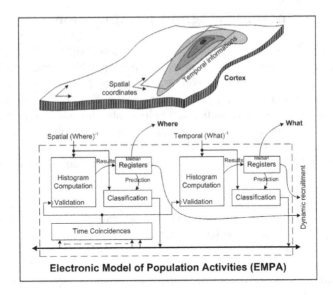

Fig. 2. Electronic Model of Population Activities (EMPA) Similarities between the cortical area of population activities (top) and the model inplemented in a silicon chip. Both architectures show a spatio-temporal arrangement with an elementary What and Where organization. The spatial reference frame, implicit in biology, is explicitly created in the model.

To do that, two other functions must be added (Fig.2). Starting from a module in pursuit state, any other module in sleeping state will first be dynamically recruited (i.e. set to attentive state) and, secondly, will inhibit the first module. In turn this recruited module waits till a second criterion appears, and the process is extended further.

A graph appears from low resolution (use of *RMax* for classification) to higher one, through dynamic recruitment. The graph can be itself annotated as a label.

This model has a bio-inspired approach. The cerebral neural population activities are emulated by our electronic model. We named this spatio-temporal histogram computation EMPA (Electronic Model of Population Activities).

It is interesting to compare our chip computation versus cerebral computation by neural population activities. In both cases, the input, the processing modes (spatio-temporal histogram versus neural population activities) and the output (histogram's medians versus elementary "What and Where" information) are similar.

5   Generic Perception and Integration

Now, we use model EMPA for generic perception. Since 1986, our various vision applications have shown that three types of perception, *Global*, *Dynamic* and *Structural* are required together. Fig.3 shows the close similarities between our

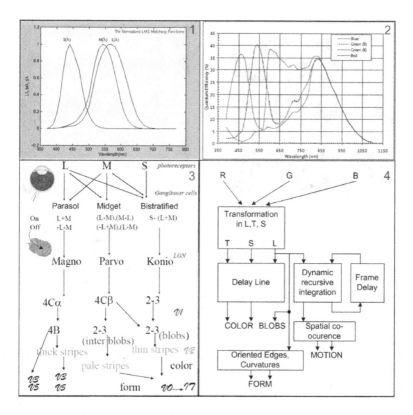

Fig. 3. Comparative synoptics in biology and in the model *top*: The color perception results from the light decomposition on three independent channels. *bottom*: A first processing is performed at the retinal level (left) and, more and more often nowadays, in the imager chip (right). Below, the computation level shows the same kind of extraction into color, motion and form perceptions.

technology and the structure of the brain. In this visual chip, we have separated the video signal into different optic flows with these three modalities (Fig.4).

Global perception (color vision). Based on neural processing for color vision in cerebral area V4 [10] [11] [13], our model uses the criteria of luminance, hue and saturation. Today, CMOS or CCD sensors send images with the industrial processes BAYER, BT656 and MIPI encoding (bio-inspired artificial eye is out of our scope [8] [9]). From this sensor, criteria of luminance Lum, hue Hue, and saturation Sat are extracted.

Dynamic perception (velocity and direction of motion) is computed by a first recursive algorithm that perceives the temporal variation DT, followed by a second algorithm that perceives this variation DT as a spatial value DX, the ratio DX/DT being the motion.

In a basic recursive function, the smoothened value at time t is equal its value at time t-1 added with the ratio of the difference between the input value and the smoothened value at time t-1 by a time constant : $In(t) = In(t-1) + ((E(t) -$

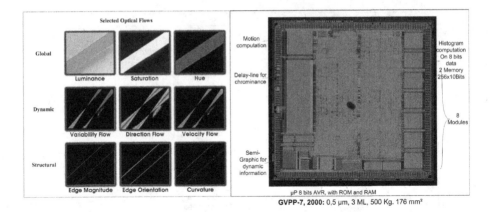

Fig. 4. Optical flows (*left*) The scene depicts an homogeneous background with a uniform red bar rotating from the center at constant speed, as the foreground. This video input is sliced into three modalities, each one being itself separated into three criteria. For each of the nine pictures, a color chart is defined to show the result for that criterion. For each pixel and at the same time, the nine values of the pictures are available for histogram computation. **EMPA implementation** (*right*) In 2000, GVPP-7 was made. For the first time, the microcontroller μP 8bits was integrated, and color was implemented. In this example, the size of each function can be appreciated. The memory for histogram computation is composed of two memories of 256×10 bits, and organized either for temporal criteria as 256×20 bits or for spatial criteria as 512×10 bits. This chip was able to process PAL 50 fps.

$In(t-1))/\tau$, the latter difference representing the error between the input and its short memory. Biologically, when an error occurs it is necessary to solve it quickly and, for that, to decrease the time constant. Otherwise, it is usual to do nothing and to increase that time constant. Integrating this biological mode into the basic recursive function, τ becomes a function of the temporal variability DT.

Three optic flows (criteria) are computed from the luminance, the temporal variation DT, the direction Dir and the velocity Vel of motion [15].

Structural Perception (Oriented Edges Oed, Curvatures Cur, and texture Tex) is computed from the luminance optical flow, a correlation with a Gaussian function in a matrix, a derivation of it in two perpendicular axes, and a CORDIC computation for extracting a module and angle. Based on the brain's cerebral processing of borders [2] [3] [6], the module determines the presence of a shape and its angle (i.e. its orientation). The perception of the spatial variation of the angle by an active module gives the curvature. The repetitivity of the oriented edges gives the texture. Three optic flows (criteria) are sent : the oriented edges Oed, the curvature Cur and the texture Tex [16]. Physical implementation is shown on Fig.4. By external programming each EMPA receives a selected criterion as input and, in its association block, a relation with other EMPAs. A complete organization on a chip is shown on Fig.5.

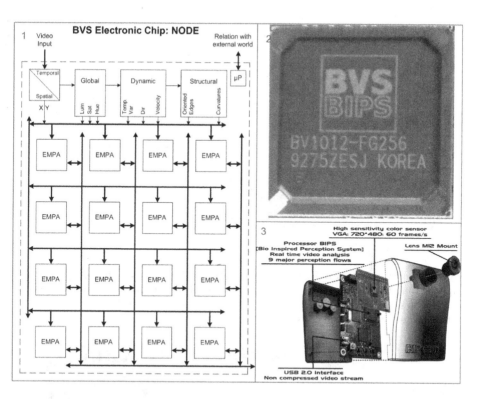

Fig. 5. NODE : Complete organization of an EMPA-based chip 1- The video
signal coming the imager is input in the chip. A first module separates this input into
data information and synchronization signal. The synchronization signal is sent to a
sequencer and delivers the spatial information (X and Y). The data information is input
to three modules that slice it into three modalities, each one being itself separated into
three criteria. the nine values of the pictures are available for a network of EMPA.
A microcontroller μP writes the registers to program EMPA. Then, it transmits the
results (medians of "What and Where") from the EMPA to the external world. 2-
Current chip in production. 3- Generic product all-in-one.

6   Application in Automatic Car Driving by Gaze Control

Driving can be considered as a phase-locked loop control between gaze and steering wheel position. Optimally the driver must gaze a position whose distance
from the car should be equal to his reaction time multiplied by the vehicle's
speed [7]. The gaze provides a global view of the road in various modalities :
globally the road is homogeneous, then *dynamically* the temporal variation is
low, and *structurally*, borders (oriented edges) define orientation of the road.

The scene is separated in three horizontal areas, a first near the vehicle up to
5 m forward, a second from 5 to 20 m and the third from 20 m up to the horizon
(Fig 7.1). In each area, the perception of the oriented edges *Oed* is automatically

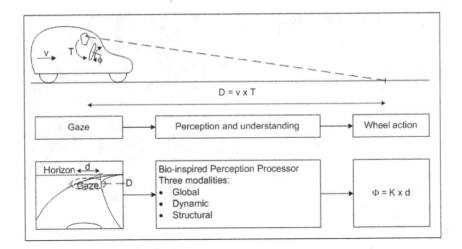

Fig. 6. Visual perception of the road during car driving An appropriate control requires that the delay T needed for perception, between gaze fixation and wheel rotation, is such that during that delay, the vehicle moves forward a distance d equal to the product of T by the car's velocity v.

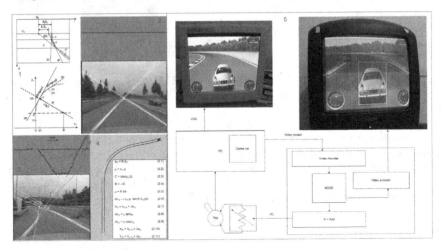

Fig. 7. Automatized car driving by gaze-like control 1- General scheme showing the different horizontal areas with border detections and their perspective. 2- Horizon line detection (*in red*) together with flow field's focus (car direction) when driving on a straight road. 3- From a video during actual driving, the perspective points (crossing of the two road's border lines) are obtained in the three zones (*in red* for near, *in green* for intermediate, *in blue* for far). The distance d between the perspective point in the far-distance area and the flow field's focus is extracted. Multiplying d by a learnt coeficient K gives the angle ϕ of the wheel. 4- Reconstruction of the road map from this previous data (formulae are described in [17]). 5- Real test with a car game (*left screen*) as a simulator using formula $\phi = K \cdot d$. The car becomes the reference (*right screen*) and the whole scene moves around it.

managed by the search for the maximum of local energy; population neuronal activities are modeled by calculations with spatiotemporal histograms as described above, the time being sampled at 50 fps. The median of every histogram calculation on oriented edges criterion Oed gives the energetic barycenter and its angular orientation ("What and Where") for each border perceived in each area. Fig. 7.2 shows six lines resulting of the previous information, each line is computed by a point (barycenter) and angle (angular orientation).

In each of the three horizontal areas, two lines coexist (borders of the road) that cross at the horizon. The three resulting points are obtained by X and Y histogram computations. In X, the optic flow's focus is computed. It can be noted that, when the road is straight (Fig 7.2), these points are close. In Y, the horizon line (in red) is determined by Y histogram values. At this step, the horizon line and the position of the optic flow's focus (car direction) are perceived. They will be used as reference for the next step.

Let us now consider the far horizontal area. The intersection of the two lines moves with road's curvature. Let us call d the distance, on the horizon line, between this point and the optic flow's focus. During driving, the time course of d (Fig. 7.3, blue curve) shows, from left to right, an horizontal line (infinite road curvature), an oblique line (clothoid road curvature), an horizontal line (constant road curvature) and another oblique line in opposite direction(clothoid road curvature). Note that, when the road curvature is constant, the steering wheel do not rotate and when the road curvature has a clothoid shape, the wheel turns at constant speed.

A proportional relation exists between the d and the angle of the wheel (ϕ) with coefficient K that must be learnt during training. In operating mode, using $\phi = K . d$ automatically drives the car.

Fig. 7.5 shows such an experimentation done with a car game in a PC controlled by a joystick. The video PC output was used as input to the electronic board, and the potentiometer of the joystick was replaced by an electronic resistor controlled by I^2C. In this configuration, the board replaces the human driver. Using only $\phi = K . d$, the car stays in place on the screen and all the scene moves around. This demonstration proves the efficiency of the model in situation.

We can see in Fig. 7.3 that the forms of the red and the blue curves differ on the right side (end of recording). The red curve is obtained with the same method as the blue one, but with the near horizontal area. It represents the position of the car on the road. By contrast, the blue curve shows the form of the road. The difference results from a difference in driving of the driver's behavior. Similarly on a straight road (left of Fig. 7.3), we can observe a deviation, an automatic control would have been more efficient.

On Fig. 7.4, we use the far and near perceptions in order to rebuilt the map of the road with, inside that road, the position of the car. All formulae shown here are described in [17].

7 Future Work

We are interested in extending this model (Fig. 8.1) to multimodal integration with a new organization named POLYNODE. To do that we tested with success the model in sensory fusion [1].

Fig. 8.3 shows the multimodality integration in a new organization, named POLYNODE. For that purpose, the size of electronic functions has to be reduced. We propose (Fig. 8.2) a shrinking process, from 220 nm to 65 nm, that has many advantages. The current board working in VGA format will be extended to HD format. Multiscale spatial processing will be implemented. The size will be reduced to 5*5 mm and the power divided by ten. So it is possible to now build a new board based on the POLYNODE organization with different sensory inputs and, as a result, a better understanding of the perceived scene.

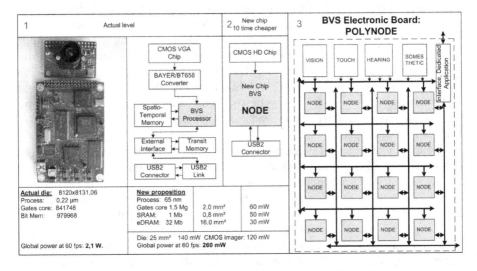

Fig. 8. Future work. 1- The actual design, around 55x80 mm, is integrated in one 4x4 mm silicon chip. 2- A new POLYNODE model allows a reduced scale for multi-modalities fusion. From left to right: the current design, its synoptic view (yellow box : our bio-inspired perception processor system, BIPS), the future implementation on one chip, and its synoptic view with POLYNODE organization.

References

1. Caluwaerts, K., Staffa, M., N'Guyen, S., Grand, C., Dollé, L., Favre-Félix, A., Girard, B., Khamassi, M.: A biologically inspired meta-control navigation system for the Psikharpax rat robot. Bioinspir. Biomim. 7(2), 025009 (2009)
2. Fournier, J., Monier, C., Pananceau, M., Frégnac, Y.: Adaptation of the simple or complex nature of V1 receptive fields to visual statistics. Nat. Neurosci. 14, 1053–1060 (2011)

3. Frégnac, Y., Pananceau, M., René, A., Huguet, N., Marre, O., Levy, M., Shulz, D.E.: A re-examination of Hebbian-covariance rules and spike timing-dependent plasticity in cat visual cortex in vivo. Front. Synaptic Neurosci. 2, 147 (2010)
4. Georgopoulos, A.P., Schwartz, A.B., Kettner, R.E.: Neuronal population coding of movement direction. Science 233, 1416–1419 (1986)
5. Guigon, E., Burnod, Y.: Modelling the acquisition of goal-directed behaviors by populations of neurons. Int. J. Psychophysiol. 19, 103–113 (1995)
6. Kandel, E.R., Schwartz, J.H., Jessel, T.M. (eds.): Principles of Neural Science, 3rd edn. Elsevier, New York (1991)
7. Land, M.F., Lee, D.N.: Where we look when we steer. Nature 369, 742–744 (1994)
8. Liu, S.C., Delbrück, T.: Neuromorphic sensory systems. Curr. Op. Neurobiol. 20, 288–295 (2010)
9. Liu, S.C., Kramer, J., Indiveri, G., Delbrück, T., Burg, T., Douglas, R.: Orientation-selective a VLSI spiking neurons. Neural Netw. 14, 629–643 (2001)
10. Rousselet, G.A., Fabre-Thorpe, M., Thorpe, S.J.: Parallel processing in high-level categorization of natural images. Nat. Neurosci. 5, 629–630 (2002)
11. Sincich, L.C., Horton, J.C.: The circuitry of V1 and V2: integration of color, form, and motion. Annu. Rev. Neurosci. 28, 303–326 (2005)
12. Yger, P., El Boustani, S., Destexhe, A., Frégnac, Y.: Topologically invariant macroscopic statistics in balanced networks of conductance-based integrate-and-fire neurons. J. Comput. Neurosci. 31, 229–245 (2011)
13. Zeki, S.: The representation of colours in the cerebral cortex. Nature 284(5755), 412–418 (1980)

Patents

14. Patent FR2611063: Method and device for real-time processing of a sequenced data flow, and application to the processing and digital video signal representing a video image (1987)
15. Patent FR2751772: Method and device for real-time detection, localisation and determination of the speed and direction of movement of an area of relative movement in a scene (1996)
16. Patent WO2005010820: Automated method and device for perception associated with determination and characterisation of borders and boundaries of an object of a space, contouring and applications (2003)
17. Patent FR2884625 (A1): Vehicle e.g. aerial vehicle, guiding method for e.g. highway, involves determining space between effective and virtual positions, and determining control signal from space, where signal directs vehicle for maintaining it in traffic lane (2006)

Embodied Simulation
Based on Autobiographical Memory

Gregoire Pointeau, Maxime Petit, and Peter Ford Dominey[*]

Robot Cognition Laboratory, INSERM U846, Bron France
{gregoire.pointeau,maxime.petit,peter.dominey}@inserm.fr

Abstract. The ability to generate and exploit internal models of the body, the environment, and their interaction is crucial for survival. Referred to as a forward model, this simulation capability plays an important role in motor control. In this context, the motor command is sent to the forward model in parallel with its actual execution. The results of the actual and simulated execution are then compared, and the consequent error signal is used to correct the movement. Here we demonstrate how the iCub robot can (a) accumulate experience in the generation of action within its Autobiographical memory (ABM), (b) consolidate this experience encoded in the ABM memory to populate a semantic memory whose content can then be used to (c) simulate the results of actions. This simulation can be used as a traditional forward model in the control sense, but it can also be used in more extended time as a mental simulation or mental image that can contribute to higher cognitive function such as planning future actions, or even imagining the mental state of another agent. We present the results of the use of such a mental imagery capability in a forward modeling for motor control task, and a classical mentalizing task. Part of the novelty of this research is that the information that is used to allow the simulation of action is purely acquired from experience. In this sense we can say that the simulation capability is embodied in the sensorimotor experience of the iCub robot.

Keywords: Humanoid robot, perception, action, mental simulation, mental imagery, forward model.

1   Introduction

One of the central capabilities that cognitive systems provide to living organisms is the ability to "travel in time," that is, to imagine the future, and recall the past, in order to better anticipate future events [1]. This can be considered in the context that one of the central functions of the brain is to allow prediction [2, 3]. One of the most classical uses of prediction in the context of control is the forward model, which allows a system to predict responses to a motor command, and then compare the predicted and actual outcome. This notion has been extensively applied in the

---

[*] Corresponding author.

N.F. Lepora et al. (Eds.): Living Machines 2013, LNAI 8064, pp. 240–250, 2013.
© Springer-Verlag Berlin Heidelberg 2013

neuroscience of motor control [4]. At a more extended time scale, a suitably detailed forward model can be used as a simulation system for allowing the system to image how things might have been, or how they might be in the future. This can allow perspective taking, as required for solving tasks in which one must take the perspective of another. In the "Sally – Anne" task, a child is shown a set-up with two dolls, Sally and Anne. Sally puts her ball in a basket, and then leaves. Meanwhile, Anne moves Sally's ball into a box. Then Sally returns, and we can ask the child "where will Sally look for her ball?" Frith and Frith demonstrated [5] that before a certain age, children will "mistakenly" indicate that Sally will look in the box, the ball's actual location, rather than in the basket, where she put it. They suggest that the ability to mentalize – to represent other's mental states – relies on a system that has evolved for representing actions and their consequences. Such a capability to compare representations of others mental states with reality can form the basis for detecting that another agent is not telling the truth [6]. In the current study, we present a capability that allows the development of an internal simulation function, based on experience acquired by the agent, which allows the generation of mental simulations that can be used both in low level motor control.

The Experimental Functional Android Assistant (EFAA) system functions in a domain of physical interaction with a human as illustrated in Figure 1. Objects are manipulated by the robot and the human in cooperative interactions, and thus it is important that the EFAA system can accurately perform these manipulations and keep track of the actual and predicted physical state of itself and the human in their shared space.

2   System Description

The system provides control for real-time human interaction with the iCub robot that is achieved by articulation of three software modules: Autobiographical Memory, abmReasoning and OPCManager. The two first modules (AutobiographicalMemory and abmReasoning) have been previously described [7, 8]. They provide the ability for the system to store the history of all interactions in the ABM, and to extract conceptual information from the ABM, including the meaning of spatial and temporal referencing terms. We will briefly describe these functions, and then focus on the OPCManager (OPCM). Our complete system is developed in the context of the emergence of a self of the robot (use of an autobiographical episodic-like memory and a reasoning based on the experience) but an important property of the self is the ability to mentally simulate and predict consequences of the actions of himself or of other.

Illustrated in Figure 1, the Objects Properties Collector (OPC), contains the world-related knowledge of the robot at the current time. Here, we use two different OPCs. One will be related to the "real" world (realOPC), and the second one to the "mental" picture of the robot and to his imagination (mentalOPC). The main purpose of the OPCmanager module will be to simulate in the mentalOPC activities previously learned through the joint action of the AutobiographicalMemory and the ambReasoning, then to observe the possible implication of these activities, and to compare this with the final state of the same activities in the real world.

In summary, we will focus on the coordinated interaction of the three modules: 1) Autobiographical Memory: Take a snapshot of the world at a given time, Store snapshots and manage them. 2) abmReasoning: Manipulate the data of the ABM, Summarize and generalize different levels of knowledge. 3) OPC Manager: Simulate action interaction in a mental OPC, and extract differences between realOPC and mentalOPC.

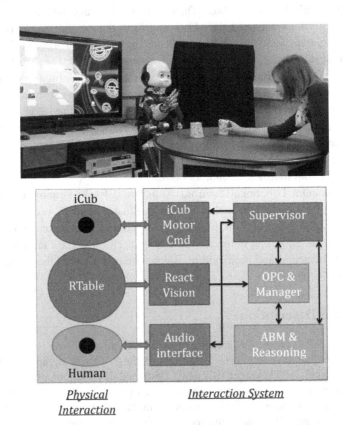

Fig. 1. Illustration of the iCub EFAA interacting with a human (Above), and System Architecture overview (Below). Human and iCub interact face-to-face across the ReacTable, which detects objects through the translucid surface, and communicates object locations via ReactVision to the Object Property Collector (OPC). The Supervisor coordinates spoken language and physical interaction with the iCub via spoken language technology in the Audio interface. The autobiographical memory ABM & reasoning system encodes world states and their transitions due to human and robot action as encoded in the OPC, and generates semantic representations. The OPC manager generates the representations of the actual and imagined states.

2.1   Autobiographical Memory and ABM Reasoning

The Autobiographical memory (ABM) is made up of an episodic memory and a semantic memory. The Episodic memory consists of 12 SQL tables, and stores the content of the OPC, with related contextual knowledge. The information will be: content of the realOPC, time, date, agent performing an action, semantic role of the argument of an action (i.e.: "ball": object, "north": spatial).

The Semantic Memory is made up of 12 SQL tables and stores the knowledge of the iCub related to different levels. Levels are: spatial, temporal, contextual, shared plan, behaviors. The semantic memory is constructed by extracting regularities from the episodic memory as human and robot actions cause changes to the states of objects on the ReacTable.

As such actions take place during the course of ongoing interactions, these events are stored in the episodic memory. The ABMreasoning function then extracts regularities that are common to experiences that are encoded in the episodic memory, to populate the semantic memory. The semantic memory thus includes the names and locations corresponding to locations taught by the human, and actions (e.g. *put* an *object* at a *location*) and their pre-conditions (e.g. that the object should be present) and post-conditions (e.g. that the *object* is now at *location*). Thus, through interaction, the system learns about the pre- and post-conditions of actions. This knowledge will be crucial in allowing the system to mentally simulate action.

2.2   OPC Management of Physical Reality and Mental Simulations

The OPC manager ensures the proper functioning of the realOPC and the mental OPC. The realOPC should maintain an accurate reflection of the physical state of the world. This state will be modified after the execution of actions. Thus, when the robot or the human perform an action of the type "put the triangle on the left", the physical state changes that result from this will be that the triangle is at the north location. For the realOPC, these changes will occur as part of the normal functioning of the OPC as it is updated by perceptual inputs from the ReacTable. This corresponds to the update of an internal model (the realOPC) via perception (ReactVision inputs to realOPC).

The novel aspect concerns the updating and maintenance of the mentalOPC. The function simulateActivity will simulate an action by retrieving its pre-conditions and post-conditions from the Semantic memory, and then "executing" this action by checking that its pre-condtions hold in the mentalOPC, and then updating the mentalOPC so that the post-condtions now hold, and the pre-conditions are removed. Thus, we emphasize that mental simulation is based on experience, initially encoded in the episodic memory and then extracted in the semantic memory.

3   Experiments

Here we report on two experimental evaluations of the use of the real and mental OPCs in different contexts.

3.1   Forward Model in Grasp Control

A current problem in robotics is the use of feedback in motor control, for example, when a robot attempts to grasp an object and the grasp fails, feedback can be used to detect the failure [9]. Such a feedback control look is illustrated in Figure 3. The motor command is sent to the forward model, and to the body, and the resulting predicted sensory feedback and actual sensory feedback are compared. If they match, the movement has been successfully completed, and if not, a failure is detected. We can use this method in the dispositive described above with the iCub. After experience producing the "put object at location" action, the system has acquired semantic information that the result of this action is that the object is now positioned at the specified location.

Functionally, the mentalOPC is used as the forward model. The "put" command is sent for execution to the ICubMotorCmd module, and it is also sent for simulation to the SimulatActivity function of the OPCmanager. The realOPC and mentalOPC can then be compared to assess the success of the action.

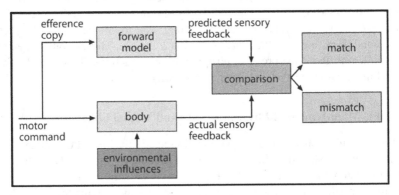

Fig. 2. Illustration of the forward model in the context of motor control. The motor command is sent to the motor command system and to the internal model. Subsequent comparison allows the system to determine if the grasp was correctly executed. Figure from [2].

Figure 4 illustrates the contents of the realOPC and mentalOPC before and after a successful "put circle left" action is executed by the iCub. The circle is indicated in the OPCs by a blue cube. In the Before panels it is at the "North" location near the robot's midline, and in the After panels it is displaced to the robot's left, to the location labeled Left. The diffOPC function produces a report indicating that there is no significant difference in the two positions, as illustrated in Table 1.

Figure 5 illustrates the mentalOPC and realOPC before and after "put cross left" action in which there is a physical disturbance during the execution by the iCub. In the lower right panel (Actual – After) we can observe that the representation of the cross object is not positioned on the localization "Left" in contrast to the predicted location that can be visualized in the upper right panel (Mental – After). During the execution a perturbation occurred and the put action resulted in a final positioning of the object that does not match with the predicted location. This mismatch is detected by the diffOPC function, as illustrated in Table 2.

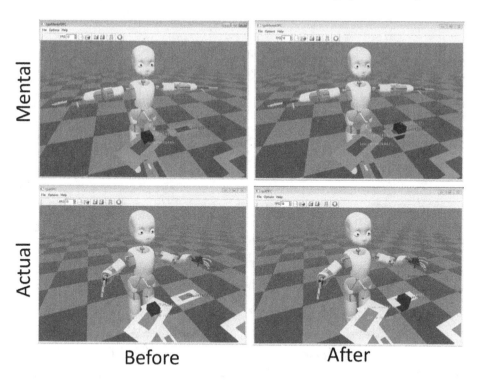

Fig. 3. Mental image and actual physical state before and after a successful grasp and move action. The action to perform is to put the circle (the blue object) to the left of the robot in the delimited location..

Table 1. Comparison from diffOPC function of realOPC and mentalOPC indicating no difference.

```
Entity : circle
        robot_position_x -0.025075
        robot_position_y 0.016605
        robot_orientation_z -0.048426
        rt_position_x 0.112762
        rt_position_y 0.180602

        No semantic differences
```

The detected mismatch can be used to allow the iCub EFAA to determine a correcive course of action. Our next step in this context will be to include experiments with this forward modeling capability integrated in the iCubMotorCMD so that failed actions can automatically initiate appropriate retrials.

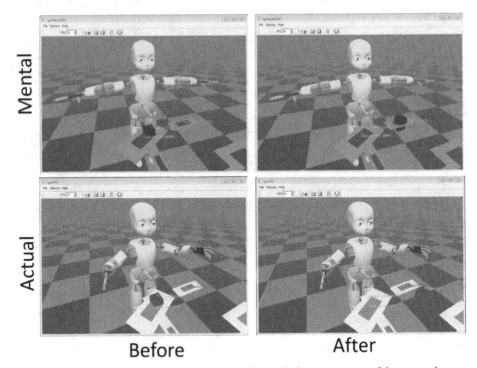

Mental

Actual

Before   **After**

Fig. 4. Mental image and actual physical state before and after an unsuccessful grasp and move action. The final state of the object in the actual condition is different from that in the mental simulation (forward model), thus indicating that the action failed. The action to perform is to put the cross (the red object) to the left of the robot in the delimited location.

3.2   Simulating Other's Beliefs in the "Sally Anne" Task

Such mental simulation can also contribute to the ability of the robot to mentalize. Mentalizing is the ability to represent the mental states of others, traditionally referred to in the context of theory of mind (ToM) tasks [5]. A classic method to assay this capability to represent false beliefs is via the "Sally – Anne" task. In this task, the

Table 2. Results returned from diffOPC comparing realOPC and mentalOPC. The comparison indicates a significant difference, corresponding to the error in the execution of the action.

```
Entity : cross
        robot_position_x -0.091195
        robot_position_y -0.035511
        robot_orientation_z 0.179689
        rt_position_x 0.194235
        rt_position_y 0.183612

Semantical differences :

Entity : icub
        Beliefs removed:cross is left (mental-predicted)
```

child is shown a small stage with two dolls, Sally and Anne, and a toy ball. Sally puts her ball in a basket, and then leaves. Meanwhile, Anne moves Sally's ball into a box. Then Sally returns, and we can ask the child "where will Sally look for her ball?"

A long history of experimental research has demonstrated that the ability to perform tasks that require such representations develops over time, though the details of precisely when young children can perform such tasks depends signficicantly on the detail of how the task protocol is implemented, and children as young as 15 months of age can be demonstrated to represent the false beliefs of others [10].

We hypothesized that the mentalOPC can allow the iCub EFAA to resolve this problem correctly. The mentalOPC can be used to represent the intial state of affairs. The key element is that after Sally leaves, whatever happens in the mentalOPC - which is intended to represent Sally's perspective - should not change and should not be subject to the results of any actions that Sally does not actually witness.

The contents of the realOPC and mentalOPC in this context are represented in Figure 6. In this case, while the agent Sally is present, the toy is placed on the left. This is represented in the mentalOPC and the realOPC. In the After column, for the

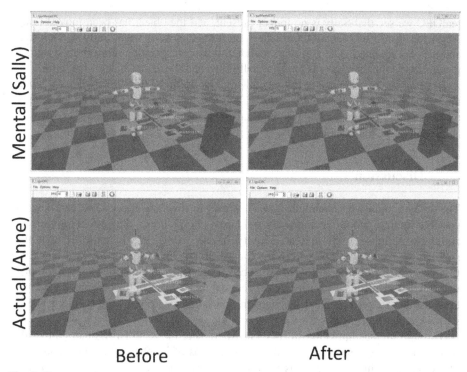

Fig. 5. Contents of mentalOPC and realOPC in the Sally-Anne task. In the "Before" column is represented the contents of both OPC when the toy has been placed at the first location. The mentalOPC is the systems representation of what it and Sally have seen. In the "After" column, the Actual situation represents the contents of the realOPC after the toy has been moved. In that same column the mentalOPC represents the what Sally observed before she left. If this is maintained in memory, then it will persist after the world has been changed, and it can be used to mentalize about where Sally would look for the toy.

Table 3. Results returned from diffOPC comparing realOPC and mentalOPC. The comparison indicates a significant difference, corresponding to difference between the "false belief" attributed to Sally in the mentalOPC and the "true beliefs" attributed to Anne in the realOPC.

```
3 entities changed :
Entity : toy
        robot_position_x -0.034557
        robot_position_y 0.276599
        robot_orientation_z -0.042716
        rt_position_x -0.118581
        rt_position_y -0.252269
Entity : icub
        Beliefs added :toy is column toy is north (after)
        Beliefs removed :toy is left, Sally is isPresent (before)
Entity : Sally
The beliefs of Sally didn't change, because she wasn't here.
Her beliefs are : toy is left.
```

realOPC we see that Sally is no longer present, and the object has been moved to the North location. The mentalOPC is the same image as seen when Sally was present, and it is not updated. The ability to maintain this representation allows the system to recognize the mismatch between what Sally saw, and the actual state of the world.

Here we see that the use of the mentalOPC allows the system to "mentalize" about the belief state of another agent. This experiment has potential impact in the context of the ongoing debate on what is required for passing false belief tasks, and will be addressed in the discussion.

4   Discussion

The human cognitive system allows us to travel in time and space – we can imagine possible futures, and relive and analyze the past [1]. To do so, the system requires the ability to simulate itself and its activity in the world. We hypothesize that this simulation capability derives from the long evolved capability for forward modeling that was crucial for the ability of advanced primates to navigate through a complex world where real-time sensorimotor was crucial to survival. In the current research we demonstrate a developmental mechanism that could contribute to the emergence of such a simulation capability.

Through the accumulation of its own experience, the iCub EFAA can extract the regularities that define the pre- and post-conditions of its physical actions, and those of the human. This knowledge is then used to drive the mental simulation of action, which can actually operate faster than real-time, and generate predictions of expected outcome before the real movement is achieved. We demonstrate the functionality of this mechanism in two settings: Forward modeling in sensory-motor control, mentalizing in a false-belief task.

Learning forward models has been successfully applied in robotics [11, 12]. In the context of forward modeling, it is important for the system to detect that inconsistent information is being provided. This can be important in the ongoing learning of the system based on experience. Thus, if the human says that it will perform an action, and then the system can detect a difference between the actual and predicted action, then it can mark this experience as suspect, and not include it in future learning, thus not contaminating experience with questionable content.

In the context of mentalizing and the false belief task, the current research has significant potential impact. There is an ongoing debate concerning the nature of the mental processes that are required to take the mental perspective of another agent. This includes discussion of whether distinct language capabilities are required [10]. Our research provides insight into this question, by illustrating how a simulation capability that is directly derived from experience can be used to provide an agent with the basic representational capabilities to perform the false belief task.

It can be considered that the mere notion of "autobiographical memory" presupposes that the system must have a first person perspective, from which that memory is situated. The notion of first person perspective is in fact a deep philosophical issue (see eg. [13]). From the perspective of the current research, we can say that the robot has taken steps towards achieving a minimal form of 1PP in that it has developed an integrated representation of itself within the peripersonal space. This is also related to the notion of ecological self as defined by Neisser, which is the individual situated in and acting on the immediate environment [14]. What is currently missing with respect to these notions of self is a reflective capability, where the system reasons on a self-model as an integrated model of the very representational system, which is currently activating it within itself, as a whole [15].

In summary, the current research makes a significant contribution to the cognitive systems research. It allows the iCub EFAA system to autonomously generate an internal simulation capability based on its own personal experience. This simulation capability can operate at the level of physical control, and at high levels of cognition including mentalizing about the belief states of others. Our current research integrates this capability in the context of simulated situations models and language comprehension [16, 17].

Acknowledgements. This research has been funded by the European Commission under grant EFAA (ICT-270490).

References

[1] Vernon, D., Metta, G., Sandini, G.: A survey of artificial cognitive systems: Implications for the autonomous development of mental capabilities in computational agents. IEEE Transactions on Evolutionary Computation 11, 151–180 (2007)

[2] Bubic, A., Von Cramon, D.Y., Schubotz, R.I.: Prediction, cognition and the brain. Frontiers in Human Neuroscience 4 (March 22, 2010)

[3] Friston, K.: A theory of cortical responses. Philos. Trans. R. Soc. Lond. B Biol. Sci. 360, 815–836 (2005)

[4] Wolpert, D.M., Ghahramani, Z., Jordan, M.I.: An internal model for sensorimotor integration (1995)

[5] Frith, C.D., Frith, U.: Interacting minds–a biological basis. Science 286, 1692–1695 (1999)

[6] Singer, T.: The neuronal basis and ontogeny of empathy and mind reading: review of literature and implications for future research. Neuroscience and Biobehavioral Reviews 30, 855–863 (2006)

[7] Petit, M., Lallee, S., Boucher, J.-D., Pointeau, G., Cheminade, P., Ognibene, D., Chinellato, E., Pattacini, U., Demiris, Y., Metta, G., Dominey, P.F.: The Coordinating Role of Language in Real-Time Multi-Modal Learning of Cooperative Tasks. IEEE Transactions on Autonomous Mental Development (2012)

[8] Pointeau, G., Petit, M., Dominey, P.F.: Robot Learning Rules of Games by Extraction of Intrinsic Properties. In: The Sixth International Conference on Advances in Computer-Human Interactions, ACHI 2013, pp. 109–116 (2013)

[9] Shirai, Y., Inoue, H.: Guiding a robot by visual feedback in assembling tasks. Pattern Recognition 5, 99–108 (1973)

[10] Onishi, K.H., Baillargeon, R.: Do 15-month-old infants understand false beliefs? Science 308, 255–258 (2005)

[11] Dearden, A., Demiris, Y.: Learning forward models for robots. In: International Joint Conference on Artificial Intelligence, p. 1440 (2005)

[12] Metta, G., Fitzpatrick, P.: Early integration of vision and manipulation. Adaptive Behavior 11, 109–128 (2003)

[13] Metzinger, T.: Empirical perspectives from the self-model theory of subjectivity: a brief summary with examples. In: Banerjee, R., Chakrabarti, B. (eds.) Models of Brain and Mind: Physical, Computational and Psychological Approaches, vol. 168, p. 215 (2008)

[14] Neisser, U.: The roots of self-knowledge: perceiving self, it, and thou. Ann. N. Y. Acad. Sci. 818, 18–33 (1997)

[15] Metzinger, T.: Being no one: The self-model theory of subjectivity. Bradford Books (2004)

[16] Barsalou, L.W.: Simulation, situated conceptualization, and prediction. Philos. Trans. R. Soc. Lond. B. Biol. Sci. 364, 1281–1289 (2009)

[17] Madden, C., Hoen, M., Dominey, P.F.: A cognitive neuroscience perspective on embodied language for human-robot cooperation. Brain Lang. 112, 180–188 (2010)

Three-Dimensional Tubular Self-assembling Structure for Bio-hybrid Actuation

Leonardo Ricotti[*], Lorenzo Vannozzi, Paolo Dario, and Arianna Menciassi

The BioRobotics Institute, Scuola Superiore Sant'Anna, Viale R. Piaggio 34,
56025 Pontedera (PI), Italy
{l.ricotti,p.dario,a.menciassi}@sssup.it,
l.vannozzi@bioroboticsinstitute.it

Abstract. This work aims at reporting an innovative approach towards the development of a three-dimensional cell-based bio-hybrid actuator. The system, made of polydimethylsiloxane and based on a stress-induced rolling membrane technique, was provided with different elastic moduli (achieved by varying the monomer/curing agent ratio), with proper surface micro-topographies and with a proper surface chemical functionalization to assure a long-term stable protein coating. Finite element modeling allowed to correlate the overall contraction of the polymeric structure along its main axis (caused by properly modeled muscle cell contraction forces) with substrate thickness and with matrix mechanical properties.

Keywords: Bio-hybrid systems, Cell-based actuators, Living machines, Engineered bio/non-bio interfaces.

1   Introduction

Actuation is a key function of any artificial or living machine, whose aim is moving. Despite the wide range of opportunities offered by traditional and recent actuation solutions [1], current actuation technologies represent a real bottleneck for many applications, especially those concerning robotics and biomimetics. Their performances, indeed, are far from those achieved by natural muscles, especially concerning inertia, back-drivability, stiffness control and power consumption [2]. This implies that current machines are not able to replicate many life-like movements (and their consequent efficacy and efficiency) typical of living animals. It has been argued that the fabrication of "soft" robotic artefacts able to mimic animal movements would be a critical step allowing a disruptive advancement in robotics and engineering [3][4].

To face these issues, an innovative approach has been recently proposed, based on the exploitation of the unique features of living cells and tissues not for mimicking muscles, but for physical integrating them in a properly engineered and controllable artificial structure, towards the development of living cell-based actuators [5].

---

[*] Corresponding author.

N.F. Lepora et al. (Eds.): Living Machines 2013, LNAI 8064, pp. 251–261, 2013.
© Springer-Verlag Berlin Heidelberg 2013

From the pioneeristic work proposed by Herr and Dennis in 2004, in which a swimming robot was actuated by two explanted frog semitendinosus muscles controlled by an embedded microcontroller [6], literature was progressively enriched by examples of autonomous contractile systems, relying on the self-beating activity of primary rat cardiac cells [7]-[9] or insect dorsal vessel tissue [10], and by more recent attempts of culture and differentiation of skeletal muscle cells (controllable by means of external electrical inputs) on properly engineered substrates [11][12].

Despite such efforts, both technological and biological advances are still needed to achieve reliable and efficient muscle cell-based three dimensional actuators resembling natural muscles, even if very recent results of Asada's group are going toward this direction [13].

In this work, we propose a strategy for fabricating three-dimensional tubular self-assembling engineered polymeric structures to be used as matrices for bio-hybrid actuators. We also show preliminary results concerning matrix fabrication and characterization, as well as finite element model (FEM)-based simulation results, predicting actuator deformation due to muscle cell contractile activity.

2   Materials and Methods

Matrix was fabricated by using polydimethylsiloxane (PDMS, Sylgard 184, Dow Corning) in different concentrations of monomer/curing agent (5:1, 10:1, 15:1, 20:1, and 25:1 w/w, respectively). The strategy followed to build a tubular structure exploited a stress-induced rolling membrane technique [14], involving two coupled PDMS layers provided with different topographical cues (Fig. 1). The top membrane was baked at 150° C for 15 min, then stretched (40% of its initial length) to induce stress and placed on top of the bottom membrane, which was semi-cured (at 80°C for 195 s). The whole bilayer was finally baked to stabilize its structure (at 80° C for 30 min).

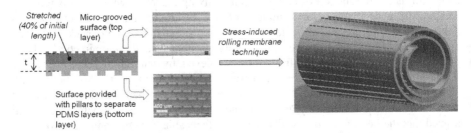

Fig. 1. Representation of the procedure followed to fabricate three-dimensional tubular PDMS structures provided with defined topographical cues

The desired topographies were achieved by means of standard photolithographic techniques and *ad hoc* designed masks. Briefly, the molds needed to fabricate micro-grooved top layers were obtained by spinning first a Microposit primer (3500 rpm for 30 s) and then a positive resist (Shipley S1813, 4500 rpm for 35 s) on Si wafers.

After baking (120° C for 1 min), the wafers were exposed for 7 s by means of a mask aligner and then developed for 30 s. The molds needed to fabricate pillars-provided bottom layers were obtained by directly spinning a negative resist (SU-8 50, MicroChem) on Si wafers (two cycles were performed: the former based on 500 rpm for 10 s, the latter based on 2000 rpm for 30 s). The wafers were then baked at 65° C for 6 min, then baked again at 95° C for 20 min, exposed for 20 s by means of a mask aligner and finally developed for 6 min.

PDMS surface was functionalized in order to provide it with a long-term stable protein coating, by following a protocol recently published by our group [15]. Samples were incubated in a solution of 60% ethanol, 20% (3-aminopropyl) triethoxysilane (APTES) and 1.2% NH3 (pH = 9) for 3 h at 60° C, and then with a 5 mg/ml genipin (Sigma) solution in pure water (Milli-Q, Millipore) for 60 min at 37° C. Surface chemistry was finally modified by incubation with a 20 μg/ml fibronectin (Sigma) solution in pure water at 37° C for 24 h.

Matrix surface was imaged by means of scanning electron microscope (SEM, EVO MA15, Zeiss) and atomic force microscope (AFM, Veeco). Substrate mechanical properties were evaluated by means of traction tests, performed with a mechanical testing system (INSTRON 4464) provided with a ± 10 N load cell. All specimens were pulled at a constant speed of 5mm/min, until reaching sample failure. Data were recorded at a frequency of 100 Hz. The stress was calculated as the ratio between the load and the cross-section area of a tensile specimen, while the strain was calculated as the ratio between its extension and its initial length. The Young modulus for each tested sample was extracted from its stress/strain curve.

Preliminary *in vitro* tests were performed to assess the ability of the functionalized substrate to induce cell alignment and to maintain cell viability. C2C12 murine skeletal myoblasts were used as cell model. They were cultured in Dulbecco's Modified Eagle Medium (DMEM) supplemented with 10% fetal bovine serum (FBS), 100 IU/mL penicillin, 100 μg/mL streptomycin and 2 mM L-glutamine. Cells were seeded on both flat and micro-grooved PDMS samples at a density of 50,000 cells/cm^2, thus reaching almost confluence on the substrates 24 h after seeding. Cell-provided samples were maintained at 37°C in a saturated humidity atmosphere containing 95% air and 5% CO_2.for 24 h, and then treated with a Live/Dead® viability/cytotoxicity assay (Invotrogen). The kit stained live cells in green and dead cells in red. Fluorescence images were acquired by means of an inverted fluorescence microscope (Eclipse Ti, FITC-TRITC filters, Nikon) equipped with a cooled CCD camera (DS-5MC USB2, Nikon) and with NIS Elements imaging software. Images were then elaborated with a free software (ImageJ), in order to quantify the preferential alignment of cells towards a specific direction.

FEM simulations were performed in ANSYS (Version no. 13, Ansys Ltd.). For each sample type we imported in the model the correspondent stress-strain curve obtained during mechanical characterization tests. The model was based on a tubular structure showing an overall thickness of 3t (three times the thickness of the PDMS bilayer (t in Fig. 1) used to build the rolled structure, assuming that it will be characterized by three turns). Cell-derived contraction forces acting on the matrix were modeled by means of a couple of forces. Each force was homogeneously distributed on the internal walls of one half of the tubular structure. The two forces had the same modulus but opposite directions, in order to induce a contraction of the polymeric structure along its longitudinal axis. Force moduli were estimated by taking into account the contraction force values reported in literature for primary cardiomyocytes (~ 10 μN per cell) [16] and for *in vitro* cultured skeletal myotubes (~ 1.2 μN per myotube) [17]. Differences in the distribution of the two cell types were also considered in the model, as depicted in Fig. 2. Typical dimensions of cardiomyocytes are 100 x 50 μm [16] and we considered a distance of 100 μm between two adjacent cells, in both X and Y directions. Typical dimensions of *in vitro* cultured myotubes are 500 x 100 μm [11] and we considered a distance of 200 μm between two adjacent myotubes in both X and Y directions.

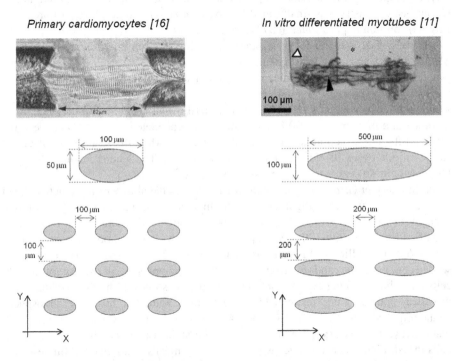

Fig. 2. Cell-related parameters considered for FEM-based simulations: typical dimensions of cardiomyocytes and *in vitro* differentiated myotubes and their topological distribution on an X-Y plane

The data collected were subjected to analysis of variance in order to evaluate the statistically significant differences among samples. A *t*-test was performed for comparison between two groups, while Holm–Sidak tests were performed for comparisons among several groups. Significance was set at 5%.

3   Results and Discussion

AFM characterization (Fig. 3a) revealed that PDMS top layer was provided with 9.53 ± 0.37 μm-wide grooves, spaced by 10.59 ± 0.39 μm and showing a height of 1.07 ± 0.02 μm. Such features have been demonstrated to be effective in aligning cells along the groove axis [18], thus achieving a strongly anisotropic muscle cell culture, that is an important pre-requisite to maximize substrate contraction [19]. The analysis of SEM images permitted to quantify the dimensions of pillars on the PDMS bottom layer (Fig. 3b): pillars were characterized by a length of 295 ± 1 μm, a width of 53 ± 1 μm and an estimated height of ~ 50 μm. This height, much larger than that of cultured cells (~ 1-2 μm), allows the pillars to operate as "spacers" between the rolled PDMS layers, thus reducing the mechanical stresses on cells, cultured on the internal walls, and keeping them alive by assuring a correct medium flow between the layers.

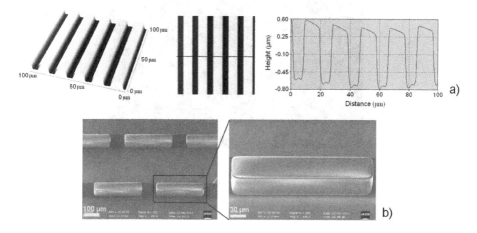

Fig. 3. (a) AFM images of the micro-grooved PDMS top layer and height profile extraction from a horizontal line on the image; (b) SEM images of the PDMS bottom layer, provided with pillars

Cell viability and orientation on PDMS samples showing different topographical cues were assessed by analyzing fluorescence images of the samples treated with the Live/Dead® assay. Results showed that both flat and micro-grooved samples assured a high cell viability (Fig. 4a,b). Furthermore, cells cultured on micro-grooved substrates showed a strong anisotropy, being highly aligned along the groove axis (Fig. 4c).

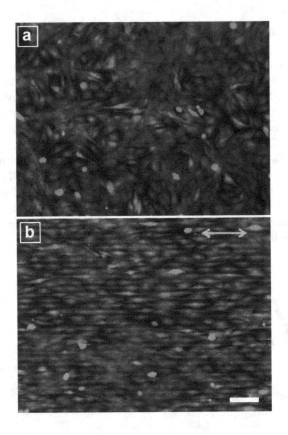

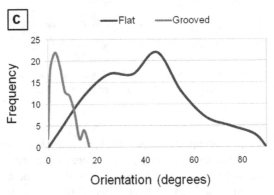

Fig. 4. Live/Dead® fluorescence images of C2C12 cells cultured on flat (a) and micro-grooved (b) PDMS substrates. Live cells are stained in green, dead cells are stained in red. The red arrow shows the groove orientation on the PDMS matrix. Scale bar is 100 μm. (c) Distribution of orientation angles for cells cultured on flat substrates (blue line) and cells cultured on micro-grooved substrates (orange line). The angles refer to an arbitrary axis in the first case and to the groove axis in the second case. Angle values close to 0° mean high alignment in the axis direction, while angle values close to 45° mean random alignment.

As known, matrix stiffness is a crucial factor for the achievement of terminally differentiated and functional skeletal myotubes [20]. At the same time, substrate flexibility is also important in order to minimize the opposition to cell-induced contraction [21]. The mechanical properties of the different samples used in this study are reported in Fig. 5.

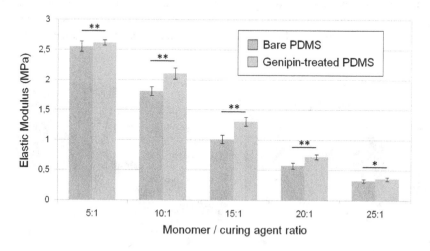

Fig. 5. Young's moduli of the different samples, assessed by means of traction tests (at least 10 specimens for each sample type). For each monomer/curing agent ratio both bare and genipin-treated [15] PDMS were tested. *=p<0.05, **=p<0.01.

For all the tested sample types, the chemical treatment used to stably cross-link fibronectin to the sample surface induced a stiffening of the matrix. Since surface functionalization is a crucial step for the achievement of stable cell cultures on PDMS substrates, we were obliged to accept as a compromise a slightly stiffer matrix, thus increasing the distance from the optimal stiffness value reported in literature for skeletal muscle maturation, namely ~ 14 kPa [20]. The option to use higher monomer – curing agent ratios in order to achieve lower Young's modulus values is hardly feasible in our case, due to manipulability issues, especially concerning the bottom PDMS layer of the structure, which is semi-cured and *per se* almost impossible to manipulate when its elastic modulus goes down below 250 kPa and when its thickness goes down below 2-3 μm. Therefore, we considered such values (25:1 for the monomer – curing agent ratio and 2.5 μm for the thickness of the bottom layer, i.e. 5 μm for the thickness (t) of the PDMS bilayer) as our internal limits. It could be argued that these mechanical characteristics would represent a strong limitation for a correct development of muscle tissue on these substrates. We envision to overcome this issue by providing the system with a feeder layer of fibroblasts, on top of which muscle cells will be seeded and differentiated. This strategy already proved to be effective in providing muscle cells both with a suitable mechanical interface and with specific cytokines, produced by fibroblasts, which help muscle tissue maturation [22],[23].

We demonstrated the feasibility of the proposed approach by developing a PDMS bilayer provided with both a micro-grooved top surface and a pillar-based bottom surface and by rolling the structure by exploiting the stress-induced technique. We were able to obtain the desired structure showing concentric PDMS layers separated by the pillars (Figure 6) and with micro-grooved internal walls (Figure 6, inset).

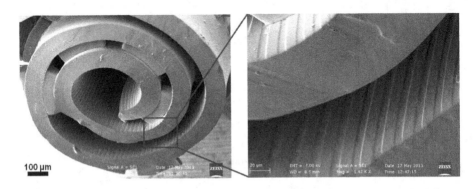

Fig. 6. SEM image of a rolled PDMS structure, showing concentric layers separated by pillars. Inset: close view of the internal walls of the PDMS structure, provided with 10 μm-wide grooves.

FEM analysis allowed to predict the overall device contraction depending on bilayer thickness and on substrate mechanical properties, by considering literature values for cell contraction forces applied to the polymeric structure. Results (Fig. 6) showed that the use of primary cardiomyocytes would theoretically permit to achieve a considerable contraction, up to ~ 55% (Fig. 7a), a value that outperform the maximum contraction of natural skeletal muscles, which is ~ 40% [24]. Results based on *in vitro* differentiated skeletal myotubes showed lower contraction values, due to current inefficiency in obtaining mature and functional (highly contractile) myotubes starting from standard immortalized cell lines. However, even in this case, the total contraction is appreciable (up to ~ 3.7%, Fig. 7b), highlighting the potential of this solution for the achievement of three-dimensional muscle-resembling bio-hybrid actuators.

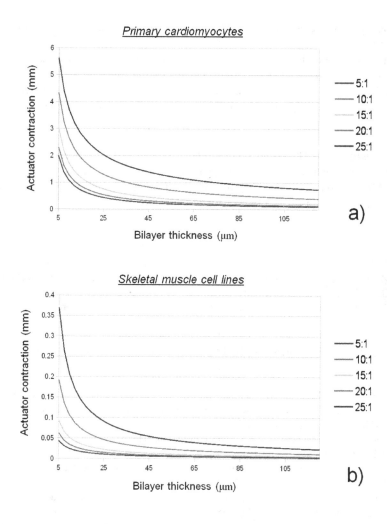

Fig. 7. FEM-based simulation results reporting the overall actuator contraction for the different sample types and considering literature values for cell contraction force concerning cardiomyocytes (a) and immortalized skeletal muscle cell lines (b). In both cases the initial actuator length is 10 mm.

4   Conclusions and Future Work

Our preliminary results highlight that PDMS substrates with different elastic moduli can be shaped in three-dimensional tubular structures by means of a stress-induced rolling membrane technique, provided with defined topographical features, functionalized with a natural cross-linker allowing a stable fibronectin coating and lately used as matrices for cardiac or skeletal muscle cell culture, with the possibility of achieving large device contractions (up to ~ 55%), thus quite closely mimicking the behavior of natural muscles.

Future work will be focused on *in vitro* validation of the proposed model, by culturing both cardiac and skeletal muscle cells on the substrate and by measuring its actual deformation. Further characterization of the bio-hybrid actuator will be also performed, in terms of overall contraction force, efficiency, power density and lifetime.

Acknowledgments. The authors would like to deeply thank Mr. Carlo Filippeschi for his unvaluable contribution to all the clean room procedures. They would also like to thank Mrs. Giada Graziana Genchi and Dr. Gianni Ciofani for their suggestions concerning the genipin-based PDMS functionalization process.

References

[1] Zupan, M., Ashby, M.F., Fleck, N.A.: Actuator classification and selection – the development of a database. Adv. Eng. Mat. 4, 933–939 (2002)

[2] Pons, J.L.: Emerging actuator technologies: a micromechatronic approach. John Wiley & Sons Ltd., Chichester (2005)

[3] Shepherd, R.F., Ilievski, F., Choi, W., Morin, S.A., Stokes, A.A., Mazzeo, A.D., Chen, X., Wang, M., Whitesides, G.M.: Multigait soft robot. Proc. Nat. Acad. Sci. 108, 20400–20403 (2011)

[4] Dario, P., Verschure, P.F.M.J., Prescott, T., Cheng, G., Sandini, G., Cingolani, R., Dillmann, R., Floreano, D., Leroux, C., MacNeil, S., Roelfsema, P., Verykios, X., Bicchi, A., Melhuish, C., Albu-Schaffer., A.: Robot companions for citizens. In: Proc. FET Conf. Exhib., Proc. Comp. Sci., vol. 7, pp. 47–51 (2011)

[5] Ricotti, L., Menciassi, A.: Bio-hybrid muscle cell-based actuators. Biomed. Microdev. 14, 987–998 (2012)

[6] Herr, H., Dennis, R.G.: A swimming robot actuated by living muscle tissue. J. Neuroeng. Rehab. 1 (2004), doi:10.1186/1743-0003-1-6

[7] Xi, A.J., Schmidt, J.J., Montemagno, C.D.: Self-assembled microdevices driven by muscle. Nat. Mat. 4, 180–184 (2005)

[8] Tanaka, A.Y., Morishima, K., Shimizu, T., Kikuchi, A., Yamato, M., Okano, T., Kitamori, T.: An actuated pump on-chip powered by cultured cardiomyocytes. Lab Chip 6, 362–368 (2006)

[9] Feinberg, A.W., Feigel, A., Shevkoplyas, S.S., Sheehy, S., Whitesides, G.M., Parker, K.K.: Muscular thin films for building actuators and powering devices. Science 317, 1366–1370 (2007)

[10] Akiyama, A., Iwabuchi, K., Furukawa, Y., Morishima, K.: Long-term and room temperature operable bioactuator powered by insect dorsal vessel tissue. Lab Chip 9, 140–144 (2009)

[11] Fujita, H., Dau, V.T., Shimizu, K., Hatsuda, R., Sugiyama, S., Nagamori, E.: Designing of a Si-MEMS device with an integrated skeletal muscle cell-based bio-actuator. Biomed. Microdev. 13, 123–129 (2011)

[12] Nagamine, K., Kawashima, T., Sekine, S., Ido, Y., Kanzaki, M., Nishizawa, M.: Spatiotemporally controlled contraction of micropatterned skeletal muscle cells on a hydrogel sheet. Lab Chip 11, 513–517 (2011)

[13] Sakar, M.S., Neal, D., Boudou, T., Borochin, M.A., Li, Y., Weiss, R., Kamm, R.D., Chen, C.S., Asada, H.H.: Formation and optogenetic control of engineered 3D skeletal muscle bioactuators. Lab Chip 12, 4976–4985 (2012)

[14] Yuan, B., Jin, Y., Sun, Y., Wang, D., Sun, J., Wang, Z., Zhang, W., Jiang, X.: A strategy for depositing different types of cells in three dimensions to mimic tubular structures in tissues. Adv. Mat. 24, 890–896 (2012)

[15] Genchi, G.G., Ciofani, G., Liakos, I., Ricotti, L., Ceseracciu, L., Athanassiou, A., Mazzolai, B., Menciassi, A., Mattoli, V.: Bio/non-bio interfaces: a straightforward method for obtaining long term PDMS/muscle cell biohybrid constructs. Coll. Surf. B: Biointerf. 105, 144–151 (2013)

[16] Lin, G., Pister, K.S.J., Roos, K.P.: Surface micromachined polysilicon heart cell force transducer. J. Micromech. Syst. 9, 9–17 (2000)

[17] Shimizu, K., Sasaki, H., Hida, H., Fujita, H., Obinata, K., Shikida, M., Nagamori, E.: Assembly of skeletal muscle cells on a Si-MEMS device and their generative force measurement. Biomed. Microdev. 12, 247–252 (2010)

[18] Charest, J.L., García, A.J., King, W.P.: Myoblast alignment and differentiation on cell culture substrates with microscale topography and model chemistries. Biomaterials 28, 2202–2210 (2007)

[19] Okano, T., Satoh, S., Oka, T., Matsuda, T.: Tissue engineering of skeletal muscle: highly dense, highly oriented hybrid muscular tissues biomimicking native tissues. ASAIO J. 43, M749–M753 (1997)

[20] Engler, A.J., Griffin, M.A., Sen, S., Bönnemann, C.G., Sweeney, H.L., Discher, D.E.: Myotubes differentiate optimally on substrates with tissue-like stiffness. J. Cell Biol. 166, 877–887 (2004)

[21] Ricotti, L., Taccola, S., Pensabene, V., Mattoli, V., Fujie, T., Takeoka, S., Men-ciassi, A., Dario, P.: Adhesion and proliferation of skeletal muscle cells on single layer poly(lactic acid) ultra-thin films. Biomed. Microdev. 12, 809–819 (2010)

[22] Cooper, S.T., Maxwell, A.L., Kizana, E., Ghoddusi, M., Hardeman, E.C., Alexander, I.E., Allen, D.G., North, K.N.: C2C12 co-culture on a fibroblast substratum enables sustained survival of contractile, highly differentiated myotubes with peripheral nuclei and adult fast myosin expression. Cytoskeleton 58, 200–211 (2004)

[23] Greco, F., Fujie, T., Ricotti, L., Taccola, S., Mazzolai, B., Mattoli, V.: Microwrinkled conducting polymer interface for anisotropic multicellular alignment. Appl. Mat. Interf. 5, 573–584 (2013)

[24] Woledge, R.C., Curtin, N.A., Homsher, E.: Energetic aspects of muscle contrac-tion. Academic Press, Bellington (1985)

Spatio-temporal Spike Pattern Classification in Neuromorphic Systems

Sadique Sheik, Michael Pfeiffer, Fabio Stefanini, and Giacomo Indiveri

Insitute of Neuroinformatics, University of Zurich and ETH Zurich,
Wintherthurerstrasse 190, 8057 Zurich, Switzerland

Abstract. Spike-based neuromorphic electronic architectures offer an attractive solution for implementing compact efficient sensory-motor neural processing systems for robotic applications. Such systems typically comprise event-based sensors and multi-neuron chips that encode, transmit, and process signals using spikes. For robotic applications, the ability to sustain real-time interactions with the environment is an essential requirement. So these neuromorphic systems need to process sensory signals continuously and instantaneously, as the input data arrives, classify the spatio-temporal information contained in the data, and produce appropriate motor outputs in real-time. In this paper we evaluate the computational approaches that have been proposed for classifying spatio-temporal sequences of spike-trains, derive the main principles and the key components that are required to build a neuromorphic system that works in robotic application scenarios, with the constraints imposed by the biologically realistic hardware implementation, and present possible system-level solutions.

1 Introduction

Spiking neural networks represent a promising computational paradigm for solving complex pattern recognition and sensory processing tasks that are difficult to tackle using standard machine vision and machine learning techniques [1, 2].

Most research on spiking neural networks is done using software simulations [3]. For large networks such approaches require high computational power and need to be run on bulky and power-hungry workstations or distributed computing facilities. Alternative approaches, for example based on full-custom implementations of spiking neural networks built using analog-digital Very Large Scale Integration (VLSI) circuits, can lead to compact and real-time solutions that are optimally suited for robotic applications. Systems built following this approach would involve multiple neuromorphic VLSI sensory and processing devices interfaced among each other to perform complex cognitive tasks. For example, they would use event-based sensors [4] for producing sensory signals in the form of continuous and asynchronous streams of spikes, and spiking multi-neuron chips [5, 6, 7] for implementing state dependent neural processing, learning mechanisms, and generation of appropriate behaviors.

N.F. Lepora et al. (Eds.): Living Machines 2013, LNAI 8064, pp. 262–273, 2013.

Within this context, one of the basic requirements for exhibiting complex behaviors is the system's ability to classify spike patterns generated by the sensors. In real-world application scenarios, these spike patterns would be typically spread across space and time reflecting the *spatio-temporal* properties of the stimuli they are encoding. Several experiments in various animal species have shown that behaviorally relevant information is encoded in spatio-temporal spike patterns, and used for decision making. Barn owls for example rely on auditory cues to capture the prey, by accurately determining the spatial location of the sound source. This is done by computing the temporal delay in the spikes originating at the two ears of the owl [8, 9]. Remarkably precise temporal codes have been found in bush crickets [10] in early auditory inter-neurons, even in the presence of strong natural background noise, and such precise temporal coding could reduce the uncertainty involved in decision making.

The timing of spikes does not necessarily have to encode only temporal features of the input signal, but can also carry information about other aspects. For example, in the human peripheral nervous system, Johansson et al. [11] found that the response of afferent projections from touch receptors in the fingertips carried significant information about pressure in the relative timing of the first spike.

To utilize information encoded in spike times, temporal codes typically require a signal that indicates the onset of the stimulus. In the cortex there is evidence for an alternative model of temporal coding in which information is encoded by the phase relative to one or several background oscillations [12], rather than relative to an external input signal. As an example, hippocampal place cells in rodents encode the spatial location of the animal by their spike timing in relation to the phase of gamma band oscillation [13]. A more detailed analysis of the role of temporal coding in biological systems is discussed e.g. in [14]. It is clear therefore that biological systems employ multiple mechanisms and resources for efficiently processing and classifying spatio-temporal spike patterns. If we want to build biologically inspired neuromorphic systems that utilize similar concepts for robotic applications, we therefore first have to answer the question: *what are the essential components that enable an electronic neuromorphic system to accomplish this task?*

Here we will address this question by evaluating the state-of-the-art spike-based spatio-temporal processing models proposed in the literature and determining their common underlying principles. We will then highlight the general principles that can be applied to design efficient spatio-temporal spike pattern classification mechanisms with electronic neuromorphic components, and present examples of possible hardware solutions for real-time robotic applications.

2 Single Neuron Models

A wide range of models has been proposed in the past that employ single neurons to classify spatio-temporal spike patterns [15, 16, 17, 18, 19]. All these models employ some form of spike based learning, such as Spike Timing Dependent

Plasticity (STDP), on the afferent input synapses that converge onto a single neuron to adjust their synaptic weights. These neurons are then typically exposed to multiple trains of input spike patterns, which contain a specific temporal pattern of spikes presented embedded in them, and presented repeatedly during the training phase. Eventually the synapses settle to a set of weights that make the neuron respond selectively to presentations of the embedded spike patterns.

A detailed analysis of what type of patterns neurons can learn through STDP is presented in [20]. The authors show that the synaptic weights converge to the *perceptron* learning rule if the input spike trains have Poisson distributions. But it does not address the cases in which the input spike trains have temporal structures that can be critical for classification tasks.

The type of features that can be detected with such types of patterns depends on the exact form of learning rule, and the underlying neuron model. We will discuss two of the most commonly used neuron models, the Poisson and the Leaky Integrate and Fire (LIF) neuron models.

2.1   Coincidence Detection

Let us first consider a Poisson neuron that has a probability of spiking proportional to its instantaneous input firing rate, which can be derived from the current membrane potential, given by the sum of all input Excitatory Post-Synaptic Potential (EPSP)s. For constant mean firing rates of input spike trains the post-synaptic neuron's membrane potential fluctuates around a mean value. In this condition synchronous inputs increase the instantaneous input firing rate, and therefore increase the probability of a post-synaptic spike. This model can therefore be used directly to detect synchronicity [21].

Coherent spiking activity of a population of neurons is easily detectable with LIF neurons [22], although the analysis of the firing properties of LIF model is not as straight forward as in the Poisson case. In [21] the authors show how the threshold and membrane time constants of a LIF neuron determine its ability to act as a coincidence detector. They demonstrate that even with constant mean input rate the mean output rate varies as a function of *temporal structure* of the input. They define "temporal structure" as the amount of spikes in the input spike trains that are phase locked to a periodic stimulus. They argue that a "small" membrane time constant is essential for a coincidence detector. The small membrane time constant should be chosen such that the mean membrane potential, under uniform input spike rate, is just below the neuron's firing threshold. Under these conditions, the neuron is sensitive to the fluctuations in the membrane potential, and can detect coincidences in its input spike trains in the same way that the Poisson neuron model described above does. It is this coincidence-detection ability of the LIF model that allows single-neuron training schemes to learn to respond selectively to specific spatio-temporal spike train sequences [23].

But *to what extent is coincidence detection in spike-patterns enough to distinguish patterns with complex spatio-temporal structure?*

To answer this question, let us consider a simple spatio-temporal pattern recognition task that involves patterns of spike-trains. Take for example those produced by an event-based (spiking) vision sensor such as the Dynamic Vision Sensor (DVS) [24], in which every pixel responds only to temporal changes in light intensity in its visual field. When observing a visual stimulus moving from left-to-right and right-to-left, under ideal conditions, the left-ward movement of the visual stimulus would be equivalent to a temporal reversal of the right-ward movement. The spike patterns produced in response to the stimulus moving in one direction would then be equivalent to the time-reversed version of the patterns produced by the stimulus moving in the opposite direction.

We can transpose this example to a more generic case: lets consider for example five specific randomly generated spatio-temporal spike patterns and their time-reversed versions, as in the case of the visual stimulus moving in two opposite directions. By using a spike-based STDP learning rule, such as the one proposed in [25], we can train six Poisson neurons arranged in a Winner-take-all (WTA) network to recognize all five patterns (presenting repeatedly only the original non-reversed patterns), as well as the background activity when none of the five patterns is present. After training, five of the six neurons fire selectively in response to one of the five input patterns, and a 6th neuron fires in response to the background activity. We trained such a network, and present the response of the neurons during presentation of their corresponding preferred spike patterns in Fig. 1. Each spike pattern and the corresponding neuronal response is marked with a distinct color.

Figure. 1 however shows also the activity of the same set of neurons in response to the time-reversed versions of the original spike patterns. As is evident from these results, the output neurons respond selectively not only to the trained spike patterns but also to their time-reversed versions. This is because even when the spike patterns are reversed the key feature the learning mechanism captures is *coincidences* (or near-coincidences) in the spike pattern. The pattern of coincidences to which a neuron responds is encoded in the vector of weights, and thus such neurons cannot distinguish between patterns that share the same sets of coincident inputs. Similar results can be obtained using analogous learning algorithms, such as the "tempotron" model [18], which is designed to capture coincidences hidden in one class of patterns and not in the other.

2.2 Beyond Coincidence Detection

In order to train neurons to distinguish more complex temporal patterns, without relying solely on coincidences it is necessary to introduce additional mechanisms that make the neuron sensitive to the causality of the input pattern, e.g., by responding differently to reversed patterns. Multi-compartmental models, conductance-based neurons [26], short-term synaptic plasticity, and dendritic computation [27, 28, 29, 30, 31] are all examples of strategies that allow such properties.

In [32], the authors demonstrate a neuromorphic hardware implementation where single neurons with short term depression can demonstrate *Stimulus-Specific*

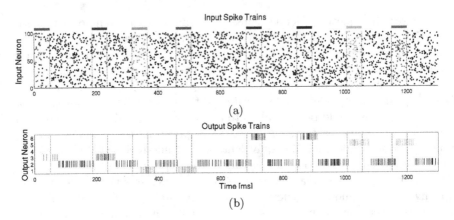

Fig. 1. a Raster plot of input spike pattern. The lighter colors, red, green, blue and yellow mark 4 of the 5 spike patterns that were used to train the network. The darker colored patterns, following each of the above colored spike patterns are time reversed spike patterns corresponding to their color. Spikes shown in black are background noise. b Response of output Poisson neurons on presentation of the input spike patterns. After training with a set of spike patterns, the neurons respond selectively to both the original and to time-reversed version of these spike patterns.

Adaptation, a phenomenon observed in neural systems. They show how short-term depression can enable neurons to adapt to specific stimuli based on the frequency of their occurrence and therefore show higher response to rare events. This behavior can be attributed to the fact that short-term plasticity lasts over hundreds of milliseconds, and carries information about synaptic inputs in the recent past. This mechanism has also been exploited in [33], where it is demonstrated that temporal information can be decoded using short-term plasticity. The author even hypothesizes that manipulations that eliminate short-term plasticity will produce deficits in temporal processing and therefore deem it is a critical computational component for temporal processing.

3   Network Models

An alternative strategy, that does not require more elaborate neuron models and dedicated circuitry, is to exploit the dynamics of a neural *network* to obtain the needed sensitivity. This approach has led to the development of several network models employed in spatio-temporal pattern recognition tasks.

Several models based on networks of recurrently coupled LIFs neurons have been proposed to classify spatio-temporal patterns. For example it has been shown how in a Liquid-State Machine (LSM) [34] that the low-pass filtered activity of a recurrently connected network (the reservoir) at any point in time reflects the temporal evolution of the input stimulus. By training a perceptron on snapshots of the neural activity in the reservoir, the liquid states, it is therefore possible to classify input patterns that have complex temporal evolution [35].

The parameters needed to obtain optimal performance are such that the input activity reverberates over a time-scale of several neuronal time-constants. With a careful choice of those parameters, the network is able to correlate events that are distant in time and that have a certain causality relationship. Neuromorphic circuits with slow dynamics for neuron and synapses can be used to exploit the reverberating properties of the network similarly to a LSM to process real-world data from spiking sensors.

An analogous approach is described in [36], where a recurrently-connected network of spiking neurons implemented in neuromorphic hardware is trained to differentiate between a sequence of spikes regularly distributed in time and its reversed version. It has been shown that this type of computation is involved in the recognition of tone sweeps in the thalamo-cortical pathway in the auditory stream [37]. The neural network consists of a rack of coincidence detectors receiving the spikes directly from the input layer and from delay lines propagating the activity of the other neurons in the network. Synapses are modified through STDP such that only the synapses capturing the temporal correlations typical of the learned stimulus survive. A·similar principle could be exploited in polychronous networks [38], where random axonal delays in a recurrent network can be used for spike pattern classification.

These models are closely related to the general concept of the Reichardt detector [39]. In a Reichardt detector two or more receptors respond at different times to the input stimuli and project to a non-linear unit, such as a neuron. Each of these projections incurs some amount of delay due to low-pass filtering. If these delays match the activation delays caused by the input stimulus, the signals from the receptors are coincident at their destination and consequently activate the nonlinear unit. Models of this kind have the advantage that they explicitly encode time within specific delay elements and lend themselves to unsupervised online learning methods. Depending on the task, such networks can be optimized to be implemented with a smaller number of neurons as compared to LSMs.

4   Neuromorphic Building Blocks for Temporal Processing

In the light of the above analysis we identified the basic computational operators required for processing of spatio-temporal spiking patterns in hardware implementations of spiking neural network. These computational operators include voltage-gating mechanisms that can make neurons sensitive to coincidences [21, 14, 28, 40], nonlinear operators such as short-term plasticity [33], low-pass filters [39] and delay elements [41].

Complex neural processing systems are likely to require large numbers of such operators. It is therefore important to design efficient hardware implementations using compact and low-power neuromorphic circuits [5]. Spiking neural networks built using such neuromorphic circuits comprise massively parallel arrays of processing nodes with memory and computation co-localized. Given their architectural constraints, these neural processing systems cannot process signals using the same strategies used by conventional von Neumann computing architectures that exploit time-domain multiplexing of small numbers of highly complex processors at high clock-rates, and operate by transferring the partial results of the computation from and to external memory banks. The synapses and neuron circuits in neuromorphic architectures have to process input spikes and produce output responses in real-time at the rate of the input data. It is not possible to virtualize time and transfer partial results in memory banks outside the architecture core, at higher rates, but it is necessary to employ resources that compute with time-constants that are well matched to those of the signals they are designed to process. Therefore, to interact with the environment and process signals with biological time-scales efficiently, hardware neuromorphic systems need to be able to compute using biologically realistic time constants. In this way, they are well matched to the signals they process, and are inherently synchronized with the real world events.

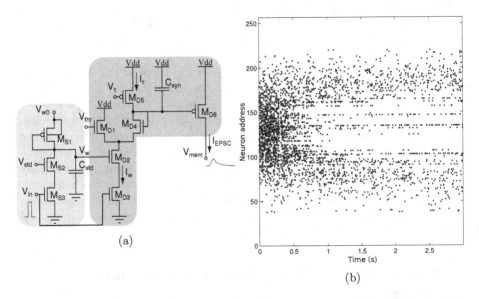

Fig. 2. a Analog synapse circuit that can implement biologically realistic synaptic dynamics and short-term depression. b Raster plot: response of a population of silicon neurons with short-term depressing synapses to a Gaussian profile of Poisson input spike trains.

Implementing biologically realistic time constants in analog VLSI technology is a non-trivial task. Conventional analog design techniques, in which typical currents are in the range of micro- or milliampere would require extremely large capacitors to implement time constants of the order of milliseconds, therefore losing the ability to integrate very large numbers of synapses in small areas. An elegant and efficient solution is that of using analog circuits that operate in the "weak inversion regime" [42], in which currents are on the order of pico-amperes.

An analog circuit of this type is the Differential-Pair Integrator (DPI), originally proposed in [43]. This circuit implements a low-pass filter which faithfully reproduces synaptic dynamics. An example of such synapse circuit is shown in Fig. 2a. It has been shown [44], by log-domain circuit analysis techniques[45, 46], that the transfer function of the part of the circuit composed by transistors $MD1 - MD6$ is:

$$\tau \frac{d}{dt} I_{EPSC} + I_{EPSC} = \frac{I_w I_{thr}}{I_\tau} \tag{1}$$

under the assumption that the input current I_w is larger than the leak current I_τ. The term τ represents the circuit time constant, defined as $\tau = CU_T/\kappa I_\tau$. The term U_T represents the thermal voltage and κ the sub-threshold slope factor [42]. The term I_{thr} represents a virtual p-type sub-threshold current that is not generated by any p-FET in the circuit, and is defined as $I_{thr} = I_0 e^{-\frac{\kappa(V_{thr} - V_{dd})}{U_T}}$. This circuit therefore implements an extremely compact and efficient model of synaptic dynamics. The transistors $MS1 - MS3$ of the DPI circuit of Fig. 2a implement a very useful non-linearity for spatio-temporal processing, that models the short-term depression properties of biological synapses [47]. This circuit is fully characterized in [48]. In [49] the authors show how this circuit can faithfully reproduce the behavior of real short-term depressing synapses, and demonstrate the equivalence with the computational model of short-term depression described in [50]. We present quantitative experimental data from this circuit in [44]. We show in Fig. 2b measurements from a population of silicon neurons that have these short-term depressing synapse circuits. The raster plot of Fig. 2b shows the response of 256 neurons, arranged along a line. The input pattern is a Gaussian profile of Poisson spike trains, with maximum firing rates centered around the address 128. As shown, high firing rate inputs, at the center of the array, are suppressed or strongly attenuated by the short-term depression mechanism of the input synapses, while low firing rate inputs at the periphery are transmitted without attenuation.

While short-term depression can be very useful for processing temporal patterns of spike trains [33, 51], additional essential elements are temporal delays and delay lines. Explicit delay circuits can be implemented in multi-compartmental neuron models to carry out spatio-temporal processing. For example Wang and Liu [28] demonstrate a VLSI neuron chip with programmable dendritic compartments and delay elements, showing how different spatio-temporal input patterns have different effects on the evoked dendritic integration. Analogous approaches have been proposed using floating-gate structures [30, 52]. But simpler Integrate and Fire (IF) neuron circuits can also implement delay elements without having to implement complex dendritic spatial structures. In [53] the authors show how slow dynamics of DPI synapses, combined with the variability and in-homogeneity "features" of neuromorphic analog circuits, can be used to generate a range of temporal delays. The low-pass filtering properties of the DPI synapses can be effectively used to create delay-elements, and these can be integrated in multi-chip networks of spiking neurons, feed-forward or recurrent, to increase their temporal retention capability. Specifically, in [53] the authors show a thalamo-cortical

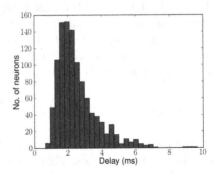

Fig. 3. Histogram of times used by post-synaptic neurons to fire after the arrival of a single pre-synaptic input spike (*delay*), across a population of neurons on an analog VLSI neuromorphic chip

neural network model that can learn to recognize specific spatio-temporal patterns of input spikes, by exploiting the mismatch present in the transistors of the DPI circuits giving rise to a wide range of delays (see also Fig. 3).

5   Conclusion

Spatio-temporal patterns produced by neuromorphic sensors reflect the time scales of events coming from the environment in real-time. In this paper we highlight the role of slow components in neural network models in order to classify spatio-temporal patterns produced by these sensors. We show a case in which IF neurons are capable of detecting coincidences and are incapable of differentiating the temporal order in the input spike pattern. We argue that this limitation is common to all single point-neuron models with first-order dynamics. We further review possible mechanisms that can be introduced to enable spatio-temporal pattern discriminability at the network level and capture the causality of the input stimuli. We present analog VLSI circuits for synaptic dynamics that have been included in recent neuromorphic implementations and can be exploited, together with the variability of the neuromorphic hardware, to classify spatio-temporal spike patterns.

Acknowledgements. This work was supported by the EU ERC Grant "neuroP" (257219).

References

[1] Maass, W., Sontag, E.: Neural systems as nonlinear filters. Neural Computation 12(8), 1743–1772 (2000)
[2] Belatreche, A., Maguire, L.P., McGinnity, M.: Advances in design and application of spiking neural networks. Soft Computing 11(3), 239–248 (2006)

[3] Brette, R., Rudolph, M., Carnevale, T., Hines, M., Beeman, D., Bower, J., Diesmann, M., Morrison, A., Goodman, P.H.J.F., Zirpe, M., Natschläger, T., Pecevski, D., Ermentrout, B., Djurfeldt, M., Lansner, A., Rochel, O., Vieville, T., Muller, E., Davison, A., El Boustani, S., Destexhe, A.: Simulation of networks of spiking neurons: A review of tools and strategies. Journal of Computational Neuroscience 23(3), 349–398 (2007)

[4] Liu, S.C., Delbruck, T.: Neuromorphic sensory systems. Current Opinion in Neurobiology 20(3), 288–295 (2010)

[5] Indiveri, G., Linares-Barranco, B., Hamilton, T., van Schaik, A., Etienne-Cummings, R., Delbruck, T., Liu, S.C., Dudek, P., Häfliger, P., Renaud, S., Schemmel, J., Cauwenberghs, G., Arthur, J., Hynna, K., Folowosele, F., Saighi, S., Serrano-Gotarredona, T., Wijekoon, J., Wang, Y., Boahen, K.: Neuromorphic silicon neuron circuits. Frontiers in Neuroscience 5, 1–23 (2011)

[6] Choudhary, S., et al.: Silicon neurons that compute. In: Villa, A.E.P., Duch, W., Érdi, P., Masulli, F., Palm, G. (eds.) ICANN 2012, Part I. LNCS, vol. 7552, pp. 121–128. Springer, Heidelberg (2012)

[7] Yu, T., Park, J., Joshi, S., Maier, C., Cauwenberghs, G.: 65k-neuron integrate-and-fire array transceiver with address-event reconfigurable synaptic routing. In: Biomedical Circuits and Systems Conference (BioCAS), pp. 21–24. IEEE (November 2012)

[8] Carr, C.E., Konishi, M.: Axonal delay lines for time measurement in the owl's brainstem. Proceedings of the National Academy of Sciences 85(21), 8311–8315 (1988)

[9] Carr, C.E., Konishi, M.: A circuit for detection of interaural time differences in the brain stem of the barn owl. The Journal of Neuroscience 10(10), 3227–3246 (1990)

[10] Pfeiffer, M., Hartbauer, M., Lang, A.B., Maass, W., Römer, H.: Probing real sensory worlds of receivers with unsupervised clustering. PloS One 7(6), e37354 (2012)

[11] Johansson, R., Birznieks, I.: First spikes in ensembles of human tactile afferents code complex spatial fingertip events. Nature Neuroscience 7(2), 170–177 (2004)

[12] Singer, W.: Time as coding space? Current Opinion in Neurobiology 9(2), 189–194 (1999)

[13] O'Keefe, J., Burgess, N.: Geometric determinants of the place fields of hippocampal neurons. Nature 381(6581), 425–428 (1996)

[14] Stiefel, K.M., Tapson, J., van Schaik, A.: Temporal order detection and coding in nervous systems. Neural Computation 25(2), 510–531 (2013)

[15] Dhoble, K., Nuntalid, N., Indiveri, G., Kasabov, N.: Online spatio-temporal pattern recognition with evolving spiking neural networks utilising address event representation, rank order, and temporal spike learning. In: International Joint Conference on Neural Networks, IJCNN 2012, pp. 554–560. IEEE (2012)

[16] Nessler, B., Pfeiffer, M., Maass, W.: Stdp enables spiking neurons to detect hidden causes of their inputs. In: Bengio, Y., Schuurmans, D., Lafferty, J., Williams, C.I., Culotta, A. (eds.) Advances in Neural Information Processing Systems, vol. 22, pp. 1357–1365 (2009)

[17] Masquelier, T., Guyonneau, R., Thorpe, S.J.: Spike timing dependent plasticity finds the start of repeating patterns in continuous spike trains. PLoS One 3(1), e1377 (2008)

[18] Gütig, R., Sompolinsky, H.: The tempotron: a neuron that learns spike timing-based decisions. Nature Neuroscience 9, 420–428 (2006)

[19] Thorpe, S., Delorme, A., Van Rullen, R., et al.: Spike-based strategies for rapid processing. Neural Networks 14(6-7), 715–725 (2001)

[20] Legenstein, R., Näger, C., Maass, W.: What can a neuron learn with spike-timing-dependent plasticity? Neural Computation 17(11), 2337–2382 (2005)

[21] Kempter, R., Gerstner, W., Van Hemmen, J.L.: How the threshold of a neuron determines its capacity for coincidence detection. Biosystems 48(1), 105–112 (1998)

[22] Gerstner, W., Kistler, W.: Spiking Neuron Models. In: Single Neurons, Populations, Plasticity. Cambridge University Press (2002)

[23] Masquelier, T., Guyonneau, R., Thorpe, S.J.: Competitive stdp-based spike pattern learning. Neural Computation 21(5), 1259–1276 (2009)

[24] Lichtsteiner, P., Posch, C., Delbruck, T.: A 128×128 120dB 30mW asynchronous vision sensor that responds to relative intensity change. In: 2006 IEEE ISSCC Digest of Technical Papers, pp. 508–509. IEEE (February 2006)

[25] Nessler, B., Pfeiffer, M., Maass, W.: Bayesian computation emerges in generic cortical microcircuits through spike-timing-dependent plasticity. PLoS Computational Biology (2013)

[26] Gütig, R., Sompolinsky, H.: Time-warp-invariant neuronal processing. PLoS Biology 7(7), e1000141 (2009)

[27] Koch, C., Poggio, T., Torre, V.: Nonlinear interactions in a dendritic tree: Localization, timing, and role in information processing. Proceedings of the National Academy of Sciences of the USA 80, 2799–2802 (1983)

[28] Wang, Y., Liu, S.C.: Multilayer processing of spatiotemporal spike patterns in a neuron with active dendrites. Neural Computation 8, 2086–2112 (2010)

[29] Arthur, J., Boahen, K.: Recurrently connected silicon neurons with active dendrites for one-shot learning. In: IEEE International Joint Conference on Neural Networks, vol. 3, pp. 1699–1704 (July 2004)

[30] Ramakrishnan, S., Wunderlich, R., Hasler, P.: Neuron array with plastic synapses and programmable dendrites. In: Biomedical Circuits and Systems Conference (BioCAS), pp. 400–403. IEEE (November 2012)

[31] Rasche, C., Douglas, R.: Forward- and backpropagation in a silicon dendrite. IEEE Transactions on Neural Networks 12, 386–393 (2001)

[32] Mill, R., Sheik, S., Indiveri, G., Denham, S.: A model of stimulus-specific adaptation in neuromorphic aVLSI. In: Biomedical Circuits and Systems Conference (BioCAS), pp. 266–269. IEEE (2010)

[33] Buonomano, D.: Decoding temporal information: A model based on short-term synaptic plasticity. The Journal of Neuroscience 20, 1129–1141 (2000)

[34] Maass, W., Natschläger, T., Markram, H.: Real-time computing without stable states: A new framework for neural computation based on perturbations. Neural Computation 14(11), 2531–2560 (2002)

[35] Maass, W., Natschläger, T., Markram, H.: Fading memory and kernel properties of generic cortical microcircuit models. Journal of Physiology – Paris 98(4-6), 315–330 (2004)

[36] Sheik, S., Coath, M., Indiveri, G., Denham, S., Wennekers, T., Chicca, E.: Emergent auditory feature tuning in a real-time neuromorphic VLSI system. Frontiers in Neuroscience 6(17) (2012)

[37] Coath, M., Mill, R., Denham, S.L., Wennekers, T.: Emergent Feature Sensitivity in a Model of the Auditory Thalamocortical System. Advances in Experimental Medicine and Biology, vol. 718, pp. 7–17. Springer, New York (2011)

[38] Izhikevich, E.M.: Polychronization: Computation with spikes. Neural Computation 18(2), 245–282 (2006)

[39] Reichardt, W.: Autocorrelation, a principle for the evaluation of sensory information by the central nervous system. Sensory Communication, 303–317 (1961)

[40] Wyss, R., König, P., Verschure, P.F.: Invariant representations of visual patterns in a temporal population code. Proceedings of the National Academy of Sciences 100(1), 324–329 (2003)

[41] Jeffress, L.A.: A place theory of sound localization. J. Comp. Physiol. Psychol. 41(1), 35–39 (1948)

[42] Liu, S.C., Kramer, J., Indiveri, G., Delbruck, T., Douglas, R.: Analog VLSI:Circuits and Principles. MIT Press (2002)

[43] Bartolozzi, C., Indiveri, G.: Synaptic dynamics in analog VLSI. Neural Computation 19(10), 2581–2603 (2007)

[44] Bartolozzi, C., Mitra, S., Indiveri, G.: An ultra low power current–mode filter for neuromorphic systems and biomedical signal processing. In: Biomedical Circuits and Systems Conference (BioCAS), pp. 130–133. IEEE (2006)

[45] Drakakis, E., Payne, A., Toumazou, C.: "Log-domain state-space": A systematic transistor-level approach for log-domain filtering. IEEE Transactions on Circuits and Systems II 46(3), 290–305 (1999)

[46] Frey, D.: Log-domain filtering: An approach to current-mode filtering. IEE Proceedings G: Circuits, Devices and Systems 140(6), 406–416 (1993)

[47] Markram, H., Tsodyks, M.: Redistribution of synaptic efficacy between neocortical pyramidal neurons. Nature 382, 807–810 (1996)

[48] Rasche, C., Hahnloser, R.: Silicon synaptic depression. Biological Cybernetics 84(1), 57–62 (2001)

[49] Boegerhausen, M., Suter, P., Liu, S.C.: Modeling short-term synaptic depression in silicon. Neural Computation 15(2), 331–348 (2003)

[50] Varela, J., Sen, K., Gibson, J., Fost, J., Abbott, L., Nelson, S.: A quantitative description of short–term plasticity at excitatory synapses in layer 2/3 of rat primary visual cortex. The Journal of Neuroscience 17, 7926–7940 (1997)

[51] Mill, R., Sheik, S., Indiveri, G., Denham, S.: A model of stimulus-specific adaptation in neuromorphic analog VLSI. Transactions on Biomedical Circuits and Systems 5(5), 413–419 (2011)

[52] Basu, A., Ramakrishnan, S., Petre, C., Koziol, S., Brink, S., Hasler, P.: Neural dynamics in reconfigurable silicon. IEEE Transactions on Biomedical Circuits and Systems 4(5), 311–319 (2010)

[53] Sheik, S., Chicca, E., Indiveri, G.: Exploiting device mismatch in neuromorphic VLSI systems to implement axonal delays. In: International Joint Conference on Neural Networks, IJCNN 2012, pp. 1940–1945. IEEE (2012)

Encoding of Stimuli in Embodied Neuronal Networks

Jacopo Tessadori[1], Daniele Venuta[1,2], Valentina Pasquale[1], Sreedhar S. Kumar[1,3], and Michela Chiappalone[1, *]

[1] Department of Neuroscience and Brain Technologies, Istituto Italiano di Tecnologia, Genova (Italy)
{jacopo.tessadori,daniele.venuta,valentina.pasquale,
michela.chiappalone}@iit.it
[2] University of Padova, Padova, Italy
[3] University of Freiburg, Freiburg, Germany
sreedhar.kumar@imtek.uni-freiburg.de

Abstract. Information coding in the central nervous system is still, under many aspects, a mystery. In this work, we made use of cortical and hippocampal cultures plated on micro-electrode arrays and embedded in a hybrid neuro-robotic platform to investigate the basis of "sensory" coding in neuronal cell assemblies. First, we asked which features of the observed spike trains (e.g. spikes, bursts, doublets) may be used to reconstruct significant portions of the input signal, coded as a low-frequency train of stimulations, through optimal linear reconstruction techniques. We also wondered whether preparations of cortical or hippocampal cells might present different coding representations. Our results pointed out that identifying specific signal structures within the spike train does not improve reconstruction performance. We found, instead, differences tied to the cell type of the preparation, with cortical cultures showing less segregated responses than hippocampal ones.

Keywords: spike train, burst, coding performance, micro-electrode array, in vitro networks.

1 Introduction

Primary neuronal cultures on Micro-Electrode Arrays (MEAs) have been used for several decades as a substrate for investigating basic mechanisms of the central nervous systems underlying more complex functions observed in living animals, such as memory, learning, processing and integration of multi-dimensional information. Such cultures can be kept healthy for significant periods of time [1] and have been proven to display activity patterns remarkably similar to those recorded in their in vivo counterparts [2]. These two observations led to the consideration that populations of dissociate cultured neurons could be used as an experimental platform to develop hybrid modules for information processing and control [3-5].

One unresolved issue in neuroscience is related to the processing of sensory information within the central nervous system. Strong evidence exists that different

---

* Coressponding author.

N.F. Lepora et al. (Eds.): Living Machines 2013, LNAI 8064, pp. 274–286, 2013.
© Springer-Verlag Berlin Heidelberg 2013

features of the spike train (e.g. bursts, isolated spikes) might be coding different aspects of the sensory signal [6-9]. In this work, we took advantage of the neuro-robotic setup developed in our lab and previously described [10], constituted by a robot with sensors and actuators able to move in an arena with obstacles, to generate time-varying input data, code it as a stimulation train and deliver it to a neural culture. Two different preparations have been used in the experiments: cortical and hippocampal neurons. Data presented in this study has been analyzed off-line (i.e. after the end of the closed-loop experiment) in order to determine the amount of information about the input signal that can be correctly reconstructed from different features of the spike train induced by the stimulation.

2   Materials and Methods

2.1   Neuronal Preparation and MEA Electrophysiology

Dissociated neuronal cultures were prepared from hippocampi and cortices of 18-day old embryonic rats (Charles River Laboratories). All experimental procedures and animal care were conducted in conformity with institutional guidelines, in accordance with the European legislation (European Communities Directive of 24 November 1986, 86/609/EEC) and with the NIH Guide for the Care and Use of Laboratory Animals. Culture preparation was performed as previously described [11]. Cells were afterwards plated onto standard MEAs (Multichannel Systems, Reutlingen, Germany) previously coated with poly-D-lysine and laminin to promote cell adhesion (final density around 1200 cells/mm^2) and maintained with 1 ml of nutrient medium (Fig. 1A). They were then placed in a humidified incubator having an atmosphere of 5% CO_2 and 95% air at 37°C. Half of the medium was changed weekly. Recordings were performed on cultures between 25 and 60 Days In Vitro (DIV).

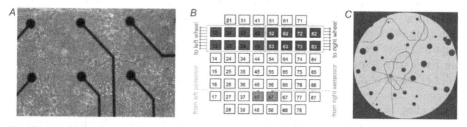

Fig. 1. A. A hippocampal culture grown on a standard MEA device. **B.** Schematical illustration of the input/output electrode configuration in the MEA device: the orange and light blue electrodes are selected as inputs, i.e. they are used to deliver stimuli as a function of the sensors readings from the robot. The red and dark blue sets of electrodes are instead used as control regions for the wheels: the speed of the left wheel will be a function of the average activity detected on the red electrodes, while the right wheel will be controlled by the blue ones. **C.** The virtual realization of the Khepera robot (pink circle) is moving within the virtual arena, while readings from its sensors (black lines departing from the robot) collect information on distances from obstacles. Each green circle represents an obstacle of different diameter. The red line highlights the trajectory followed by the robot.

The electrophysiological activity of a culture was recorded through the MEA, which consists of 60TiN/SiN planar round electrodes (30 μm diameter; 200 μm center-to-center inter-electrode distance) arranged in an 8×8 square grid excluding corners. Each electrode provides information on the activity of the neural network in its immediate area. The amplified 60-channel data (amplifier gain = 1100) is conveyed to the data acquisition card which samples them at 10 kHz per channel and converts them into 12 bit data.

2.2   Sensory Stimuli Production and Experimental Protocol

In accordance with Cozzi et al.[12], we named signals presented to the neural network 'sensory', even though this is a misnomer: there is, in fact, no actual sensory structure, since our biological networks grow randomly on a planar substrate. In our case, instead, 'sensory area' is defined simply as the region of the network neighboring the electrodes delivering input signals.

A typical MEA configuration is illustrated in Fig. 1B: two electrodes are used to deliver stimulation as a function of the readings of the robot sensors, one for the left and one for the right sensor set of the robot (orange and light blue, in Fig. 1B). The speed of each wheel is determined as a function of the activity recorded over a predefined set of electrodes (i.e. red electrodes control the left wheel, while blue electrodes the right wheel). Sensory information consisted in recordings obtained from the sensors of a simulated robot (i.e. the virtual representation of a Khepera robot, by KTeam, Switzerland) moving in a virtual 2-D environment. The robot is constantly acquiring information about its surroundings, providing data on the distance of the closest objects in different directions, while the speed and direction of the robot are computed from this data in order to steer the platform away from obstacles while keeping moving forward, i.e. the robot behaves as a Braitenberg vehicle [13]. The virtual environment consists of 400x400 pixels circular arena containing several round dark green obstacles in random positions. A typical experiment with the virtual robot is shown in Fig. 1C.The robot (small pink circle) is collecting information about its environment through its six sensors: each black line departing from the robot represents the line of sight of a different sensor; their angles are fixed with respect to the robot heading (in this case, 30°, 45° and 90° on both sides of the robot direction), while the length of each line is equal to the distance from the robot center to the closest obstacle in the sensor direction. This distance defines the reading of the sensor: the output is 0 if the robot is in direct contact with an obstacle, 1 if the closest obstacle is at the maximum distance possible (the diameter of the arena, in this case). The sensory signals used in the experiments are the averaged recordings of the three sensors on the left or right side of the robot. These inputs are provided to the neural preparation as a series of electrical stimulations: one channel of the MEA is selected to code information from the left set of sensors and another one is selected for the right set (Fig. 1B). Information is linearly coded in the frequency of these stimulations, ranging from 0.5 Hz for a sensor reading of 0 to a maximum of 2 Hz for a reading of 1.

Before the navigation phase, each sensory channel is associated with up to eight recording channels, according to a stimulus-response test (for details, see [10]. A sample MEA configuration is shown in Fig. 1B). Therefore, in the ideal scenario, the activity observed in the recording channels should carry all the information of the associated sensory input channel (i.e. 'same' in the rest of the text) and none of the opposite one (i.e. 'opposite in the rest of the text):

2.3   Signal Processing

Spike Detection & Blanking of Stimulus Artifact. A custom spike detection algorithm was used to discriminate spike events and to isolate them from noise [14]. Briefly, the algorithm is based on the use of three parameters: 1) a differential threshold (DT) set independently for each channel and computed as 6 or 7-fold the standard deviation (SD) of the noise of the signal; 2) a peak lifetime period (PLP) set to 2 ms; 3) a refractory period set to 1 ms. The algorithm scans the raw data to discriminate the relative minimum or maximum points. Once a relative minimum point is found, the nearest maximum point is searched within the following PLP window (or vice versa). If the difference between the two points is larger than DT, a spike is identified and its timestamp saved. The result of the spike detection procedure is a 'point process', i.e. a sequence of time stamps each of them indicating the occurrence of a spike. Given the extreme difficulty to identify single neuron's waveforms in our dense dissociated cultures [15-17], we did not sort the detected spikes.

Stimulus artifacts are detected when the recorded signal exceeds a defined threshold much higher than the one used for spike detection. The artifact is then suppressed by blanking the first 4 ms after each stimulus delivery.

Bursts and Doublets Detection. Neuronal networks from both hippocampus and cortex show both random spiking activity and bursting behavior. Bursts consist of packages of spikes distributed over a range of a few milliseconds, which generally last from hundreds of milliseconds up to seconds, and are separated by long quiescent periods. A previously developed algorithm [18, 19] was used to detect the presence of bursts. Briefly, a burst was detected when a sequence of at least 5 spikes with an inter-spike interval less than or equal to 80 ms was found. For each recognized burst, we stored the timing of the first spike within the burst (i.e. Burst Event, BE). Besides bursts, also doublets were detected. A doublet consists of a couple of spikes separated by less than 10 ms.

Sensory Signal Recontruction. In order to evaluate the amount of signal information encoded in the neural activity, we estimated the sensory input that can be recovered from the spike train recorded at each site through an optimal linear filtering approach [12, 20, 21]. More specifically, the linear filter $h(\cdot)$ was computed through Welch's averaged periodogram method with windows of 2 s, a 75% overlap between windows and the use of a Hamming window [22]. The sensory signal $s(t)$ is reconstructed from the train of stimuli $y(t)$ as:

$$s(t) = s_{est}(t) + n(t) = \int_{-\infty}^{+\infty} h(\tau)y(t-\tau)\,d\tau + n(t)$$

where $s_{est}(t)$ represents the reconstructed signal, $s(t)$ the original one and $n(t)$ is the residual, i.e. unmodeled portion of $s(t)$. $h(\cdot)$ is constructed to be the average of the linear filters that minimize the variance of $n(t)$ in each of the 2 s observation windows. An identical approach has been used to reconstruct signals from delivered stimulation, using stimuli trains instead of spike trains.

Coding Performances. Coding performance of each reconstruction has been evaluated by comparing the variance of the original signal with that of the residual. Namely, we used as parameter for comparison the coding fraction, defined as in [12]:

$$CF = 1 - \left(\frac{\sigma_n}{\sigma_s}\right)^2$$

Simulated Data. Simulated data has been generated to validate the reconstruction process and test its robustness to different kinds of 'disturb'. We generated a sensory signal through the convolution of one of the recorded stimulation trains with a Gaussian kernel of a width compatible with the filter generation (the value of the kernel has to be zero or close to zero at the end of the considered window), thus granting perfect match (CF>99%). This train-signal pairing will be referred to as 'mixed data' in the following, since the event train is recorded, but the signal is simulated. A second train was generated as a point-process in which each time sample has a probability of being an event. This probability is computed so that the average number of events per time unit is the same as that in the recorded train. The matching signal has also been generated through convolution as described above (this train-signal pair will be referred to as 'simulated'). The different kinds of disturb introduced either a temporal jitter or changed the number of stimuli, by adding or removing to the original signal. In the first case, the timing of each stimulus was moved forward or backward in time by a random number of samples selected from a uniform distribution with a given maximum. Removal of stimuli was performed in a similar way, with each stimulus presenting a fixed probability of being removed. In the third kind of disturb, each non-stimulus time point in the stimulation train had a fixed probability of being converted to a stimulus. We introduced these simulated signals to estimate the amount of information lost by the reconstruction algorithm under different scenarios. In particular, in the "mixed signal" we recorded a stimulation sequence which therefore presented inter-stimuli intervals between 0.5 and 2s, as implied by the stimulation protocol during close-loop experiments. In this specific test, the distance between stimuli fell in the 0.5-2s range, which is the ideal range the algorithm has been set to perform on. In the simulated signal, the stimulation train is not recorded, but rather stochastically generated. This implies that there is no fixed inter-stimulation interval range, but rather any value can be observed. We ran this scenario to understand what could be the reconstruction performances under more realistic conditions.

Reconstructions. The first aim of the analysis on spike trains was that of defining which features provide the largest amount of information about the input sensory signal. Furthermore, we wanted to find out whether significant differences could be found in the information content of different firing modalities of cortical and hippocampal cultures. For each analyzed channel, one reconstruction has been generated from each of several observed features. In particular, the following reconstructions have been performed:

1. **all spikes**: the entire spike train of a channel was used to reconstruct the relative sensory signal;

2. **doublets**: the train used for reconstruction consisted of the first spike in each detected doublet;
3. **bursts**: the train used for reconstruction consisted of all the spikes belonging to all detected bursts;
4. **burst events**: the train used for reconstruction consisted of the first spike in each detected burst;
5. **isolated spikes**: the train used for reconstruction consisted of all the spikes not belonging to any burst;
6. **burst events + isolated spikes**: the train used for reconstruction consisted of the first spike in each burst and all those not belonging to any burst;
7. **other spikes in burst**: the train used for reconstruction consisted of all the spikes following the first within each burst;
8. **linear reconstruction:** this reconstruction was obtained as a linear combination of the reconstructions from the points 4, 5 and 7 (further details are provided below).

During the close-loop experiments, the robot collects information through its left-side and right-side sensors. This led to two different input signals, one for the left and one for the right side of the robot, being delivered to the culture through two different stimulating electrodes. At the beginning of the experiment, all the recording channels are paired with either the left- or right-side input, according to the observed responses from single pulses delivered through one stimulating electrode at a time. All the reconstructions mentioned above try to recover the same side input signal (i.e. left input signal for electrodes paired with the left input, right input signals for electrodes paired with the right one). Two more reconstructions have been generated starting from the complete spike train, in order to match a different sensory signal: in the first case, reconstruction of the 'opposite' sensory channel is attempted (i.e. the sensory input not associated with the recording channel); in the second case, the considered sensory signal is the sum of the two actual inputs. Furthermore, for each experiment, a reconstruction of the sensory signal was generated starting from the corresponding stimulation train.

Linear Combination. In order to evaluate whether different features of the spike train code for different characteristics of the input signal, the reconstructions obtained from complete spike trains have been compared against a linear combination of reconstructions obtained from three subsets of the spike train, namely burst events, following spikes within bursts and spikes outside of bursts. The multiplication coefficients have been chosen so as to minimize the reconstruction error of the linear combination. If bursts were reliably generated as a response to a specific signal feature, the reconstruction error should be significantly smaller on linear combination reconstructions rather than that observed on reconstructions from the complete spike train.

2.4   Database and Statistics

We performed experiments on 15 different cell cultures (4 cortical and 11 hippocampal). In each experiment, 2 electrodes were selected to deliver the stimulation train, while 16 channels were monitored and recorded. Half of these were associated with the left input channel, the other half with the right one. Datasets were thinned by removing experiments displaying a CF of sensory signal from stimuli train less than 80% on either input channel or an average CF computed on all the channels less than 10%. The remaining dataset (i.e. the one actually used in this work) included 3 experiments on cortical and 3 experiments on hippocampal preparations, for a grand total of 96 analyzed spike trains. Furthermore, single spike trains have been removed if they obtain CFs less than 5% when reconstructing either sensory signal.

Data within the text are expressed as mean ± standard error of the mean (se). Statistical tests were employed to assess the significant difference among different signal's features. The normal distribution of experimental data was assessed using the Kolmogorov-Smirnov normality test. We then performed the two sample t-test or the one-way ANOVA test for multiple comparison and p values < 0.05 (indicated with an '*' in the figures) were considered significant. The mean comparison was performed through the post-hoc Bonferroni test. Statistical analysis was carried out by using OriginPro (OriginLab Corporation, Northampton, MA, USA).

3   Results

In order to evaluate the performances in reconstructing the input sensory signal of cortical and hippocampal neurons, we used a neuro-robotic set-up constituted by a virtual robot bi-directionally connected to a MEA coupled to a neuronal assembly. Raw data were first acquired and then a spike detection procedure was applied. The result of this algorithm produced a spike train (second plot from the top in Fig. 2), which was in turn subdivided into several different features. The elements extracted from each spike train were doublets and bursts (cf. section 2.3). Bursts were in turn divided into burst events (the first spike of each burst) and subsequent spikes. Isolated spikes are here defined as all those spikes that are not part of a burst. An example of the features just described is shown in Fig. 2.

First, we tested the reconstruction algorithm on mixed, simulated and recorded (cf. see Section 2.3) train-signal pairings under different conditions in order to evaluate its

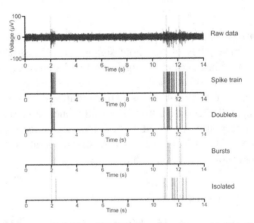

Fig. 2. Several different features of the spike train have been considered: the original spike train itself, doublets (two spikes whose distance is less than 10 ms), bursts (a group of five or more spikes with maximum inter-spike interval less than 80 ms), isolated spikes (spikes not belonging to any burst).

robustness. As it is possible to observe from the graphs in Fig. 3, the coding fraction is above 80% in the case of recorded stimulation train and signal ('recorded' trace in the graphs), while it gets virtually to 100% in the mixed data pairing. On the other hand, the reconstruction based on the simulated pairing is missing most of the information present in the original signal, with the maximum CF around 30%: although the mean rate of events is the same in all three cases, in the simulated train, the positioning of the events is random. This leads to imperfect reconstruction of the linear filter, as events closer than the length of the window of the periodogram will lead to partly overlapping responses.

Reconstruction performances are very robust to jitter (Fig. 3A), since displacements under 10 ms (100 samples) cause a loss of less than 10% of CF, while maximum jitter has to be increased to more than 100 ms (1000 samples) to effectively destroy the information content of the recorded signal and even more than that in the simulation from recorded data. On the other hand, addition or removal of events has a very different effect on the reconstruction of a simulated or recorded signal (Fig. 3B and C): in the first case addition or removal of around 10% of the events has an almost negligible impact on the computed CF, while in the latter less than 3% of events changed result in the impossibility of recovering the correct sensory signal.

In Fig 4, for a sample channel, most of the reconstructions described above are shown in the case of a hippocampal and a cortical culture. Panels B and C of Fig. 4 show the performances obtained from the experiments that satisfied the conditions described in the Methods section. Spike trains usually provide the best reconstruction performances among all the considered features and, in general, the CF for spike trains and linear reconstructions is roughly the same, both for cortical and hippocampal cells. This implies that the division of the spike train into burst events, other burst spikes and isolated spikes is not helpful in recovering the information content of the spike train.

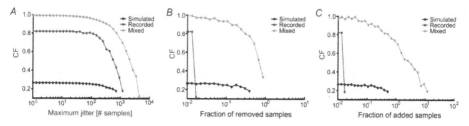

Fig. 3. A. Effect of jitter on coding fraction. For each point, 25 reconstructions have been computed for each signal-stimulation train pairing, in which each event of the train was moved from its original position a random number of samples in either direction. Shown on the graph are the median values of the 25 repetitions. **B.** Effect of event removal on coding fraction. For each point, 25 reconstructions have been computed for each signal-stimulation train pairing, with each event in the train having the same fixed chance of being removed. Median values of the 25 repetitions are shown in the graph. **C.** Effect of event addition on coding fraction. For each point, 25 reconstructions have been computed. Each non-event sample of the train had the same probability of being changed into an event (10% fraction of added samples indicates that each non-event point had a probability of being changed into an event so that, on average, the output train would have 10% more events than the input train). Median values of the 25 repetitions are shown in the graph.

The reconstruction performance has also been used to evaluate the ability of different preparations to process the information of two spatially separated input channels at the same time. In the setup of the neuro-robotic experiments, eight recording channels were associated with one of the two sensory inputs (see Methods and [10] for details). In this work, we are evaluating a posteriori the selectivity of the

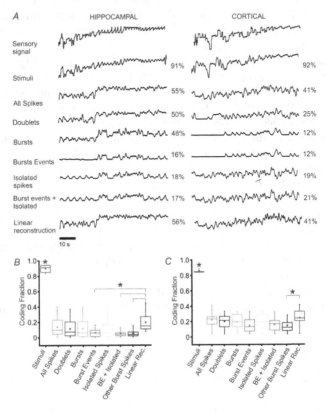

Fig. 4. A. The two set of graphs show, from top to bottom, one of the two input sensory signals delivered to the culture; reconstruction performed from the corresponding train of stimuli; from the complete set of spikes on a channel; from the train of doublets; from all the spikes belonging to bursts; from spike events (i.e. only the first spike in a burst); from spikes not belonging to bursts; from burst events and isolated spikes; from the best linear combination of reconstructions from burst events, other spikes within bursts and isolated spikes. The left set of graphs is generated from data recorded on a hippocampal culture, while the data from the right set is from a cortical culture. On the right side of each graph, the corresponding CF is indicated as a percentage. **B.** Box plots of reconstruction performances, expressed as coding fractions, on all channels recorded from hippocampal cultures. Thresholds for exclusion of a channel from the dataset are explained in detail in the Materials and Methods section. **C.** Box plots of reconstruction performances, expressed as coding fractions, on all channels recorded from cortical cultures. Thresholds for exclusion of a channel from the dataset are the same as for plot B. Statistics presented in the lower panels was done by using the one way ANOVA with Bonferroni's method for multiple comparison, $*p<0.05$. The 'stimuli' condition is statistically different from all the others, the other significance are indicated by a line connecting the statistically different conditions.

representation of each recording channel by scoring the reconstructions of the associated sensory input, the opposite sensory input and that of a signal equal to the sum of both inputs from the spike train observed.

Fig. 5 shows these results. Specifically, we reported the performance of each channel in reconstructing its associate stimulus and the opposite one (left top panel), a comparison in the performances obtained in the reconstruction of the associated/opposite stimuli against those of the sum of stimuli (middle/right top panels). While results from hippocampal cultures are rather spread in the parameter space, data relative to cortical cultures tend to cluster along the bisector of both graphs. A quantitative analysis of the selectivity of responses is shown in Fig. 5, where, for each considered condition and cell type, a box plot is represented, showing the distribution of distances of each point in the top graphs from the bisector. Points lying on the bisector (i.e. distance close to 0) represent activity from an electrode whose activity allows reconstruction of both considered signals, while points lying closer to the axes represent electrodes whose activity may be used to reconstruct only one of the input traces (or their sum).

Spike trains from hippocampal cultures present better performances in the reconstruction of both associate and opposite signals, whereas recordings on cortical cultures tend to code more precisely the sum signal than either the associate or opposite ones.

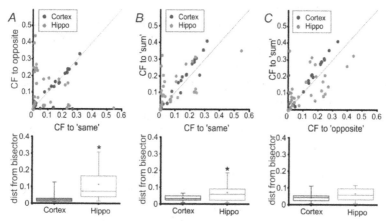

Fig. 5. Comparison of CFs computed from the complete spike trains of channels in the reconstruction of the two input signals. The coordinates of each point represent the CF obtained when reconstructing one input signal (A: 'same'-'opposite' signals; B: 'same'-'sum' signals; C: 'opposite' – 'sum' signals). Color codes the cell type of the culture each recording is from: red for cortical cultures, green for hippocampal ones. 'same' is the input signal the recording channel is associated with (see text for details), 'opposite' is the other input signal and 'sum' is the sum of both. The lower panels show box plots of the Euclidean distances of each point in the graphs above from the bisector, separated according to cell type. Statistics presented in the lower panels was done by using two-sample t-test, *p<0.05.

4   Discussion

In this paper, we made use of a hybrid neuro-robotic platform to investigate the basis of sensory coding of cortical and hippocampal cultures plated on micro-electrode arrays.

First, we tested the chosen reconstruction technique on simulated data and we further asked whether and how reconstruction performances are affected by different kinds of data loss and/or corruption (see Fig. 3). For all datasets (i.e. recorded, mixed and simulated data) we found that the exact positioning of events within the train is not critical for signal reconstruction, as the addition of jitter to the event train does not cause significant loss of performance until very large values. This result was partially expected, since the employed coding scheme is rate-based. On the other hand, removal of events from the train causes a very quick degradation of performance, as does the addition of events in random positions or the loss of spectral information, as the simulated dataset indicates (Fig. 3). Translating these observations to spike trains, it turns out that the single most important property of the neural preparation is the reliable generation of spikes following stimulation and the absence of interfering spontaneous activity, as even a low number of spikes uncorrelated with the sensory signal are sufficient to disrupt reconstructions: if a stimulus does not generate responses, no estimation of sensory data can be performed in the corresponding interval. On the other hand, uncorrelated spontaneous activity will degrade the quality of the linear filter.

These observations are indicative of the findings from the other experiments performed: subdividing the spike train into finer features (in contrast to what reported by Cozzi and co-workers [12]) does not improve the reconstruction performance. In our preparation, there does not seem to be any relationship between features of the sensory signal (such has sharp changes or high values for long periods of time) and the chance of occurrence of bursts: burst events score constantly lower than other trains and even linear reconstructions are not significantly different from those obtained from the complete spike train (Fig. 4). It is worth noting, though, that in Cozzi et al. bursts were defined as a group of at least 2 spikes closer than 100 ms, whereas in our work we considered bursts groups of at least 5 spikes, the commonly accepted definition of burst [18, 19, 23]. Doublets were introduced in this work to provide a more direct comparison to the work of Cozzi and co-workers [12]. It is possible to observe, furthermore, that even the best reconstructions from spike trains rarely exceed a CF of 40%, but are closer to the performance obtained on simulated data in Fig. 4. This is ascribable to the fact that recorded spike trains do not have the same peaked spectral content as stimulation trains (data not shown), therefore they constitute suboptimal candidates for reconstruction. In particular, a better reconstruction could be obtained by designing suitable, different windows for the computation of the linear filter in each experiment. On the other hand, in our work, a difference between the behavior of cortical and hippocampal neurons can be noted. When preparations were subjected to two different sensory inputs at the same time, cortical preparations were generally unable to differentiate between the two, but the activity recorded on different channels was rather a function of the sum of the two channels (i.e. responses are not segregated). Hippocampal cultures showed a much more varied behavior, ranging from very selective responses to a behavior close to that observed for cortical preparations (Fig. 5). In other works [10, 24], it was shown that the use of a confinement mask to create interconnected sub-populations of neurons improves the segregation of responses to different input signals, thus correlating structural modularity to functional segregation. Here, we can speculate that our hippocampal cultures spontaneously tended to cluster, thus reproducing a modular architecture, whereas cortical ones are likely to reproduce in vitro a more homogeneous network.

Acknowledgments. The research leading to these results has received funding from the European Union's FP7 (FET Young Explorers scheme) under grant agreement n° 284772 BRAIN BOW (www.brainbowproject.eu). The authors wish to thank Dr Marina Nanni and Claudia Chiabrera for the technical assistance in cell culturing; Prof. Sergio Martinoia and Prof. Vittorio Sanguineti, for their useful comments and discussion.

References

[1] Potter, S.M., DeMarse, T.B.: A new approach to neural cell culture for long-term studies. J. Neurosci. Methods 110, 17–24 (2001)

[2] Kriegstein, A.R., Dichter, M.A.: Morphological classification of rat cortical neurons in cell culture. J. Neurosci. 3, 1634–1647 (1983)

[3] Bakkum, D.J., Shkolnik, A.C., Ben-Ary, G., Gamblen, P., DeMarse, T.B., Potter, S.M.: Removing Some 'A' from AI: Embodied Cultured Networks. In: Iida, F., Pfeifer, R., Steels, L., Kuniyoshi, Y. (eds.) Embodied Artificial Intelligence. LNCS (LNAI), vol. 3139, pp. 130–145. Springer, Heidelberg (2004)

[4] Mulas, M., Massobrio, P., Martinoia, S., Chiappalone, M.: A simulated neuro-robotic environment for bi-directional closed-loop experiments. Paladyn. Journal of Behavioral Robotics 1, 179–186 (2010)

[5] Warwick, D., Xydas, S.J., Nasuto, V.M., Becerra, M.W., Hammond, J.H., Downes, S., Marshall, B.J.: Controlling a Mobile Robot with a Biological Brain. Defence Sci. J. 60, 5–14 (2010)

[6] Metzner, W., Koch, C., Wessel, R., Gabbiani, F.: Feature extraction by burst-like spike patterns in multiple sensory maps. J. Neurosci. 18, 2283–2300 (1998)

[7] Reinagel, P., Godwin, D., Sherman, S.M., Koch, C.: Encoding of visual information by LGN bursts. Journal Neurophys. 81, 2558–2569 (1999)

[8] Martinez-Conde, S., Macknik, S.L., Hubel, D.H.: The function of bursts of spikes during visual fixation in the awake primate lateral geniculate nucleus and primary visual cortex. PNAS 99, 13920–13925 (2002)

[9] Krahe, R., Gabbiani, F.: Burst firing in sensory systems. Nat. Rev. Neurosci. 5, 13–23 (2004)

[10] Tessadori, J., Bisio, M., Martinoia, S., Chiappalone, M.: Modular neuronal assemblies embodied in a closed-loop environment: toward future integration of brains and machines. Front Neural Circuits 6, 99 (2012)

[11] Frega, M., Pasquale, V., Tedesco, M., Marcoli, M., Contestabile, A., Nanni, M., Bonzano, L., Maura, G., Chiappalone, M.: Cortical cultures coupled to Micro-Electrode Arrays: a novel approach to perform in vitro excitotoxity testing. Neurotoxicol Teratol 34, 116–127 (2012)

[12] Cozzi, L., D'Angelo, P., Sanguineti, V.: Encoding of time-varying stimuli in populations of cultured neurons. Biol. Cybern. 94, 335–349 (2006)

[13] Braitenberg, V.: Vehicles - Experiments in synthetic psychology. The MIT Press, Cambridge (1984)

[14] Maccione, A., Gandolfo, M., Massobrio, P., Novellino, A., Martinoia, S., Chiappalone, M.: A novel algorithm for precise identification of spikes in extracellularly recorded neuronal signals. J. Neurosci. Methods 177, 241–249 (2009)

[15] Rolston, J.D., Wagenaar, D.A., Potter, S.M.: Precisely timed spatiotemporal patterns of neural activity in dissociated cortical cultures. Neuroscience 148, 294–303 (2007)

[16] Shahaf, G., Marom, S.: Learning in networks of cortical neurons. J. Neurosci. 21, 8782–8788 (2001)

[17] Eytan, D., Marom, S.: Dynamics and effective topology underlying synchronization in networks of cortical neurons. J. Neurosci. 26, 8465–8476 (2006)

[18] Chiappalone, M., Novellino, A., Vajda, I., Vato, A., Martinoia, S., Van Pelt, J.: Burst detection algorithms for the analysis of spatio-temporal patterns in cortical networks of neurons. Neurocomputing 65, 653–662 (2005)

[19] Turnbull, L., Dian, E., Gross, G.: The string method of burst identification in neuronal spike trains. J. Neurosci. Methods 145, 23–35 (2005)

[20] Gabbiani, F., Metzner, W., Wessel, R., Koch, C.: From stimulus encoding to feature extraction in weakly electric fish. Nature 384, 564–567 (1996)

[21] Rieke, F., Warland, D., De Ruyter van Steveninck, R., Spikes, W.B.: Exploring the neural code. MIT press, Cambridge (1997)

[22] Welch, P.: The use of fast Fourier transform for the estimation of power spectra: a method based on time averaging over short, modified periodograms. IEEE Transactions on Audio and Electroacoustics 15, 70–73 (1967)

[23] Leondopulos, S.S., Boehler, M.D., Wheeler, B.C., Brewer, G.J.: Chronic stimulation of cultured neuronal networks boosts low-frequency oscillatory activity at theta and gamma with spikes phase-locked to gamma frequencies. J. Neural Eng. 9, 026015 (2012)

[24] Levy, O., Ziv, N.E., Marom, S.: Enhancement of neural representation capacity by modular architecture in networks of cortical neurons. Eu. J. Neurosci. 35, 1753–1760 (2012)

Modulating Behaviors Using Allostatic Control

Vasiliki Vouloutsi[1], Stéphane Lallée[1], and Paul F.M.J. Verschure[2]

[1] Universitat Pompeu Fabra (UPF). Synthetic, Perceptive, Emotive and Cognitive
Systems group (SPECS)
http://specs.upf.edu
[2] Institució Catalana de Recerca i Estudis Avançats (ICREA)
Passeig Llus Companys 23, 08010 Barcelona, Spain
http://www.icrea.cat

Abstract. Robots will be part of our society in the future. It is therefore
important that they are able to interact with humans in a natural way.
This requires the ability to display social competence and behavior that
will promote such interactions. Here we present the details of modeling
the emergence of emotional states, adaptive internal needs and motivational drives. We explain how this model is enriched by the usage of a
homeostatic and allostatic control that aim at regulating its behavior.
We evaluate the model during a human-robot interaction and we show
how this model is able to produce meaningful and complex behaviors.

Keywords: human-robot interaction, behavioral modulation, allostatic
control.

1 Introduction

As the introduction of robots into our society is slowly coming closer to reality,
their ability to be able to interact with humans in a meaningful and intuitive
way gains importance. Traditionally, the usage of robots is constrained in situations where little interaction with humans is required, such as environmental
monitoring [1], looking for hazardous substances [2] etc. Recently, we observe a
change in paradigm as robots with more social character seem to gain ground.
Such robots vary from museum tour guides [3][4], robotic nurse maids [5], for kids
with autism [6] to sociable partners [7]. It is therefore essential that we start developing robots that are not just tools for automated processes but rather social
agents that are able to interact with humans.

Humans have evolved to be experts in social interaction, attributing causality
to entities, provided that they obey specific regularities consistent with the necessary contingency and contiguity conditions of causality [8]. In fact there is a
large body of work that shows the propensity of humans to make social inferences
and judgements even to shapes on a screen to explain their behavior [9]. Hence,
if there is a social model that humans can attribute to the robot's behavior, the
robot can be considered socially competent [10] [11]. Such operationalization of
social competence nonetheless seems to exclude both the mechanisms that underlie such competence as well as a broader range of non-human social behaviors.

N.F. Lepora et al. (Eds.): Living Machines 2013, LNAI 8064, pp. 287–298, 2013.
© Springer-Verlag Berlin Heidelberg 2013

Our goal is to address this set of social skills starting from the communicative and interactive non-anthropomorphic artifact Ada [12].

2   Making Social Robots

Studying social behavior cannot be uncoupled from the ability to socially perceive, which in turn requires a self. To support social perception a number of issues need to be addressed such as what is the analogy between self and other. We propose that the minimal requirements for a functional robot that can act as a social agent are: (i) intrinsic needs to socially engage; (ii) have an action repertoire that supports communication and social interaction; (iii) the ability to distinguish between self and non-self by realizing a "Phenomenal Model of the Intentionality-Relation" (PMIR) [13]; (iv) the ability to evaluate how the self is situated in the world by assessing if the self's needs and goals are satisfied; and finally (v) the ability to infer mental states of other social agents and use this information to modify the self's behaviors and actions. Here, we report our results of point (i) and set the framework for further development of the rest of the points.

We propose an Experimental Functional Android Assistant (EFAA) that has the following requirements: (i) intrinsic needs to socially engage, as successful interaction requires an agent that is socially motivated; (ii) action repertoire that supports communication and interaction, in a way that the agent is able to perform actions such as object manipulation, produce linguistic responses, recognize and identify a social agent, establish and maintain interaction etc. and finally (iii) the core ingredients of social competence: actions, goals and drives. We define as drives the intrinsic needs of the robot. Goals define the functional ontology of the robot and depend on the drives whereas actions are generated to satisfy goals.

A socially competent android requires a combination of drives and goals coupled together with an emotional system. Drives and goals motivate the robot's behavior and evaluate action outcomes and emotions aim at appraising situations (epistemic emotions) and define communicative signals (utilitarian emotions). Although emotions are considered a controversial subject and a general consensus is needed [14] on the definition of emotions, Ekman [15] has defined six basic emotions that are considered universal and can be found in most cultures: happiness, surprise, fear, anger, disgust and surprise. How and why emotions arise is still under discussion with many different views. Recently, the mechanisms in the neural circuitry of emotion have gained increasing interest and attention [16] [17]. An interesting work is that of [18], where they present the following basic emotions: SEEKING, FEAR, RAGE, LUST, CARE, PANIC and PLAY. The authors provide a detailed explanation of the neural mechanisms that serve these emotions, supporting the continuity of animal and human emotions, as similar neural mechanisms are found across mammalian species, shedding light to behavioral and physiological expressions associated with these emotions. Thus emotions exist for both assisting communication by expressing one's internal state (external/utilitarian) and for organizing behaviors (internal/episthemic) [19] [16].

An organism is also endowed with internal drives. For Hull [20], a drive triggers behavior, but there is the belief that drives also maintain behaviors and direct them [21]. Like emotions, there are various opinions regarding the nature of a drive, however, it is generally accepted that an organism has multiple drives [22]. According to Cannon [23] and Seward [24] , drives are part of a homeostatic mechanism that aims at preserving their basic needs in steady states. Animals are able to perform real-world tasks by combining the homeostatic mechanism with an allostatic control, where stability is achieved through physiological or behavioral change. This way animals are able to adjust their internal state and at the same time achieve stability in dynamic environments [25].

We propose an affective framework for a socially competent robot that uses an allostatic control model as a first level of motivational drive and behavior selection combined with an emotion system. In the following sections we present the model in detail and display how complex behaviors emerge through human-robot interaction.

3   EFAA Agent: A Humanoid Robot That Promotes Interaction

In our framework, the robot's behavior is guided by its internal drives and goals in order to satisfy its needs. Drives set the robot' goals and contribute to the process of action-selection. The overall system is based on the Distributive Adaptive Control (DAC) architecture [26] which consists of four coupled layers: soma, reactive, adaptive and contextual. Each level's organization is increasingly more complex starting from the soma which designates the body itself. The reactive layer is the first level of behavior control and it is modeled in terms of an allostatic process; in this level stimuli are hard-wired with specific actions. The adaptive layer is predicated on the reactive layer; here adaptive mechanisms are deployed to deal with the unpredictability of the environment. Finally the contextual layer develops the state space that was previously acquired by the adaptive layer to generate behavioral plans and policies. Our model mainly focuses on the reactive and adaptive layer of the DAC architecture setting a framework for higher cognitive processes such as state space learning. To validate our model we propose a setup in which consists a humanoid robot, namely iCub [27], a human partner and the tabletop tangible interface Reactable [28]. The interaction defined by this setup involves a human communicating with the iCub. In the beginning of the scenario, the robot is sleeping and the human has to wake it up in order to start the interaction. Once the robot is awake, it engages in different set of activities aimed at satisfying its needs, such as play games with the human using objects placed on the Reactable. An example of the proposed setup is illustrated at figure 1. We implemented the model of drives and emotions using IQR, an open-source multilevel neuronal simulation environment [29] that is able to simulate biological nervous systems by using standard neural models.

Fig. 1. Example of the proposed scenario where the humanoid robot iCub interacts with a human and uses the Reactable objects as means of playing a game

3.1 Emotions and Drives

The behavior of the robot is highly affected by its internal drives. Inspired by the intelligent space Ada [12], an interactive entertainment space that promotes interactions with several people, the robot has the following goals that it aims at optimizing: *Be social*: the robots goal is to interact with people and regulate its behavior accordingly. *Exploration*: the need to be constantly stimulated. *Survival*: consists of two parts: physical and cognitive survival. As physical survival we define the need of the robot to occasionally rest, whereas cognitive survival is the need to reduce complexity so as to not get confused. *Play*: the robot's need to engage the human with different games in order to form a more pleasant and interesting interaction. *Security*: the need to protect itself and avoid unwanted stimuli or events.

The goal of the EFAA agent is to socially engage with humans and its drives and emotions are designed to propel such a social interaction. The main goal of the robot is to maximize its happiness by keeping its drives in a homeostatic level. A homeostatic control is applied at each drive and on top of each subsystem we employ an allostatic control that aims at maintaining balance through behavioral change.

The emotions that emerge through the agent's interaction with a human and the environment are the following: happiness, anger, sadness, fear, disgust and surprise. These emotions are compliant with Ekman's emotions [15] that are considered to be basic from evolutionary, developmental and cross-cultural studies. The emotional system is responsible for exhibiting emotional responses that are consistent with the agent's internal state and are expressed through facial expressions. The emergence of emotions depends on two main factors: the satisfaction of the drives and external stimuli such as different tactile contacts (poke, caress, grab) which affect poke, happiness and fear respectively. At a neuronal level, each emotion is expressed by a single neuron whose activity varies from 0 to 1. At an expression level, this number determines the intensity of each emotion and sets the facial expression of the robot. An example of two different intensities of the same emotion, namely happiness is depicted at figure 2.

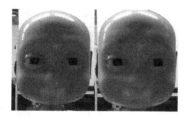

Fig. 2. Example of the emotional expression of happiness. On the left, the intensity is set to 0.5 whereas on the right the intensity is set to 1.

3.2   Homeostatic and Allostatic Control

We propose three main states that define each drive: *under homeostasis*, *homeostasis* and *over homeostasis*. A homeostatic control is applied on each drive to classify it in homeostatic terms. A drive is in *homeostasis* when it is encountering the appropriate stimulus to satisfy its needs. The absence of a stimulus leads the drive in *under homeostasis* whereas the presence of an extensive stimulus leads the corresponding drive to a *over homeostasis* state. The allostatic control aims at achieving consistency and balance in the satisfaction of the drives through behavioral change. It is responsible for choosing which action to take, what behavior to trigger and avoid cases of conflict, like the case when two drives need to be satisfied at the same time and contradict each other (e.g. when energy and play need to be satisfied), by setting priorities. The allostatic control is constantly monitoring the environment and the drives in parallel, assessing only the relevant stimuli for each drive, for example the presence of a human for the social drive, the presence of objects on the table for the exploration drive or the presence of both objects on the table and the presence of a human for the play drive.

The implementation of a combined homeostatic and allostatic control that runs in parallel, contradicts the paradigm of state machines, as the proposed system allows the robot to display more complex behaviors. The dynamics of the model are depicted in figure 3.

3.3   Behavioral Modulation

The EFAA agent has to perform different actions in order to satisfy its drives. Such behavior is considered adaptive since it allows the system to achieve specific goals, like the satisfaction of a specific drive in a dynamic environment. We have employed the following behaviors: *Wake up*: the procedure in which the robot transits from inactivity to being "awake" and ready to interact. Waking up behavior also initializes its drives and emotions. *Explore*: the robot interacts with objects on the table. *Look around*: the robot is looking around in an explorative way in order to find relevant and salient stimuli. *Track*: once a salient stimulus is found, the robot shifts its attention focus to the salient stimulus. *Play*: the

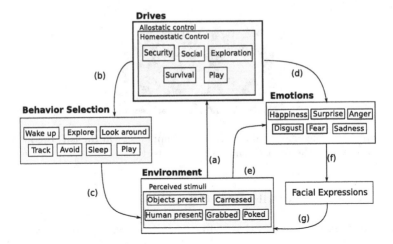

Fig. 3. Overview of the parts involved at behavioral level. Inputs from the environment are fed into the drives control mechanism (a) where there is an assessment of the homeostatic value of each drive and on top we have the allostatic control that is monitoring the drives and the related stimuli. Depending of value of each drive, an appropriate behavior is being selected (b) and executed (c). At the same time, the level of satisfaction of each drive affects the emotions of the system (d) and in combination of the assessment of certain stimuli (e) emotions emerge in the emotion system. The most dominant emotion (f) is expressed (g) through the facial expressions of the EFAA.

robot engages the human in an interactive game. The play behavior has two sub-scenarios: toss a dice and play a sequence game. *Avoid*: the robot informs the human that certain actions, objects or events are unwanted. *Sleep*: the robots drives and emotions stop. The robot will not try to satisfy its drives nor express its emotional state. During sleep, the robot's drive's are reset.

Currently, most of these behaviors are at a single level, i.e. they do not underlie a set of behaviors to choose from with the exception of the *play* behavior. However this is setting the ground for a more thorough implementation of behavior selection where the EFAA agent can learn to pick the optimal behavior. Table 1 illustrates the interaction between drives, emotions, perceived stimuli and behavioral processes. Some of the suggested behaviors are considered reflexive, such as the waking up of the robot when it is touched while asleep. However certain behaviors are employed not to satisfy a drive, but rather to create the appropriate conditions for the satisfaction of a drive. A typical example of the adaptive control is satisfaction of the socialize drive: it requires a human to interact with. In case a human is already present and tracked, the robot enters in a social behavior (dialog, game). However, in case there is no human present, the robot will seek one by either looking around or by verbally expressing its need to have someone to play with. The look around behavior in this case is considered adaptive as it does not aim at directly satisfying the social drive but rather aims at meeting the preconditions that will satisfy it.

Table 1. The perceived stimuli column refers to the presence or absence of certain stimuli that affect the drives and emotions system of EFAA. The drive column refers to the current drive that is affected by the inputs, emotion column refers to the emotions that emerge from a given situation and the behavior column denotes the kind of behavior that is triggered.

Perceived stimuli	Drive	Emotion	Behaviour
No human present	Social	Sadness	Look around
Human present	Social	Happiness	Track
No objects present on table	Exploration	Sadness	Look around
Objects on table	Exploration	Happiness	Explore
Too many objects	Cognitive survival	disgust	Avoid
Human caresses the iCub	-	Happiness	-
Human pokes the iCub	-	Anger	Avoid
Human grabs the iCub	Security	Fear	Avoid
Human leaves unexpectedly	Social	Surprise	Look around
Human touches the iCub when asleep	(drive initialization)	-	Wake up
Human present and objects on table	Play	Happiness	Play
Robot interacts too long with human	Physical survival	-	Sleep

4 System Assessment

During a human-robot interaction, the robot proceeds in action-selection and triggers behaviors that aim at satisfying its internal drives. In figure 4 we present data obtained using the model described previously during a real-time human-robot interaction.

Our results show the interplay between drives, emotions, perceived stimuli and actions while they display some key features of the overall system. The emotions panel indicates the levels of each emotion over time. The red line represents the overall happiness of the robot during interaction. On approximately the 1000th cycle we observe the presence of many objects on the table (b) which in turn promotes the emotion of disgust. At the same time, too many objects on the table cause the cognitive survival drive to rise and trigger the "avoid" behavior. On approximately the 3200th cycle (c) the robot perceived that it was grabbed which gave rise to fear. Poke rises the security drive which in turn also triggers the avoid behavior. At certain moments in the simulation, more than a single emotion emerges, however only one is dominant. This emotion is the one with highest value and is the one displayed on the facial expressions of the robot.

Another stimulus that affects the drives of the robot is that of the presence of a human. In deed we see that until the more or less 600th cycle, there is no human present. This causes the social drive to fall and rise once the human appears (a). This is a good example to show how certain behaviors cannot be triggered unless certain conditions are met. For example, to initiate the play activity, the robot needs a human to play with and play is triggered in the "actions" panel once a human appears; a drive cannot be satisfied if the appropriate conditions are not met. An example where the human participant leaves the interaction scene

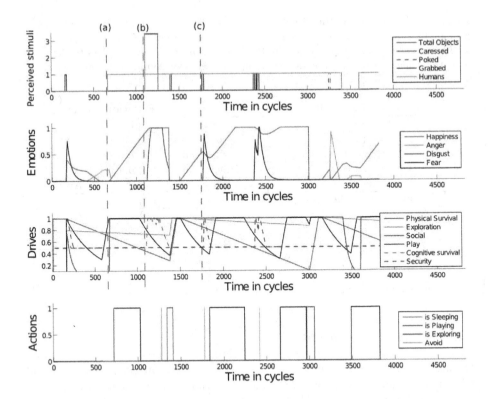

Fig. 4. Overview of the drives and emotions system over time. On the upper panel we can see the stimuli that are perceived from the environment (the number of people present, the number of objects on the table and the input from the skin of the robot: if it has been caressed, poked or crabbed). The "emotions" panel illustrates the emergence of different emotions (happiness, anger, surprise, sadness, disgust and fear). The next panel displays the drives values for survival (cognitive and physical), exploration, play, social and security whereas the actions panel indicates the emergence of the behaviors triggered in order to maintain the system in homeostasis.

(a) is depicted in figure 5. The play drive from the "drives" panel constantly decreases as conditions are not met (human is not present). Nonetheless, the robot proceeds in an explorative behavior (b) and satisfies its exploration drive displaying a more adaptive behavior. Only once the human returns, the robot is able to satisfy its play drive and trigger the appropriate behavior.

Part of the role of the allostatic control is to make sure that certain actions do not collide. A robot can look or track a stimulus while talking or playing a game however, it cannot play a game while sleeping. In the presented interaction, the physical survival of the robot gets low and needs to be satisfied at approximately the 2500th cycle (see the "drives" panel in figure 4), however at that time the robot is already playing with the human and cannot go to sleep. Once the play action is finished, the robot is free to proceed into the sleep behavior.

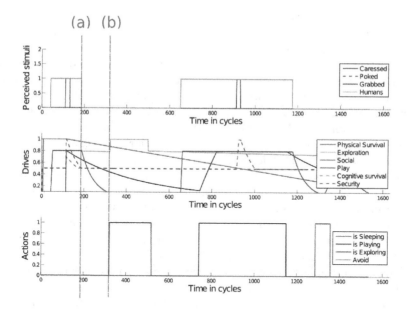

Fig. 5. Example of the robot's behavior during an interaction where the human leaves the scene (a). With the absence of humans, the robot starts exploring (b) objects on the table.

Our results indicate how emotions and drives are affected by certain perceptions. Although behaviors are triggered with the scope of maintaining the drives in homeostasis, however, they are bound to these perceptions. Nonetheless, we can observe the dynamics of the proposed system through the interplay of the external percepts, the robot's emotions and drives and the emergence of certain behaviors in the attempt to keep the balance in the agent's internal state.

5   Discussion

Nowadays there is an increased interest in developing social robots, that is robots that are able to interact and communicate with humans in a natural and intuitive way. However this raises an important question: what are the minimum prerequisites a robot should have in order to be considered a social agent? In this paper we argue that the minimal requirements for a functional robot that can act as a social agent are: (i) the need to be social, (ii) have a repertoire of actions that support communication and interaction, (iii) the ability to distinguish between self and others, (iv) the ability to infer mental states of other social agents and (v) the ability to assess its needs and goals and therefore evaluate how the self is situated in the world.

Here we propose a system that has the intrinsic need to socially engage and interact with humans and is equipped with an action repertoire that can support communication and interaction. This system includes drives that help satisfy the robot's intrinsic needs, emotions that assist the robot express its internal state (utilitarian) and organize behaviors (episthemic) and a set of actions that aims at satisfying its needs. In the proposed model we have defined the following drives: sociality, exploration, survival, security and play. Each of these drives is monitored by a homeostatic control that classifies the level of each drive into the following stages: *under homeostasis, homeostasis* and *over homeostasis*. On top of homeostasis we apply an allostatic control that is responsible for the maintenance of the system in balance by behavior selection and priority assignation in order to satisfy its needs. The model's design in based on the reactive and adaptive layers of the Distributed Adaptive Control (DAC). The reactive layer is responsible for producing reflexive almost hard-wired responses while in the adaptive layer deals with the unpredictability of the world. However, the satisfaction of its needs highly depends on the environment and the current state of the world. As the allostatic control switches from a reactive to an adaptive level, it is not anymore motivated by direct drive satisfaction but it is aiming for matching requirements so that an action leading to a given goal (that is the final drive satisfaction) will be available.

The satisfaction level of each drive defines the emotional state of the robot as well as certain external stimuli such as the robot being caressed, poked or grabbed by the human. The robot is able to exhibit six emotions: happiness, anger, sadness, disgust, surprise and fear, emotions that are considered to be basic from evolutionary and cross-cultural studies. The main goal of the robot is to maximize its happiness by keeping its drives in homeostasis. To do so, it is equipped with a set of different behaviors that it can trigger in order to satisfy its needs: wake up, explore, look around, track, play, avoid and sleep. Most of these behaviors are considered reflexive (such as wake up) and single layered, however there are also more complex behaviors such as play that trigger 2 sub-scenarios: play a dice game or play a memory task game.

The suggested scenario involves the interaction of a humanoid robot, the iCub, with a human, using the tangible interface Reactable as means of playing games. The robot's actions are triggered based on the suggested model. The data collected during a human-robot interaction suggest that there is a guided behavior emergence based on the satisfaction level of each drive and the perceptions of the environment. By monitoring drives in parallel (allostatic control) and trying to keep them in a homeostatic state(homeostatic control) we are able to produce different sets of behaviors. Although there is similar work, using emotional and motivational models applying the "homeostatic regulation rule" for action selection [30], our model of homeostatic and allostatic control can act as the first level of the motivational engine and regulate the robots internal needs and drives via behavioral modulation opening the way for a more adaptive behavior.

The allostatic control focuses on actions that could satisfy a drive, but the preconditions of which can be easily satisfied by direct execution of another behavior. This leads to a better adaptation and manipulation of the environment while still being able to satisfy only short-term goals. The long-term global satisfaction of drives, or within contexts that need reasoning about past experience are still to be investigated. Initial attempts to achieve such capabilities are to be linked tightly with cognitive components responsible for the different memory types (episodic, autobiographic) which implementation are described in [31].

Acknowledgments. This work is supported by the EU FP7 project EFAA (FP7-ICT- 270490).

References

1. Trincavelli, M., Reggente, M., Coradeschi, S., Loutfi, A., Ishida, H., Lilienthal, A.J.: Towards environmental monitoring with mobile robots. In: IEEE/RSJ International Conference on Intelligent Robots and Systems, IROS 2008, pp. 2210–2215 (2008)
2. Distante, C., Indiveri, G., Reina, G.: An application of mobile robotics for olfactory monitoring of hazardous industrial sites. Industrial Robot: An International Journal 36(1), 51–59 (2009)
3. Thrun, S., Beetz, M., Bennewitz, M., Burgard, W., Cremers, A.B., Dellaert, F., Fox, D., Haehnel, D., Rosenberg, C., Roy, N., et al.: Probabilistic algorithms and the interactive museum tour-guide robot minerva. The International Journal of Robotics Research 19(11), 972–999 (2000)
4. Bennewitz, M., Faber, F., Joho, D., Schreiber, M., Behnke, S.: Towards a humanoid museum guide robot that interacts with multiple persons. In: 2005 5th IEEE-RAS International Conference on Humanoid Robots, pp. 418–423 (2005)
5. Tapus, A., Ţăpuş, C., Matarić, M.J.: User robot personality matching and assistive robot behavior adaptation for post-stroke rehabilitation therapy. Intelligent Service Robotics 1(2), 169–183 (2008)
6. Robins, B., Dautenhahn, K., Te Boekhorst, R., Billard, A.: Robotic assistants in therapy and education of children with autism: Can a small humanoid robot help encourage social interaction skills? Universal Access in the Information Society 4(2), 105–120 (2005)
7. Breazeal, C.: Toward sociable robots. Robotics and Autonomous Systems 42(3), 167–175 (2003)
8. Michotte, A.: The perception of causality (1963)
9. Premack, D., Premack, A.J.: Origins of human social competence (1995)
10. Breazeal, C.L.: Designing sociable robots. The MIT Press (2004)
11. Reeves, B.: The media equation: how people treat computers, television, and new media (1997)
12. Eng, K., Douglas, R.J., Verschure, P.F.: An interactive space that learns to influence human behavior. IEEE Transactions on Systems, Man and Cybernetics, Part A: Systems and Humans 35(1), 66–77 (2005)
13. Gallese, V., Metzinger, T.: Motor ontology: the representational reality of goals, actions and selves. Philosophical Psychology 16(3), 365–388 (2003)
14. Griffiths, P.E.: What emotions really are: The problem of psychological categories. University of Chicago Press (1997)
15. Ekman, P.: An argument for basic emotions. Cognition & Emotion 6(3-4), 169–200 (1992)

16. Scherer, K.R.: Neuroscience projections to current debates in emotion psychology. Cognition & Emotion 7(1), 1–41 (1993)
17. LeDoux, J.: Rethinking the emotional brain. Neuron 73(4), 653–676 (2012)
18. Panksepp, J., Biven, L.: The archaeology of mind (2011)
19. Arbib, M.A., Fellous, J.M.: Emotions: from brain to robot. Trends in Cognitive Sciences 8, 554–561 (2004)
20. Hull, C.: Principles of behavior (1943)
21. Duffy, E.: The concept of energy mobilization. Psychological Review 58(1), 30 (1951)
22. McFarland, D.: Experimental investigation of motivational state. Motivational Control Systems Analysis, 251–282 (1974)
23. Cannon, W.B.: The wisdom of the body. The American Journal of the Medical Sciences 184(6), 864 (1932)
24. Seward, J.P.: Drive, incentive, and reinforcement. Psychological Review 63(3), 195 (1956)
25. Sanchez-Fibla, M., Bernardet, U., Wasserman, E., Pelc, T., Mintz, M., Jackson, J.C., Lansink, C., Pennartz, C., Verschure, P.F.: Allostatic control for robot behavior regulation: a comparative rodent-robot study. Advances in Complex Systems 13(03), 377–403 (2010)
26. Verschure, P.F.: Distributed adaptive control: A theory of the mind, brain, body nexus. Biologically Inspired Cognitive Architectures (2012)
27. Metta, G., Sandini, G., Vernon, D., Natale, L., Nori, F.: The icub humanoid robot: an open platform for research in embodied cognition. In: Proceedings of the 8th Workshop on Performance Metrics for Intelligent Systems, pp. 50–56. ACM (2008)
28. Geiger, G., Alber, N., Jordà, S., Alonso, M.: The reactable: A collaborative musical instrument for playing and understanding music. Her&Mus. Heritage & Museography (4), 36–43 (2010)
29. Bernardet, U., Verschure, P.F.: iqr: A tool for the construction of multi-level simulations of brain and behaviour. Neuroinformatics 8(2), 113–134 (2010)
30. Arkin, R.C., Fujita, M., Takagi, T., Hasegawa, R.: An ethological and emotional basis for human–robot interaction. Robotics and Autonomous Systems 42(3), 191–201 (2003)
31. Pointeau, G., Petit, M., Dominey, P.: Successive developmental levels of autobiographical memory for learning through social interaction (2013) (manuscript sumbitted for publication)

A Biomimetic Neuronal Network-Based Controller for Guided Helicopter Flight*

Anthony Westphal, Daniel Blustein, and Joseph Ayers

Depts. of Marine and Environmental Sciences, Biology, and Marine Science Center,
Northeastern University, Nahant MA 01908, USA
lobster@neu.edu
http://www.neurotechnology.neu.edu/

Abstract. As part of the Robobee project, we have modified a coaxial helicopter to operate using a discrete time map-based neuronal network for the control of heading, altitude, yaw, and odometry. Two concepts are presented: 1. A model for the integration of sensory data into the neural network. 2. A function for transferring the instantaneous spike frequency of motor neurons to a pulse width modulated signal required to drive motors and other types of actuators. The helicopter is provided with a flight vector and distance to emulate the information conveyed by the honeybee's waggle dance. This platform allows for the testing of proposed networks for adaptive navigation in an effort to simulate honeybee foraging on a flying robot.

Keywords: robotic flight, honeybee, neuronal control, helicopter.

1 Introduction

Autonomous and adaptive flight as performed by insects has long served as an inspiration to the development of micro air vehicles (MAVs). Efforts to date have focused on mimicking different biological characteristics of flying insects including sensors, behaviors, and biomechanics. Here we present a holistic biomimetic platform that combines and advances robotic implementation of neuromorphic sensors, electronic nervous system control, and motor neuron activation of effectors.

The control rules for foraging insect flight are well established [33]. What is less understood is how these rules are realized by neuronal networks in the central nervous system [34]. Given the methodological difficulties in recording from neurons in behaving animals, we have taken an alternative approach to circuit tracing: embodied simulation through robotics [2,39,40]. Comparative physiology and biomimetics provide numerous examples of how many aspects of behavior are mediated by neuronal networks [28,37]. The command neuron, coordinating neuron, central pattern generator model appears generalizable for the control of innate behavior throughout the animal kingdom and has withstood critical review for over 30 years [20,21,26].

---

* Supported by NSF ITR Expeditions Grant 0925751.

N.F. Lepora et al. (Eds.): Living Machines 2013, LNAI 8064, pp. 299–310, 2013.
© Springer-Verlag Berlin Heidelberg 2013

Some of the neural components underlying arthropod behavior have been identified including optical flow sensitive neurons [in Drosophila [25], in honeybees [17], in locusts [4], in crayfish [44]]. Although recent efforts have begun to investigate the neural pathways involved in optical flow responses [18], the neural circuitry underlying visually mediated flight remains a mystery. Insect responses to optical flow, the movement of the world across the visual field, have been shown to control and stabilize flight [36] and have been mimicked on various robot platforms. Other bioinspired sensory systems well-studied in animals and mimicked on robots include magnetic compasses in animals [7] and robots [42], inertial sensors in animals [16] and robots [22], and air velocity sensors in animals [46] and robots [31]. Previous efforts to engineer autonomous flight have adopted such biologically inspired sensors but neglect neuronal control principles [8,9,45]. As the details of the dynamics of the mechanical system of Robobee emerge [15,24] we have been addressing the possiblity of adapting neuronal network control [2] to the control of the steering system [11] for directed foraging.

Hybrid analog/digital systems provide efficiencies essential to micromechanical systems [32]. Advances in nonlinear dynamics [29,30] have yielded technologies that allow for the control of robots by synaptic networks in real time [1,23]. Controlling robots with neuronal and synaptic networks provides a powerful heuristic framework to compare and evaluate different hypothetical network topologies and candidate mechanisms for the control of behavior [2,43]. However, this approach requires the development of an autonomous flying biomimetic platform that can interface with sensors and actuators. Recent advances as part of the Robobee project have lead to the engineering of a flapping wing MAV demonstrating passive stability [38] and optical flow sensing [12]. Robobee flight currently relies on an off-board motion tracking system for control [24] but current efforts are focused on achieving autonomy using onboard sensors, processors, and power. The helicopter-based model we describe here serves as a platform to establish and verify basic neuronal network functionality as an alternative to traditional algorithmic control prior to miniaturization.

The coaxial helicopter platform mediates lift and steering with a reasonable approximation to that of the flight system of asynchronous insects [10,19]. Here we use it to explore neuronal network models. Interfacing electrical components with an electronic nervous system simulation requires the translation of digital data to and from the language of neurons, i.e. action and synaptic potentials. To accomplish this we present an approach to convert sensory information into the pulse-interval code of biological networks and to convert neuronal activity into a pulse width-modulated (PWM) signal for the control of motors and servos.

2   Helicopter Platform

A custom electronics and software package implemented on a Blade MX2 coaxial helicopter serves as the foundation for the biomimetic platform. The electronics package is divided into three parts: sensory integration, neural processing, and motor output. These parts are modular and can be expanded and customized based on the robot's requirements. A programmed phenomenological neural model [30] allows for rapid prototyping and simulation of neural network hypotheses in LabVIEW^{TM} and their subsequent implementation in C on an 8bit microprocessor.

Our studies have shown [1,43] that numerical simulations of networks needed to control a biomimetic robot can be implemented in real time using standard microprocessors. The hardware layout for the helicopter control system is shown in Figure 1. There are two ATMega 2560 microprocessors at the heart of the helicopter control board. The sensory microprocessor (SP) is responsible for reading data from the gyroscope, ultrasonic range finder, and optical flow sensors. It also sends data to a computer logger for analysis via a Bluetooth link, sends commands to the rotors, and controls pitch and roll servomotors. User control override via a 5 channel RC receiver is also enabled. The nervous system microprocessor (NS) runs the discrete time map-based (DTM) neurons and synapses [30] that are the building blocks of the nervous system simulation. The sensory information passes from the first processor to the second where it modulates neuron network activity. Motor neuron activity is then passed back to the first processor to drive the motors and servos.

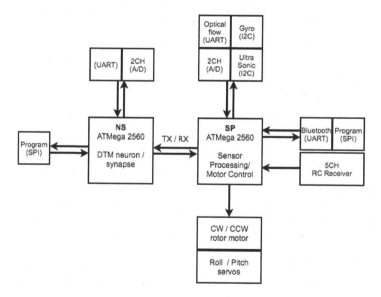

Fig. 1. *Helicopter nervous system processor hardware.* The ATMega 2560 sensory microprocessor (SP) is used to gather sensory information, transmit data to an off-board logging device and send motor signals to the main rotor motors and to the roll and pitch servomotors. The nervous system microprocessor (NS) is passed filtered sensory data that modulates the neural networks which produce motor neuron outputs that are passed back to control the motors and servos. The operator can resume control of the motor output via the 5ch RC receiver.

A custom PC board 30 mm wide and 50 mm long was built. The top of the PC board holds the NS microprocessor (Fig. 2A) while the SP processor is located on the bottom of the board (Fig. 2B).

Fig. 2. *Top and bottom views of the nervous system microprocessor board.* Two ATMega 2560 microprocessors on a custom DSP to operate a DTM nervous system on the helicopter. **A.** The NS IC runs the DTM neurons and synapses. The connectors are outlined by boxes: [1] 5ch RC receiver; [2] output to rotors and servos; [3] two UART lines to processor SP; [4] SPI programming jack for processor SP; [5] programming jack for processor NS; [6] I2C bus for SP processor/UART to NS processor. **B.** The SP IC is responsible for motor control and sensor processing.

3 Sensory System

A variety of analog sensors are mounted to the sensor array board (Fig. 3) and connected via a series of wires between the individual sensors and the helicopter SP microprocessor. It is well established that bees use a sun compass for navigation [14]. To simulate this we have integrated an ITG-3200 triple-axis digital gyro capable of 2000 deg/sec measurements. It is used to calculate the current heading since vibration and magnetic interference generate inaccurate readings from a standard magnetic compass module. The ultrasonic range finder is a Devantech SRF08 capable of cm resolution up to a height of 6 m. The optical flow sensor is a Centeye Ardueye Aphid Vision Sensor (v1.2), which produces optical flow readings in both the x and y axes. A 4 mm gel pad is used to isolate the sensor mount from helicopter vibrations. Additional vibration dampening in the form of antistatic foam was used to encase the ultrasonic sensor. The sensor mount, printed on a 3D printer (MakerBot Replicator 2) using PLA, clips onto the helicopter rails via 4 mounting tabs. The fully assembled helicopter with all of its components is shown in Figure 4.

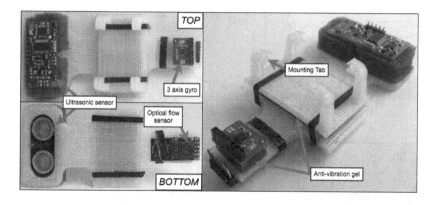

Fig. 3. *Sensor array board.* The 3D-printed mounting board holds the 3 axis digital gyroscope, ultrasonic range finder, and optical flow sensor.

4 Neuron Network Development

Based on known rules for optical flow modulation of flight [35] and sun compass operation [14], we have developed a hypothetical network that integrates a heading deviation sense along with optical flow modulation of yaw, and optical flow-mediated odometry to control the search phase of foraging (Fig. 5). To initiate a search behavior, a target heading and distance are communicated over the Bluetooth module as an analogue to the search vector communicated by the honeybee's waggle dance [13]. The target heading is conveyed as a set point for the compass network and the target distance as the strength of synaptic connections between optical flow sensory neurons and the odometer neuron. The compass sensory neurons mediate yaw towards the target heading while the helicopter climbs to the desired altitude based on the ultrasonic sensor's set point [2,42]. Once heading error has been eliminated, inhibition on the Forward Translation command is released and the helicopter pitches forward resulting in forward flight. Translation proceeds with yaw compensation and odometry mediated by optical flow [2,6]. Once the desired distance has been traversed, the odometer neuron fires resulting in inhibition of Forward Translation.

5 Neuron Network Implementation

The implementation of a nervous system simulation on the helicopter platform is a three-step process that has been developed to allow for rapid prototyping and implementation in C on a variety of processors. The DTM model is capable of modeling both spiking and bursting activity [30]. Here we simplified the model to achieve robust integration of synaptic input without higher order bursting. A nonlinear function (1) is responsible for spiking behavior and generates a new current value x_{n+1} when passed the present current x_n, the previous current

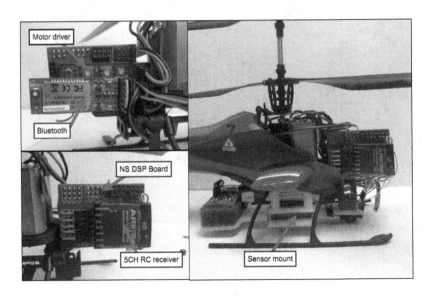

Fig. 4. *Helicopter platform.* Custom Blade MX2 helicopter platform with Motor Driver, Bluetooth module, NS/SP DSP board, 5ch RC receiver and mounted sensor array.

x_{n-1}, and a synaptic current input u. The variables α and σ determine a baseline spiking or bursting activity profile.

$$x_{n+1} = f_n(x_n, x_n - 1, u) = \begin{cases} \alpha/(1 - x_n) + u, & x \leq 0, \\ \alpha + u, & 0 < x_n < \alpha + u \ \& \ x_{n-1} \leq 0, \\ -1, & x_n \geq \alpha + u \text{ or } x_{n-1} > 0, \end{cases} \quad (1)$$

Function (2) sums σ with the excitatory synaptic currents cIe, the inhibitory current cIi, and the exogenous injected current Idc, and scales the cell's transient response to them based on the scaling factors, β_E, β_I, and β_{DC}, respectively.

$$u = \beta_E(cIe) + \beta_I(cIi) + \beta_{DC}(Idc) + \sigma \quad (2)$$

The synaptic current (I) is calculated by:

$$I = \gamma(I) - \gamma_{Inh}(spike(x_n^{post} - x_{rp})) \quad (3)$$

where γ = synaptic strength, γ_{Inh} = relaxation rate, x_n^{post} = postsynaptic current and x_{rp} = reversal potential. The variable *spike* signifies when an action potential has occured, it is either 1 or 0 based on the state of equation 1.

Based on the causal network of Figure 5, the parameters of the network are tuned using LabVIEW^{TM} software. Virtual instruments (VIs) for neurons (equations 1 and 2) and synapses (equation 3) are constructed. Neuron VIs are connected together with synapse VIs based on the proposed network layout. Neuron behaviors are set by varying α and σ. Synapses are characterized as excitatory

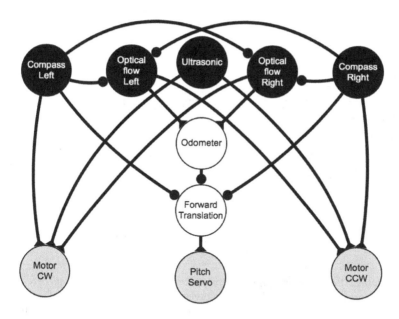

Fig. 5. *Helicopter sensory network layout.* Circles represent modeled neurons and lines represent synapse topology: filled triangles represent excitatory synapses, filled circles represent inhibitory. Black circles are sensory neurons, white circles are interneurons and grey circles are motor neurons. Compass Right and Compass Left neurons respond to heading error to either the right or left, respectively. Optical flow Right and Optical flow Left respond to translational optical flow from separate sensors facing laterally outwards. The Ultrasonic neuron responds to the error between the target and current altitude.

or inhibitory by setting x_{rp} to either a positive or negative number, respectively. Once the network settings have been tuned for optimal performance in software simulation, the network hypothesis is instantiated in C code for robot operation. A function for each type of neuron and synapse is created in a structure with the same variables to mimic the LabVIEW^{TM} architecture.

6   Transforming Sensor Data to Neuronal Activity

Sensory information can be transformed into a scaled injected current that drives sensory neurons. The sensor outputs are converted to neuronal spike trains by a variety of mechanisms such as raw value scaling, differential error scaling, and range fractionation [43]. For example, the optical flow sensor generates a scalar representing instantaneous optical flow. This scalar serves as an input to an optical flow interneuron that encodes optical flow as a spiking neuronal output (Fig. 6). As instantaneous optical flow increases, the raw sensor value increases leading to an increase in spike frequency in the corresponding sensory neuron.

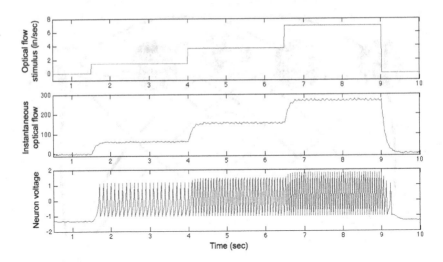

Fig. 6. *Optical flow sensory input is transformed to a neuronal pulse code.* An optical flow sensor was presented a moving visual stimulus consisting of a sinusoidal black/white contrast pattern with a period of 0.5 in [27]. The top trace shows the speed of the optical flow stimulus profile (1.42 in/sec from 1.5 to 4.0 sec, 3.70 in/sec from 4.0 to 6.5 sec, and 6.94 in/sec from 6.5 to 9.0 sec). The middle trace shows the reported instantaneous optical flow values from the sensor. The bottom trace shows the response of a sensory neuron coding this sensory input ($\alpha = 4.05$, $\sigma = $ -3.10). The sensor was mounted 5 inches away from a computer monitor displaying the stimulus.

The control parameters α and σ can be adjusted to modify the neuron's response to sensory input.

7   Controlling Motors with Neuronal Activity

Translating neuronal activity to motor actuation is the final step necessary to establish a complete robotic implementation of a nervous system simulation. While generally not an issue using robots with shape memory alloy actuators such as RoboLamprey [43] since neuronal spikes can directly activate such effectors, a method to transform spikes to a PWM signal is needed for the helicopter platform. The transformation from spike frequency to pulse width modulated duty cycle is governed by the following formula:

$$PWM_{DC} = \begin{cases} (-100 * (\frac{(\Delta_{iterations})}{Max_{iterations}}) - 1)), & \Delta_{iterations} < Max_{iterations} \\ 0, & \Delta_{iterations} > Max_{iterations} \end{cases} \quad (4)$$

$PWM_{DC}=$ Duty cycle used to drive motors. $\Delta_{iterations}=$Number of iterations since last spike. $Max_{iterations}=$Maximum number of iterations between each spike before PWM_{DC} can be considered 0.

PWM_{DC} is calculated when the motor neuron fires and is a representation of the motor neuron spike frequency. $Max_{iterations}$ is set to a value that corresponds to the maximum number iterations of the code that can occur before the motor neuron that is being transformed to a PWM signal can be considered silent. $\Delta_{iterations}$ is increased by 1 every iteration of the code. The ratio of $\Delta_{iterations}$ to $Max_{iterations}$ determines PWM_{DC}. As spike frequency increases across its range of values, the output PWM signal increases proportionally allowing for control of motors and servos (Fig. 7).

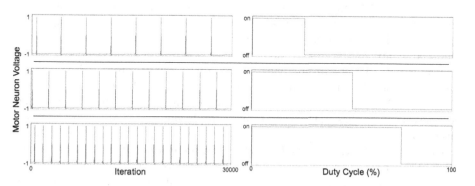

Fig. 7. *Transformation of spike frequency to duty cycle.* As the spike frequency of a motor neuron increases, the PWM output to the associated motor increases. Top panel: When a spike occurs every 3754 iterations, the injected current (Idc) equals .2836 which corresponds to a 25% duty cycle. Middle panel: 2521 iterations per spike, $Idc = 1.1.2852$, 50% duty cycle. Bottom panel: 1288 iterations per spike, $Idc = 1.2932$, 75% duty cycle.

8   Discussion

Here we have shown the development of a flying biomimetic robot platform that can be used to test embodied simulations of neuronal network function for the control of flight behavior. The custom electronics platform presented allows for the use of a variety of sensors including an ultrasonic range finder, optical flow sensors and a three-axis gyroscope. Using a discrete time map-based neuron and synapse model, neural network hypotheses can be simulated and tested on the flying robot. At this stage in the development of a neuronal network-based controller for helicopter flight, there exist two main challenges in approaching biomimetic control. The first challenge is to transform sensory information to the pulse-interval code of biological networks. We have shown that by using biologically relevant sensors, sensory data can serve as input to interneurons and motor neurons to drive reflexive behaviors. The second challenge was to translate motor neuron spikes into PWM signals to drive DC motors and servos.

By using an iteratively updating equation to convert spike frequency to PWM, the activity of a modeled neuron can drive a motor as shown in Figure 7.

This complete biomimetic platform allows for the testing of nervous system hypotheses related to network connectivity and neuronal principles in general, such as varied basal neuron activity levels and the role of efferent copy in multimodal sensory fusion [41]. Forthcoming work will use this platform to test a neuronal compass [42] and a neuron-based optical flow odometer [6] in the field. Testing of sensor fusion hypotheses demonstrating complex behaviors will follow.

While this platform is a powerful tool for the testing of nervous system hypotheses related to insect flight control, the general framework presented can be adapted for a wide range of experimentation. The technological capability to run real-time neural network simulations on-board autonomous robots has remarkable implications for the advancement of synthetic neuroscience research. The approach to software modeling of neuronal activity presented can be accomplished on almost any microprocessor including Arduino boards and the $LEGO^{TM}$ NXT brick [5]. Neuroscientists can use these robotic tools to test their own network hypotheses stemming from neurophysiological and neuroethological investigations.

We have previously developed controllers for underwater walking [3] and swimming [43] robots. The work presented here extends this implementation to a simulation of asynchronous flight and confirms that the command neuron, coordinating neuron, central pattern generator model can be implemented on a range of robotic platforms with varied sensors and actuators. We plan to adapt and miniaturize this technology for the control of Robobee.

Biomimetic embodied nervous system simulations show great promise for investigating neuron operation in vivo. Comparative testing between animal and robot simulation can serve to elucidate deficiencies in hypotheses and inform further study. The development of a biomimetic framework adaptable to a flying platform shows that the approach can be used to address research questions in a broad range of biological systems and offer new control techniques for autonomous robotics.

References

1. Ayers, J., Rulkov, N., Knudsen, D., Kim, Y.-B., Volkovskii, A., Selverston, A.: Controlling Underwater Robots with Electronic Nervous Systems. Appied Bionics and Biomimetics 7, 57–67 (2010)
2. Ayers, J., Blustein, D., Westphal, A.: A Conserved Biomimetic Control Architecture for Walking, Swimming and Flying Robots. In: Prescott, T.J., Lepora, N.F., Mura, A., Verschure, P.F.M.J. (eds.) Living Machines 2012. LNCS, vol. 7375, pp. 1–12. Springer, Heidelberg (2012)
3. Ayers, J., Witting, J.: Biomimetic Approaches to the Control of Underwater Walking Machines. Phil. Trans. R. Soc. Lond. A 365, 273–295 (2007)
4. Baader, A., Schfer, M.: The perception of the visual flow field by flying locusts: A behavioural and neuronal analysis. J. Exp. Biol. 165, 137–160 (1992)
5. Blustein, D., Rosenthal, N., Ayers, J.: Designing and implementing nervous system simulations on LEGO robots. J of Visualized Experiments (in press, 2013)

6. Blustein, D., Westphal, A., Ayers, J.: Optical flow mediates biomimetic odometry on an autonomous helicopter (in preparation, 2013)
7. Boles, L.C., Lohmann, K.J.: True navigation and magnetic maps in spiny lobsters. Nature 421(6918), 60–63 (2003)
8. Chahl, J., Rosser, K., Mizutani, A.: Bioinspired optical sensors for unmanned aerial systems. In: Proceedings of SPIE: Bioinspiration, Biomimetics, and Bioreplication, vol. 7975, pp. 0301–0311 (2011)
9. Conroy, J., Gremillion, G., Ranganathan, B., Humbert, J.S.: Implementation of wide-field integration of optic flow for autonomous quadrotor navigation. Auton. Robot. 27(3), 189–198 (2009)
10. Dantu, K., Kate, B., Waterman, J., Bailis, P., Welsh, M.: Programming micro-aerial vehicle swarms with karma. In: Proceedings of the 9th ACM Conference on Embedded Networked Sensor Systems. ACM (2011)
11. Dickinson, M.H., Tu, M.S.: The function of dipteran flight muscle. Comparative Biochemistry and Physiology Part A: Physiology 116(3), 223–238 (1997)
12. Duhamel, P.-E.J., Perez-Arancibia, N.O., Barrows, G.L., Wood, R.J.: Biologically Inspired Optical-Flow Sensing for Altitude Control of Flapping-Wing Microrobots. IEEE/ASME Trans Mechatron 18(2), 556–568 (2013)
13. Dyer, F.C.: The biology of the dance language. Annual Review of Entomology 47(1), 917–949 (2002)
14. Dyer, F.C., Dickinson, J.A.: Sun-compass learning in insects: Representation in a simple mind. Current Directions in Psychological Science 5(3), 67–72 (1996)
15. Finio, B.M., Wood, R.J.: Open-loop roll, pitch and yaw torques for a robotic bee. In: IEEE/RSJ International Conf. on Intelligent Robots and Systems, IROS (2012)
16. Fraser, P.J.: Statocysts in Crabs: Short-Term Control of Locomotion and Long-Term Monitoring of Hydrostatic Pressure. Biol. Bull. 200(2), 155–159 (2001)
17. Ibbotson, M.: Wide-field motion-sensitive neurons tuned to horizontal movement in the honeybee, Apis mellifera. J. Comp. Physiol. A: Neuroethology, Sensory, Neural, and Behavioral Physiology 168(1), 91–102 (1991)
18. Joesch, M., Weber, F., Eichner, H., Borst, A.: Functional Specialization of Parallel Motion Detection Circuits in the Fly. J.Neuroscience 33(3), 902–905 (2013)
19. Kate, B., Waterman, J., Dantu, K., Welsh, M.: Simbeeotic: A simulator and testbed for micro-aerial vehicle swarm experiments. In: Proceedings of the 11th International Conference on Information Processing in Sensor Networks. ACM (2012)
20. Kennedy, D., Davis, W.J.: Organization of Invertebrate Motor Systems. Handbook of Physiology. The organization of invertebrate motor systems. In: Geiger, S.R., Kandel, E.R., Brookhart, J.M., Mountcastle, V.B. (eds.) Handbook of Physiology, sec. I, vol. I, part 2., pp. 1023–1087. Amer. Physiol. Soc, Bethesda (1977)
21. Kiehn, O.: Development and functional organization of spinal locomotor circuits. Current Opinion in Neurobiology 21(1), 100–109 (2011)
22. Lobo, J., Ferreira, J.F., Dias, J.: Bioinspired visuo-vestibular artificial perception system for independent motion segmentation. In: Second International Cognitive Vision Workshop, ECCV 9th European Conference on Computer Vision, Graz, Austria (2006)
23. Lu, J., Yang, J., Kim, Y.B., Ayers, J.: Low Power, High PVT Variation Tolerant Central Pattern Generator Design for a Bio-hybrid Micro Robot. In: IEEE International Midwest Symposium on Circuits and Systems, vol. 55, pp. 782–785 (2012)
24. Ma, K.Y., Chirarattananon, P., Fuller, S.B., Wood, R.J.: Controlled Flight of a Biologically Inspired. Insect-Scale Robot. Science 340(6132), 603–607 (2013)

25. Paulk, A., Millard, S.S., van Swinderen, B.: Vision in Drosophila: Seeing the World Through a Model's Eyes. Annual Review of Entomology 58, 313–332 (2013)
26. Pearson, K.G.: Common principles of motor control in vertebrates and invertebrates. Annu.Rev.Neurosci. 16, 265–297 (1993)
27. Peirce, J.: PsychoPy - Psychophysics software in Python. J. Neurosci. Methods 162, 8–13 (2007)
28. Prescott, T.J., Lepora, N.F., Mura, A., Verschure, P.F.M.J. (eds.): Living Machines 2012. LNCS, vol. 7375. Springer, Heidelberg (2012)
29. Rabinovich, M.I., Selverston, A., Abarbanel, H.D.I.: Dynamical principles in neuroscience. Reviews of Modern Physics 78(4), 1213–1265 (2006)
30. Rulkov, N.F.: Modeling of spiking-bursting neural behavior using two-dimensional map. Phys.Rev. E 65, 041922 (2002)
31. Rutkowski, A.J., Miller, M.M., Quinn, R.D., Willis, M.A.: Egomotion estimation with optic flow and air velocity sensors. Biol. Cybern. 104(6), 351–367 (2011)
32. Sarpeshkar, R.: Analog versus digital: extrapolating from electronics to neurobiology. Neural Computation 10(7), 1601–1638 (1998)
33. Srinivasan, M.V.: Honey bees as a model for vision, perception, and cognition. Annual Review of Entomology 55, 267–284 (2010)
34. Srinivasan, M.V.: Honeybees as a model for the study of visually guided flight, navigation, and biologically inspired robotics. Physiol Reviews 91(2), 413–460 (2011)
35. Srinivasan, M.V.: Visual control of navigation in insects and its relevance for robotics. Current Opinion in Neurobiology 21(4), 535–543 (2011)
36. Srinivasan, M., Zhang, S., Lehrer, M., Collett, T.: Honeybee navigation en route to the goal: visual flight control and odometry. J. Exp. Biol. 199, 237–244 (1996)
37. Stein, P.S.G., Grillner, S., Selverston, A.I., Stuart, D.: Neurons, Networks and Motor Behavior. MIT Press, Cambridge (1997)
38. Teoh, Z.E., Fuller, S.B., Chirarattananon, P.: A hovering flapping-wing microrobot with altitude control and passive upright stability. In: 2012 IEEE/RSJ International Conference on Intelligent Robots and Systems (IROS), pp. 3209–3216 (2012)
39. Webb, B.: Can robots make good models of biological behaviour? Behav. Brain Sci. 24(6), 1033–1050 (2001)
40. Webb, B.: Robots in invertebrate neuroscience. Nature 417(6886), 359–363 (2002)
41. Webb, B., Reeve, R.: Reafferent or redundant: integration of phonotaxis and optomotor behavior in crickets and robots. Adaptive Behavior 11(3), 137–158 (2003)
42. Westphal, A., Ayers, J.: A neuronal compass for autonomous biomimetic robots (in preparation, 2013)
43. Westphal, A., Rulkov, N., Ayers, J., Brady, D., Hunt, M.: Controlling a lamprey-based robot with an electronic nervous system. Smart Struct. Sys. 8(1), 37–54 (2011)
44. Wiersma, C.A., Yamaguchi, T.: Integration of visual stimuli by the crayfish central nervous system. J. Exp. Biol. 47(3), 409–431 (1967)
45. Wood, R.J., Avadhanula, S., Steltz, E., Seeman, M., Entwistle, J., Bachrach, A., Barrows, G., Sanders, S.: An autonomous palm-sized gliding micro air vehicle. IEEE Robotics and Automation Magazine 14(2), 82–91 (2007)
46. Yorozu, S., Wong, A., Fischer, B., Dankert, H., Kernan, M., Kamikouchi, A., Ito, K., Anderson, D.: Distinct sensory representations of wind and near-field sound in the Drosophila brain. Nature 458, 201–205 (2009)

Bioinspired Adaptive Control
for Artificial Muscles

Emma D. Wilson[1], Tareq Assaf[2], Martin J. Pearson[2], Jonathan M. Rossiter[2],
Sean R. Anderson[1], and John Porrill[1]

[1] Sheffield Centre for Robotics (SCentro), University of Sheffield, UK
[2] Bristol Robotics Laboratory (BRL), University of the West of England
and University of Bristol, UK

Abstract. The new field of soft robotics offers the prospect of replacing existing hard actuator technologies by artificial muscles more suited to human-centred robotics. It is natural to apply biomimetic control strategies to the control of these actuators. In this paper a cerebellar-inspired controller is successfully applied to the real-time control of a dielectric electroactive actuator. To analyse the performance of the algorithm in detail we identified a time-varying plant model which accurately described actuator properties over the length of the experiment. Using synthetic data generated by this model we compared the performance of the cerebellar-inspired controller with that of a conventional adaptive control scheme (filtered-x LMS). Both the cerebellar and conventional algorithms were able to control displacement for short periods, however the cerebellar-inspired algorithm significantly outperformed the conventional algorithm over longer duration runs where actuator characteristics changed significantly. This work confirms the promise of biomimetic control strategies for soft-robotics applications.

1 Introduction

The scalability, high mechanical compliance, structural simplicity, high versatility, proprioceptive feedback and large strain capability of biological muscle make it an excellent multi-purpose actuator [1]. As a result, there has been a major effort to develop actuator technologies capable of mimicking the performance of biological muscle [2]. Such artificial muscle actuators would be ideal for use in robots working in unstructured environments and for applications where safe human-robot interaction is required [3].

Electroactive polymers (EAPs) form a class of materials that are capable of undergoing large deformations in response to suitable electrical stimuli. They possess many of the desirable properties of biological muscle [4, 5]. There are two major classes of EAPs: dielectric and ionic. Here we consider the control of dielectric electroactive polymers (DEAs). DEAs can be used to manufacture compliant actuators with high energy density, large strain capability, a relatively fast response, and which have the capacity for self sensing [6–8].

N.F. Lepora et al. (Eds.): Living Machines 2013, LNAI 8064, pp. 311–322, 2013.

Current research into EAPs has focused on the development of new actuator configurations and materials to provide better actuation properties [9, 10]. Some effort has been made to investigate control strategies for ionic EAPs using adaptive inverse control [11, 12], model reference adaptive control [13], and PID feedback control [14]. There has been less focus on the control of DEAs, and much of what there is has used non-adaptive schemes such as PID controllers [6, 10]. In fact DEAs present a number of interesting control challenges, for example they are manufactured with wide tolerances, and are subject to creep and time related ageing [6]. Although these characteristics of EAPs would seem to be well-suited to adaptive control methods there has been a gap in the development suitable techniques.

In this investigation we address the adaptive control problem for a DEA using a bioinspired approach. Although EAPs are commonly termed artificial muscles, as far as we are aware bioinspired adaptive control schemes have not yet been applied to EAPs. This is in spite of the clear similarity between EAPs and muscle in many of their control challenges [4]. Taking inspiration from the neuromuscular control system would seem to be a natural strategy since it has evolved to be well suited to adapting and tuning the control of a wide range of motor behaviours in unstructured, changing environments [3].

In this contribution we apply a bio-inspired framework based on the known properties of the cerebellum to the adaptive control of a DEA actuator. The cerebellum is a region of the brain strongly associated with adaptive control and skilled movement [20]. The basic cerebellar microcircuit has similar structure to an adaptive filter, which has been well established over numerous investigations [17–21]. Furthermore, cerebellar based control algorithms have previously been successfully applied to other robotic control applications [22–25]. In order to evaluate the performance of the cerebellar algorithm we benchmark it against a conventional adaptive control scheme (filtered-x LMS – FXLMS) [15] which has previously been applied to vibration isolation using a DEA actuator.

The paper is organised as follows. The control algorithms are described in Section 2 and experiment details are given in Section 3. Section 4 presents the results from experiments. These results include real time control of the DEA behaviour and the comparison with the performance of a conventional controller using a simulated DEA plant. Finally, the results are discussed in Section 5.

2 Adaptive Control Algorithms

A bio-inspired control scheme, based on the adaptive filter model of the cerebellum, will be compared to the FXLMS algorithm. The FXLMS algorithm was chosen as a benchmark because of its previous application to the adaptive control of a DEA [15]. The bio-inspired scheme uses an architecture originally studied in the context of the vestibular ocular reflex (VOR) [26]. This section describes the two algorithms, both of which are adaptive control schemes that use LMS to adjust the weights of an adaptive filter.

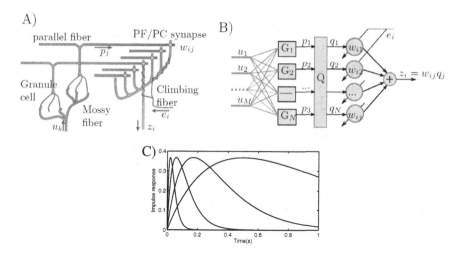

Fig. 1. A) Schematic of the basic cerebellar microcircuit showing granule cells, parallel fibres and a Purkinje cell (PC). B) Interpretation of the cerebellar microcircuit as an adaptive filter (including a Q-matrix optimisation in the granular layer to enable fast learning). Processing by the granule cells is modelled as a filter $p_j = G_j[u_1, u_2...]$, followed by signal decorrelation into signals q_j using a matrix Q. The Purkinje cell output takes a weighted sum of the inputs, q_j, where the filter weights are adjusted using the covariance learning rule. C) Normalised impulse responses of the four log-spaced α-function basis filters.

2.1   Cerebellar Algorithm

The cerebellar microcircuit can be modelled as an adaptive filter [17, 21], as illustrated in Figure 1. In the adaptive filter model of the cerebellum, initial processing in the granule cell layer is modelled as a bank of basis filters. In this contribution we model these basis as α-function filters, with Laplace transforms

$$G_k(s) = \frac{1}{(T_k s + 1)^2} \tag{1}$$

where T_k is the filter time constant. To control the DEA we use four log-spaced time constants between 0.02s and 0.5s (see Figure 1C). A constant filter implementing a bias term is also included. This α-basis replaces the tapped delay lines (TDLs) commonly used in engineering applications. This basis set provides a biologically plausible way of representing timing in which signals with larger delay are more dispersed in time, allowing a compact representation of phenomena on both fast and slow time-scales. It has been shown to greatly reduce the number of adaptable parameters required in comparison to using conventional TDLs [27, 28].

The filter outputs p_j are highly correlated, therefore a matrix Q is used to decorrelate these signals and speed up learning [29, 30], giving the signals q_j which are weighted and summed to give the cerebellar output. The weights w_{ij}

A)   B)

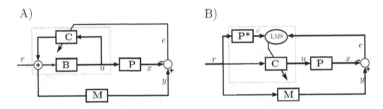

Fig. 2. The two adaptive control schemes (each controller is contained in a dotted box) used to learn to drive the output (x) of the DEA plant P to match a desired trajectory y. A) The cerebellar control architecture. The motor command is generated by a fixed element B, and an adaptive element C in a recurrent loop. B) Filtered-x LMS control. The motor command is generated by an adaptive filter C. The algorithm requires a model $P*$ of the plant P to provide a suitable teaching signal.

are learnt using the covariance learning rule $\delta w_{ij} = -\beta e_i q_j$, where e_i is the difference between the desired and actual output at each step (i) and β is the learning rate.

The cerebellar control scheme is shown in the flow diagram in Figure 2A where the cerebellar filter C is embedded in a recurrent loop. Recurrent connectivity is a characteristic feature of the cerebellum [34], possibly because in this architecture output error is a suitable training signal and no prior model of the inverse plant is required [31]. The brainstem B is an approximate fixed feedforward controller for the plant P. A reference model M is included in the control scheme (an extension to the original computational model of the VOR [26]) so that exact compensation is achieved when $B = MP^{-1}$. The behaviour of the controlled plant then matches that of the reference model M which specifies a realistic response for the controlled plant; the use of a reference model also ensures that the estimated controller is proper [32]. In this contribution the reference model used is defined in continuous time as

$$M = \frac{1}{(\tau s + 1)^n} \tag{2}$$

where $\tau = 0.1$s, and n is the difference between the number of plant poles and zeros.

2.2   Filtered X-LMS Algorithm

The FXLMS algorithm is a feedforward adaptive filter control algorithm widely used for active control applications [16, 33]. A schematic of the control scheme using FXLMS is given in Figure 2B. For the adaptive filter controller C, we used the same bank of α-function filters and bias term as the cerebellar adaptive filter (described above). The filter outputs were again passed through a Q-matrix to decorrelate the signals.

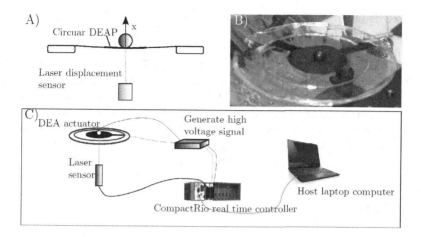

Fig. 3. Summary of the experimental set up. A) A laser displacement sensor is used to measure the vertical displacement of a mass on a circular DEA. B) DEA actuator stretched on circular perspex frame supporting spherical load. C) Summary of experimental real-time control connectivity.

3   Experimental Details

This section details the experimental set-up used to test one degree-of-freedom control of a DEA. Our DEA actuator is comprised of a thin passive elastomer film, sandwiched between two compliant electrodes. In response to an applied voltage the electrodes squeeze the film in the thickness direction, resulting in biaxial expansion. In order to constrain the controlled variable to one degree of freedom a spherical load was placed at the centre of a circular DEA and its motion in the vertical plane (i.e. vertical displacement) was measured. A summary of the experimental setup is provided in Figure 3.

The DEAs are constituted of acrylic elastomer (3M VHB 4905) with an initial thickness of 0.5mm. A conductive layer of carbon grease (MG chemicals) constitutes the capacitor plates. The elastomer was pre-stretched biaxially by 350% (where 100% was the unstretched length) prior to being fixed on a rigid perspex frame with inner and outer diameters of 80mm and 120mm respectively. The electrodes were brushed on both sides of the VHB membrane as circles with a diameter of approximately 35mm. The load used during experiments was a sphere weighing 3g. The control hardware was a CompactRio (CRIO-9014, National Instruments) platform, with input module NI-9144 (National Instruments) and output module NI-9264 (National Instruments) used in combination with a host laptop computer. LabView was run on the host laptop computer, with communication between the host laptop and CompactRio (CRio) carried out using the LabView shared variable engine. This real-time setup enabled accurate timing of control inputs and outputs with input and output signals being sampled simultaneously at 50Hz.

A laser displacement sensor (Keyence LK-G152, repeatability - 0.02mm) was used to measure the vertical movement of the mass sitting on a circular DEA. This signal was supplied to the input module of the CRio. From the output module of the CRio, voltages in the range 1.1V-3.75V were passed through a potentiometer (HA-151A HD Hokuto Denko) and amplified (EMCO F-121 high voltage module) with a ratio of 15V:12kV and applied to the DEA. During the real-time experiments the cerebellar controller (described in the previous section) was used to adjust the voltage input to the DEA to track a desired coloured-noise reference signal. The reference signal was low pass filtered white noise with frequency range 0-1Hz, and the amplitude constrained to 0.2-0.65.

4   Results

This section presents the results for measurement and control of the vertical displacement of the DEA. The results are divided into the following sub-sections: real-time control of the DEA, identification of a model of the plant input-output response, and comparison in simulation of the biomimetic and conventional algorithms.

4.1   Real-Time Control of DEA Using the Biomimetic Algorithm

In this experiment the cerebellar algorithm was applied to real-time control of the DEA actuator. In the experiments described here the reference signal was constrained to lie in the linear range of the plant. The brainstem, B was an approximate feedforward controller, modelled as in Eq. (6) using $0.087, 0.331, -0.317$ for the identified parameters a_0, b_0, c_0 respectively (see below for details of model identification). Initial cerebellar weights were set to zero.

The algorithm was tested for its ability to learn the dynamics in order to track the vertical displacement of the DEA. Figure 4 shows the results from a displacement tracking run. The algorithm quickly learnt the filter weights which allowed accurate tracking and tracking quality remained stable over the length of the experiment. The learning rate was chosen to give robust and stable learning on a time-scale which allowed tracking of variations in model parameters.

4.2   Identifying a Plant Model of the Actuator

A model of the DEA plant was identified to allow detailed comparison of the biomimetic algorithm with a conventional algorithm. This allowed us to use exactly the same data set for both algorithms in control simulations. The fitted model also allowed us to quantify the time dependence of actuator properties. The dynamics of the DEA, and their change over-time, were measured by recording the displacement response (Fig. 5B) to an applied voltage (Fig. 5A) over a 30 minute period. A non-linear dynamic model, with Hammerstein structure, of the response was estimated as

$$a\dot{x} + x = \begin{cases} bu + c, & \text{if } u < e \\ bu + c + d(u - e)^2, & \text{otherwise} \end{cases} \qquad (3)$$

where x is the vertical displacement of the EAP, u the voltage input (prior to amplification), and a, b, c, d, and e are the model parameters. These parameters were estimated by fitting to 1 minute of data. An example of the resulting fit is given in Figure 5C. The response characteristics changed over time so model parameters were estimated by fitting to 1min segments of data every 3mins over the 30mins of data. These parameter estimates and a linear least squares fit to their changes over time are shown in Figure 5D. The suitability of this identified model for investigating control performance in simulation is illustrated in Figure 4, where the performance of the cerebellar algorithm in the real-time control situation and in simulation are shown to be very similar.

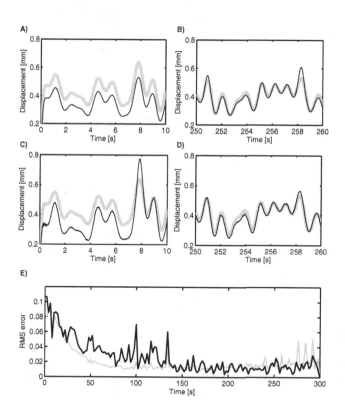

Fig. 4. Experimental adaptive control of DEA using cerebellar-inspired control. Cerebellar weights are initially set to zero and the brainstem is an approximate fixed compensator. A) Desired (__) and actual (__) response of simulated plant before learning. B) Desired (__) and actual (__) response of simulated plant after learning. C) Desired (__) and actual (__) response of experimental plant before learning. D) Desired (__) and actual (__) response of experimental plant after learning. E) Windowed rms errors during learning, for simulated (__)and experimental (__) plants.

318   E.D. Wilson et al.

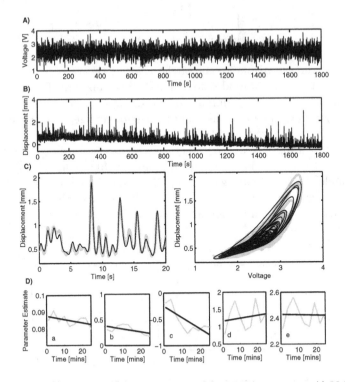

Fig. 5. Dynamic response and dynamic modelling of DEA actuator. A) Voltage input prior to amplification. The voltage input is 0-1Hz band limited white noise, constrained to range from 1.1-3.75V, this is passed through a high voltage amplifier with ratio of 15V:12kV and applied to the DEA. B) Vertical displacement response of a mass on a circular DEA during 30 minutes of actuation. C) Measured (___) and modelled (—) vertical displacement, fitting to 60 seconds of data, modelled using Eq. (3). D) Parameters of model given in Eq. (3) (___) estimated for the 30mins of data. Parameter estimates were updated every 3mins by fitting to 1min of data. The least squares linear fit to the change is parameters over time (—) is also plotted.

4.3   Control of Simulated DEA Plant

The cerebellar control algorithm was compared to the FXLMS algorithm by applying both to the control of the simulated plant described by the model given in Equation (3), with time varying model parameters

$$a(t) = 0.085$$
$$b(t) = 0.316 - 0.006t$$
$$c(t) = -0.193 - 0.020t \qquad (4)$$
$$d(t) = 0.784 + 0.008t$$
$$e(t) = 2.316$$

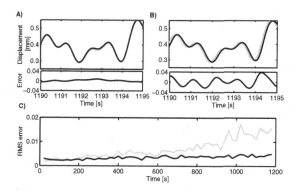

Fig. 6. Simulated adaptive control of EAP actuator using FXLMS and cerebellar-inspired control. A simulated plant based on the model in Eq. (3), with coefficients that vary slowly over time is tracked for 20mins. A) The desired (—) and actual (—) displacement response using the cerebellar algorithm, with associated error plotted underneath. B) The desired (—) and actual (—) displacement response using the FXLMS algorithm, with associated error plotted underneath. C) Comparison of windowed RMS errors in displacement for the FXLMS algorithm (—), and the Cerebellar algorithm (—).

where t is the time in minutes. Initial values are estimated as the fit to the first minute of data, and the gradients in Equations (4) come from the straight line least squares estimates presented in Figure 5D.

The actuator model has nonlinear dynamics (Figure 5C left), so to test the linear cerebellar and FXLMS algorithms described here the reference input was constrained to a range where the plant response was approximately linear. The brainstem B and estimated plant P^* for the cerebellar and FXLMS algorithms respectively were set to the ideal values for a linear approximation of the initial plant model. The input e to output v model describing P^* is

$$a_0\dot{v} + v = b_0 e + c_0 \tag{5}$$

The input (v) to output (u) model describing $\mathbf{B}$ is

$$\tau\dot{u}_t + u_t = a_0\dot{v} + v \tag{6}$$

$$u = (u_t - c_0)/b_0$$

where a_0, b_0, c_0 are the parameters given in Equations (4) at time t=0, and τ is the time constant of reference model M. The reference model used is given by Equation (2) with $n = 1$. Initial weights of the adaptive controllers were set to provide optimum compensation for the initial plant. Figure 6 shows the results when tracking the simulated plant for 20mins. Both algorithms are able to accurately track the dynamics of the plant initially, however tracking errors build up over time when using the FXLMS algorithm due to variations in the plant parameters.

5   Discussion

As expected the dynamic characteristics of the DEA show considerable change over time. These changes are likely to be due to creep and highlight the need for a control algorithm capable of compensating for secular changes in the plant dynamics. Although both the FXLMS and cerebellar control algorithms were shown to perform effectively when tracking the simulated plant for a short time period, uncompensated tracking errors eventually developed when using the FXLMS algorithm over a longer period (Fig. 6C). The same increase in error is not seen in the bioinspired cerebellar algorithm. The probable cause of the increasing error when using the FXLMS algorithm is that, over time, the estimated plant model P^* used to provide a suitable teaching signal becomes less accurate, which affects the convergence properties of the learning rule. In contrast the cerebellar algorithm is embedded in a recurrent loop in which output error itself is a suitable training signal [26], meaning an accurate estimate of the plant is not required.

In each algorithm a compact basis (with four filters, plus a bias term) was used to represent a wide range of time scales. This basis provides a biologically plausible way of representing timing. Using a set of basis functions to replace TDLs can also greatly reduce the number of model parameters. In engineering applications Laguerre, or Kautz functions are often used to replace TDLs. They form an orthonormal basis for white noise (as do TDLs), and are insensitive to the sampling frequency (unlike TDLs) [27].

In this contribution we have focused on the control of a DEA over its linear operating range. The cerebellar algorithm could be extended to the control of non-linear systems by including an approximate plant linearization stage in the brainstem, or by using a non-linear adaptive filter which includes non-linear basis, or combining the two schemes.

The presented control task is used to provide a test for the control algorithms. The cerebellar algorithm described here can be applied to generic control tasks. For example, it could be used to control the impedance, or force response. In applications to robotic systems it is likely that the movement of multiple actuators will be coupled (perhaps including agonist-antagonist configurations). It has been show multiple actuators can produce useful behaviours [9]. The recurrent cerebellar control architecture presented here has previously been shown to be well adapted to the control of such multi degree of freedom systems [19].

6   Conclusions

In this investigation we have developed an adaptive control algorithm for DEA actuators inspired by the function of the cerebellum. The adaptive cerebellar-inspired algorithm, in combination with a simple fixed controller, learns to represent the inverse dynamics of the DEA actuator and therefore forms a feed-forward inverse plant compensation control scheme. We have demonstrated the suitability of the cerebellar-inspired algorithm for real-time tracking control of a DEA actuator over its linear range of operation. The cerebellar-inspired control algorithm successfully adapted to time-variation in the dynamics of the DEA actuator. We further benchmarked the performance of the cerebellar inspired algorithm in simulation, using an identified nonlinear dynamic model of the DEA actuator, against a conventional adaptive control scheme (filtered-XLMS). The cerebellar-inspired scheme was shown to outperform filtered-XLMS over long time periods of operation.

Acknowledgements: This work was supported by EPSRC grant no. EP/IO32533/1, *Bioinspired Control Of Electro-Active Polymers for Next Generation Soft Robots.*

References

1. Carpi, F., Kornbluh, R., Sommer-Larsen, P., Alici, G.: Electroactive polymer actuators as artificial muscles: are they ready for bioinspired applications? Bioinspir. Biomim. 6(4), 045006 (2011)
2. Carpi, F., Raspopovic, S., Frediani, G., De Rossi, D.: Real-time control of dielectric elastomer actuators via bioelectric and biomechanical signals. Polym. Int. 59(3), 422–429 (2009)
3. Van Ham, R., Sugar, T.G., Vanderborght, B., Hollander, K.W., Lefeber, D.: Review of Actuators with Passive Adjustable Compliance/Controllable Stiffness for Robotic Applications. IEEE Robot. Autom. Mag., 81–94 (2009)
4. Bar-Cohen, Y.: Electroactive polymer (EAP) actuators as artificial muscles: reality, potential, and challenges. SPIE Press (2001)
5. Meijer, K., Rosenthal, M.S., Full, R.J.: Muscle-like actuators? A comparison between three electroactive polymers. In: Proc. SPIE, vol. 4329, pp. 7–15 (2001)
6. Xie, S., Ramson, P., Graaf, D., Calius, E., Anderson, I.: An Adaptive Control System for Dielectric Elastomers. In: 2005 IEEE International Conference on Industrial Technology, pp. 335–340 (2005)
7. Pelrine, R., Kornbluh, R.D., Pei, Q., Stanford, S., Oh, S., Eckerle, J., Full, R.J., Rosenthal, M.A., Meijer, K.: Dielectric elastomer artificial muscle actuators: toward biomimetic motion. In: Proc. SPIE, vol. 4695, pp. 126–137 (2002)
8. OHalloran, A., OMalley, F., McHugh, P.: A review on dielectric elastomer actuators, technology, applications, and challenges. J. Appl. Phys. 104(7), 071101 (2008)
9. Conn, A.T., Rossiter, J.: Towards holonomic electro-elastomer actuators with six degrees of freedom. Smart Mater. Struct. 21(3), 035012 (2012)
10. Ozsecen, M.Y., Mavroidis, C.: Nonlinear force control of dielectric electroactive polymer actuators. In: Proc. SPIE, vol. 7642(1) (2010)
11. Hao, L., Li, Z.: Modeling and adaptive inverse control of hysteresis and creep in ionic polymer metal composite actuators. Smart Mater. Struct. 19(2), 025014 (2010)
12. Dong, R., Tan, X.: Modeling and open-loop control of IPMC actuators under changing ambient temperature. Smart Mater. Struct. 21(6), 065014 (2012)
13. Brufau-Penella, J., Tsiakmakis, K., Laopoulos, T., Puig-Vidal, M.: Model reference adaptive control for an ionic polymer metal composite in underwater applications. Smart Mater. Struct. 17(4), 045020 (2008)
14. Yun, K., Kim, W.J.: Microscale position control of an electroactive polymer using an anti-windup scheme. Smart Mater. Struct. 15(4), 924–930 (2006)

15. Sarban, R., Jones, R.W.: Physical model-based active vibration control using a dielectric elastomer actuator. J. Intel. Mat. Syst. Str. 23(4), 473–483 (2012)
16. Widrow, B., Walach, E.: Adaptive Inverse Control A Signal Processing Approach. Reissue edn. John Wiley & Sons, Inc. (2008)
17. Dean, P., Porrill, J., Ekerot, C.F., Jörntell, H.: The cerebellar microcircuit as an adaptive filter: experimental and computational evidence. Nat. Rev. Neurosci. 11(1), 30–43 (2010)
18. Porrill, J., Dean, P., Anderson, S. R.: Adaptive filters and internal models: Multilevel description of cerebellar function. Neural Networks (December 28, 2012), http://dx.doi.org/10.1016/j.neunet.2012.12.005
19. Porrill, J., Dean, P.: Recurrent cerebellar loops simplify adaptive control of redundant and nonlinear motor systems. Neural Computation 19(1), 170–193 (2007)
20. Ito, M.: The Cerebellum and Neural Control New York, Raven (1984)
21. Fujita, M.: Adaptive Filter Model of the Cerebellum. Biol. Cybern. 206, 195–206 (1982)
22. Lenz, A., Anderson, S.R., Pipe, A.G., Melhuish, C., Dean, P., Porrill, J.: Cerebellar-inspired adaptive control of a robot eye actuated by pneumatic artificial muscles. IEEE T. Syst. Man. Cy. B 39(6), 1420–1422 (2009)
23. Miller III, W.T.: Real-Time Application of Neural Networks for Sensor-Based Control of Robots with Vision. IEEE T. Syst. Man. Cyb. 19(4), 825–831 (1989)
24. Spoelstra, J., Arbib, A.A., Schweighofer, N.: Cerebellar adpative control of a biomimetic manipulator. Neurocomputing 26-27, 881–889 (1999)
25. Smagt, P.: van der: Cerebellar control of robot arms. Connection Science 10, 301–320 (1998)
26. Dean, P., Porrill, J., Stone, J.V.: Decorrelation control by the cerebellum achieves oculomotor plant compensation in simulated vestibulo-ocular reflex. Proc. R. Soc. B 269(1503), 1895–1904 (2002)
27. Anderson, S.R., Pearson, M.J., Pipe, A.G., Prescott, T.J., Dean, P., Porrill, J.: Adaptive Cancelation of Self-Generated Sensory Signals in a Whisking Robot. IEEE T. Robot. 26(6), 1065–1076 (2010)
28. Ljung, L.: System Identification - Theory for the User, 2nd edn. Prentice Hall, Upper Saddle River (1999)
29. Schweighofer, N., Doya, K., Lay, F.: Unsupervised Learning of Granule Cell Sparse Codes Enhances Cerebellar Adaptive Control. Neuroscience 103(1), 35–50 (2001)
30. Coenen, O.J.D., Arnold, M.P., Sejnowski, T.J.: Parallel Fiber Coding in the Cerebellum for Life-Long Learning. Auton. Robot. 11, 291–297 (2001)
31. Porrill, J., Dean, P.: Recurrent cerebellar loops simplify adaptive control of redundant and nonlinear motor systems. Neural Comput. 19(1), 170–193 (2007)
32. Sastry, S., Bodson, M.: Adaptive Control Stability, Convergence and Robustness. Prentice Hall, Englewood Cliffs (1989)
33. Elliott, S.J., Nelson, P.A.: Active noise control. IEEE Signal Proc. Mag, 12–35 (1993)
34. Kelly, R.M., Strick, P.L.: Cerebellar loops with motor cortex and prefrontal cortex of a nonhuman primate. Journal of Neuroscience 23(23), 8432–8444 (2003)

TACTIP - Tactile Fingertip Device, Texture Analysis through Optical Tracking of Skin Features

Benjamin Winstone[1], Gareth Griffiths[1], Tony Pipe[1],
Chris Melhuish[1], and Jonathon Rossiter[2]

[1] Bristol Robotics Laboratory, University of the West of England
Benjamin.Winstone@brl.ac.uk
[2] Engineering Mathematics, University of Bristol
Jonathan.Rossiter@bristol.ac.uk

Abstract. In this paper we present texture analysis results for TACTIP, a versatile tactile sensor and artificial fingertip which exploits compliant materials and optical tracking. In comparison to previous MEMS sensors, the TACTIP device is especially suited to tasks for which humans use their fingertips; examples include object manipulation, contact sensing, pressure sensing and shear force detection. This is achieved whilst maintaining a high level of robustness. Previous development of the TACTIP device has proven the device's capability to measure force interaction and identify shape through edge detection. Here we present experimental results which confirm the ability to also identify textures. This is achieved by measuring the vibration of the in-built human-like skin features in relation to textured profiles. Modifications to the mechanical design of the TACTIP are explored to increase the sensitivity to finer textured profiles. The results show that a contoured outer skin, similar to a finger print, increases the sensitivity of the device.

1 Introduction

TACTIP is a biologically-inspired sensing device, based upon the deformation of the epidermal layers of the human skin. Deformation from device-object interaction is measured optically by tracking the movement of internal papillae pins on the inside of the device skin. These papillae pins are representative of the intermediate epidermal ridges of the skin, whose static and dynamic displacement are normally detected through the skin's mechanoreceptors, see Fig.1.

In the work presented in [1, 2, 3, 4] they presented the first oversize prototype TACTIP device as a 40mm diameter probe. The mechanics are inspired by the human fingertip but at more than twice the size. In more recent work we have miniaturised the size of the TACTIP design closer to the size of a human fingertip (20mm) in order to integrate it with a robotic hand, [5]. Our work in [5] explores comparison between the two sizes of TACTIP. The TACTIP device lends itself well to human-like fingertip applications due to its similarity and range of sensing capabilities. There is growing interest due to the need to place

N.F. Lepora et al. (Eds.): Living Machines 2013, LNAI 8064, pp. 323–334, 2013.

robots in the workplace along side humans. For robots to function effectively in these environments, they need to be able to perform the same complex functions of sensing and manipulation as humans [6]. Traditional industrial grippers equipped with contact and pressure sensors are not capable of the diverse tasks humans undertake, where objects have different sizes and require a wide range of pressures and grips. A compliant sensing gripper, such as the human fingertip, is far more attractive.

2   Texture Sensation

Texture recognition is a prominent and sophisticated characteristic of the human fingertip, useful in distinguishing objects, particularly when visual information is not available [7]. Tactile information is communicated through the human fingertip via the four mechanoreceptors embedded within glabrous skin: Pacinian corpuscles (fast adaption), Meissner's corpuscles (moderate adaption), Merkel's discs (slow adaption), and Ruffini endings (slow adaption).

Recent research has suggested that at least two encoding methods for roughness of texture are employed by the body [8]. These separate textures into those with elements less than $200\mu m$ in size and those with elements greater than $200\mu m$. These encodings are dependent on signals from different mechanoreceptors, where finer vibro-tactile surfaces activate the fast adaption mechanoreceptor, the Pacinian corpuscle, and more coarse surfaces activate the moderate adaption Meissner's corpuscle. Stimulation of the Pacinian corpuscle occurs through vibration where optimal sensitivity is 250 Hz, whilst stimulation of the Meissner's corpuscle occurs at 50 Hz and below, [9]. The Pacinian corpuscle is situated deep within the dermis region of the skin on the boundary with the subcutis region, as shown in Fig.1, whilst the Meissner's corpuscle is located in glabrous skin just beneath the epidermis within the dermal papillae.

There have been a number of studies focussing on discriminating texture information with various sensor technologies. In [10] an electret microphone was

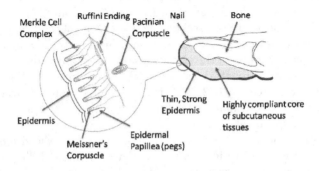

Fig. 1. Cross section of human finger

integrated with a silicone moulded finger-shaped probe to measure surface texture. That approach measures vibro-tactile stimulation of the finger, mirroring the role of the fast Pacinian corpuscle, which is stimulated by vibration. In [11] and [12] active whisking, inspired by rats, shrews and other whiskered animals, to obtain tactile texture information. That approach mirrors the role of the moderate speed Meissner's corpuscle or hair follicle receptor. The work in [13] also mimics the role of the Meissner's corpuscle through the use of strain gauges embedded close to the surface of a silicone moulded robot finger.

3 Device Description

[1] introduced the TACTIP device as a biologically-inspired sensing device, based around the deformation of the epidermal layers of the human skin. In particular the relationship between the papillae ridges/pins that form between the dermis and epidermis regions of the skin and the Pacinian corpuscle and Meissner's corpuscle mechanoreceptors. Similar approaches to replication of a natural fingertip sensor can be seen in [14] and a comparison between soft and rigid sensors for sensitive manipulation are given in [15]. The TACTIP device works by visual observation of biomimetic sub-dermal structures which replicate the papillae in human skin. In this scenario the embedded camera adopts the role of the mechanoreceptor, being activated by movement in the papillae pins. The nature of the device provides an enclosed and constrained image processing task that can identify a range of contextual data, including pressure force, shear force, edge detection and shape detection. As a deformable contact device, the TACTIP shares characteristics with the human fingertip that make it not only a sensing device but also an efficient gripping device. With greater applied pressure, the contact surface area naturally increases, providing a varying level of friction, aiding in gripping. This separates the TACTIP device from conventional rigid sensing devices.

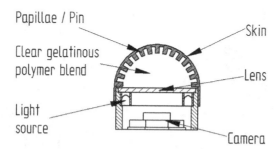

Fig. 2. Cross section of device design

Fig.2 shows the main components that make up the TACTIP device; a silicone outer skin, a compliant optically clear gelatinous polymer inner membrane (similar to the consistency of human flesh), a transparent window to contain the membrane, a light source and a camera. The silicone outer skin is matt black in colour, and the tips of the internal papillae pins are painted white, providing a high contrast image. The camera looks through the window to monitor the skin deformation through movement and interaction of the papillae pins. Optical-based approaches have been utilised in other sensors including [16] and [17] where both monitor two dimensional patterns printed on a skin surface. The novelty of the TACTIP resides in the raised three dimensional papillae pins and optical tracking of their free movement. These raised papillae are a mechanical representation of the biological epidermal papillae shown in Fig.1. Fig.3a shows the original oversized TACTIP device next to the new small TACTIP device, along side their associated moulds. The moulds clearly show the difference in pin arrangement and pin density between the two sizes of sensor.

4 Previous Work

The potential of the TACTIP device to measure force interaction through pin displacement and also to characterise contact shape was presented in [1]. For example, edge detection is possible using a simple image processing approach for blob detection [4]. When contact with the device is made, at the point of contact, the adjacent pins separate in relation to the skin deformation, revealing a greater area of the black background. This large region of background is representative of an edge.

Further applications with this device have been presented in [3] within a haptic tactile feedback system where the TACTIP assumes the role of a remote tactile sensor in the context of a surgery robot. Deformation of the TACTIP skin is translated to linear movement of sixteen linear actuators that make up a haptic feedback device attached to the user's fingertip.

Current research on this device focuses on integration with a robot hand, as shown in Fig.3b. The device's oversized sensor design in [1, 2, 4, 3] is unsuitable

(a) Photograph of the large 40mm and small 20mm TACTIP devices next to their corresponding casting moulds

(b) TACTIP attached to ELU-2 Elumotion hand

Fig. 3. Photographs of the TACTIP device

for use on a humanoid robot hand due to its large size (40mm diameter, see Fig.3a). In response a smaller, 20mm diameter, TACTIP sensor was fabricated [5]. This is much closer to the size of the human fingertip (16-22mm [18]).

The TACTIP uses a standard manual focus web camera. For fair comparison between different designs of the TACTIP, images taken from the camera are cropped to the region of interest and scaled to 100 pixel by 100 pixels. This enables a consistent benchmark comparison between different sensor configurations. For experiments presented in this paper, the image processing task initially measures pixel change with respect to a rest state where the device is not in contact with any object.

Fig.4a presents a raw image from the smaller 20mm device. The image is initially converted to grayscale then binary thresholded. This creates the image shown in Fig.4b. A region of interest is then extracted around the linear path of any chosen pin. This is shown to the right of Fig.4b where a pin is shown moving linearly up and down. The distance of pixel movement is related to applied pressure or shear force at the skin surface.

In the next section we will describe experiments and give results for texture analysis using TACTIP.

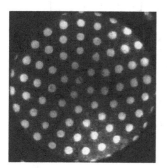

(a) Example raw image from the 20mm TACTIP device

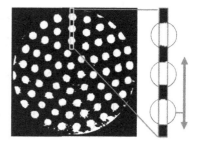

(b) Demonstration of an extracted region of interest

Fig. 4. Internal image of TACTIP before and after conditioning

5   Experiment Design

In order to evaluate the capabilities of TACTIP a series of textures were exposed to the sensor and a texture analysis system was developed. This approach to texture sensing is based upon monitoring the internal papillae vibratory movement during interaction with artificial textured surfaces. By demonstrating that the TACTIP is capable of texture discrimination we will add a fourth sensing capability to the device in addition to pressure, shear and edge detection.

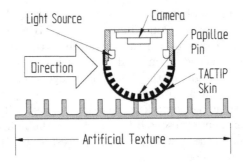

Fig. 5. Artificial texture plate

A series of artificial surfaces were designed as shown in Fig.5 and Fig.6. Here a series of square or triangular ridges with a range of pitches were fabricated using a Stratasys Titan fused deposition modelling 3D printer.

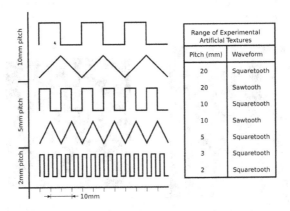

Fig. 6. Profile of artificial textures

The ability of TACTIP to distinguish edge types aids in characterisation of both surface shape and texture. The smaller 20mm TACTIP now used in all experiments presented below, is mounted on a 6-axis industrial arm. The arm moves the device along each surface at one of three velocities; 10mm/s, 20mm/s and 30mm/s. Given that the Pacinian corpuscle is stimulated at around 250 Hz and a conventional web camera cannot operate at more than 60Hz, an alternative imaging solution is required. For these texture experiments an alternative CASIO EX-ZR200 high speed camera has been used with an image sensor which records at 1000 fps at an image size of 224 x 64 pixels. This is the raw sample rate for any data recorded in these experiments. Using a 1KHz sample rate provides sufficient opportunity to capture papillae movement at multiple stages of a vibration cycle. The nature of high speed recording means the exposure time of the camera is greatly reduced. To compensate for a shortened exposure time, an increase in environmental lighting is required so that the subject is visible in each frame. This is addressed by using six super bright (3200 mcd) LEDs integrated in the assembly around the internal circumference of the device.

In the typical configuration of TACTIP a gelatinous polymer is used between the device skin and image sensor. For this experiment the polymer has been removed in order to isolate papillae movement from the polymer dynamics. However, future work would involve tests with the polymer. To further constrain the texture recognition task the algorithm only tracks one papillae pin as shown in Fig.7. It is expected that when the TACTIP is moved across a textured surface, pins at the focal point of the area of contact and some neighbouring pins will move or vibrate in relation to the textured roughness. As with previous development of the TACTIP, [2, 3, 5], the image is preprocessed with a greyscale conversion and then thresholding. From this state, the central papillae pin is easily identified against its black surrounding environment.

Fig. 7. Left; Profile of TACTIP with fingerprint, Middle: TACTIP mounted on to 6-axis arm, being pushed along artificial textures. Right; internal view of TACTIP presented to camera showing one papillae pin painted with white marker.

The first generation oversized TACTIP, and initial miniaturised TACTIP, employ a smooth outer skin surface. This differs from a human fingertip in which the fingerprint is integral to the skin structure and sensing mechanisms. [19] investigated the benefit of fingerprint-like surface ridges on a tactile array for detecting surface shape with promising results. [20] proposed that fingerprint ridges may enhance tactile sensitivity by magnifying subsurface strains. To investigate this characteristic a fingerprint design has been incorporated into the skin of the new miniaturised TACTIP. This fingerprint is embodied as raised bumps positioned directly over the papillae pins, on the outer skin surface. It is expected that these bumps will act as amplifiers to the movement of the papillae pins. The experiments undertaken involved moving the TACTIP over textured surfaces with both a smooth TACTIP with no fingerprint, and a TACTIP with fingerprint. Results show a comparison of performance, highlighting the benefits of a fingerprint in texture recognition.

6   Results of Experiments

6.1   Without Fingerprint

Fig.8 and Fig.9 present example wave-forms produced from tracking the central papillae pin whilst the TACTIP is moved over each textured profile. These figures show comparison between a TACTIP without fingerprint and TACTIP with fingerprint. With wider pitch textures (5mm or greater) the profiles of the textures have clearly been captured, see Fig.8. However a finer pitch becomes undetectable with a smooth-skinned TACTIP. This is shown most clearly in Fig.9 where the TACTIP with no fingerprint shows minimal papillae pin movement.

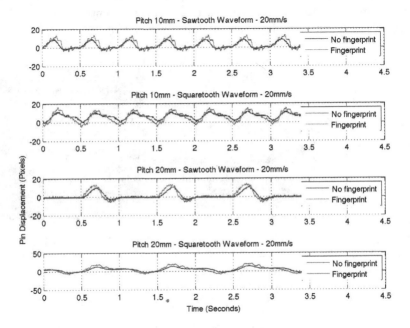

Fig. 8. Example waveforms sensed with the TACTIP moved over artificial textures with pitch from 20mm spacing to 10mm, with both squaretooth and sawtooth profiles. TACTIP with and without fingerprint.

6.2   With Fingerprint

Actually, both Fig.8 and Fig.9 show that the fingerprint enables much higher resolution texture sensing. Not only do the signals from the larger textures in Fig.8 show much greater detail, but the fingerprinted sensor is clearly effective for the smaller scale textures in Fig.9 where the fingerprint-less TACTIP fails. The bumps that form the fingerprint are approximately 1mm in diameter, which is smaller than the pitch of the artificial test pieces used. More realistic surfaces,

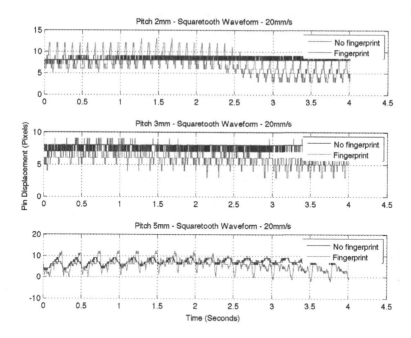

Fig. 9. Example waveforms sensed with the TACTIP moved over artificial textures with pitch from 5mm spacing to 2mm, with both squaretooth and sawtooth profiles. TACTIP with and without fingerprint.

such as rough fabric or sand paper will offer a greater challenge. In these cases higher frequencies and associated harmonics will need to be analysed.

6.3 Frequency Analysis

A practical application of the fingerprinted TACTIP would require a frequency-based classification method capable of operating in real time in order to identify different surface textures. However, in order to ease development, analysis of data is performed offline, after the experiment has been performed. A Fast Fourier Transform can be used to identify dominant frequencies within a waveform such as those shown in Fig.8 and Fig.9. Fig.10 presents examples of FFT results from TACTIP output, in the form of a power spectrum, where dominant frequencies have been identified. Table 1 shows the full results in tabular form. Note that the TACTIP with no fingerprint cannot capture the correct frequencies for textures with pitch under 5mm.

The FFT analysis confirms that the TACTIP is capable of identifying dominant frequencies of clean consistent textured surfaces, however most common

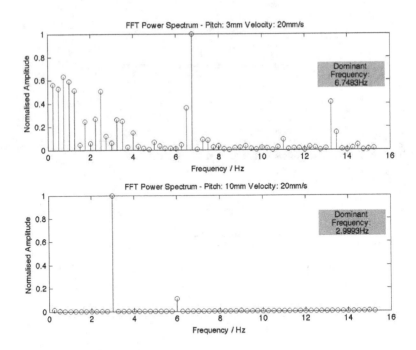

Fig. 10. Examples of FFTs used to identify the dominant frequency sensed through the TACTIP with fingerprint over square waveform artificial texture.

Table 1. FFT identified dominant frequencies (Hz) over range of artificial textures with comparison between use and no use of fingerprint. Highlighted cells indicate errorneous sensor reading.

	Velocity mm/s					
	Without fingerprint			With fingerprint		
Pitch	10	20	30	10	20	30
20mm Sq	0.5	1	1.5	0.5	1	1.5
20mm Sw	0.5	1	1.5	0.5	1	1.5
10mm Sq	1	2	3	1	2	3
10mm Sw	1	2	3	1	2	3
5mm Sq	2	4	6	2	4	6
3mm Sq	0.25	0.25	6.7	3.3	6.6	9.9
2mm Sq	0.25	0.25	0.2	5	10	15

surfaces in a human-centric environment are not clean consistent textures. Common surfaces such as carpet, wood, velcro and foam have much more complex frequency spectra.

7 Conclusion

Here we have presented an exploration of using the TACTIP device to recognise surface texture through monitoring internal papillae pin vibration. We have identified that a relationship exists between papillae pin vibration and the exploration of the contours formed on an object surface. First we have recorded data presented by the TACTIP over time whilst interacting with a series of rough 3D printed surfaces with a smooth artificial skin without fingerprint. This has produced a time based waveform applicable for signal processing algorithms.

The TACTIP skin without fingerprint showed promise but presented limited sensitivity as the test surfaces became less rough, and incapable of reading finer textures. Taking further inspiration from biology and also previous work presented by [19] and [20], an artificial fingerprint was added to the TACTIP skin in an attempt to increase the sensitivity to texture. This addition of a fingerprint increased the sensitivity of the TACTIP such that it was capable of presenting a waveform for each test surface, a significant improvement on the smooth skin TACTIP.

Diversity in sensing of object contact such as pressure, shear, stress, strain and texture provide crucial information for many robotic tasks that involve gripping, manipulation and recognition of objects in a human centric environment. TACTIP is a robust device with a large working range in relation to hand manipulation tasks. Typical robotic hand gripping activities require compliant contact surfaces, the fingertip and palm, coupled with good force sensing capabilities. The TACTIP offers both of these attributes in one device. These results have proven the ability to analyse and textured surfaces, providing the grounding for further development towards a realtime algorithm that could be integrated with the current force sensing algorithms used with the TACTIP.

Acknowledgement. Bristol Robotics Laboratory gratefully acknowledge that this work was supported by the the European Union FP7 project ECHORD - Call 2, www.echord.info, grant number 231143.

References

[1] Chorley, C., Melhuish, C., Pipe, T., Rossiter, J.: Development of a Tactile Sensor Based on Biologically Inspired Edge Encoding. Design (2008)

[2] Chorley, C., Melhuish, C., Pipe, T., Rossiter, J.: Tactile Edge Detection. Sensors, 2593–2598 (2010)

[3] Roke, C., Melhuish, C., Pipe, T., Drury, D., Chorley, C.: Deformation-Based Tactile Feedback Using a Biologically-Inspired Sensor and a Modified Display. Technology, 114–124 (2011)

[4] Assaf, T., Chorley, C., Rossiter, J., Pipe, T., Stefanini, C., Melhuish, C.: Realtime Processing of a Biologically Inspired Tactile Sensor for Edge Following and Shape Recognition. In: TAROS (2010)

[5] Winstone, B., Griffiths, G., Melhuish, C., Pipe, T.: TACTIP - Tactile Fingertip Device, Challenges in reduction of size to ready for robot hand integration. In: ROBIO, pp. 160–166 (2012)

[6] Brooks, R., Aryananda, L., Edsinger, A., Fitzpatrick, P., Kemp, C., Torres-jara, E., Varshavskaya, P., Weber, J.: Sensing and manipulating built-for-human environments. International Journal of Humanoid Robotics, 1–28 (2004)

[7] Cole, J.: Pride and a Daily Marathon. 1 edn. The MIT Press (1995)

[8] Hollins, M.: Somesthetic senses. Annual Review of Psychology 61, 243–271 (2010)

[9] Hollins, M., Bensmaïa, S.J.: The coding of roughness. Canadian Journal of Experimental Psychology 61(3), 184–195 (2007)

[10] Mayol-Cuevas, W.W., Juarez-Guerrero, J., Munoz-Gutierrez, S.: A First Approach to Tactile Texture Recognition. In: SMC. Number 1993, pp. 4246–4250 (1998)

[11] Diamond, M.E., von Heimendahl, M., Arabzadeh, E.: Whisker-mediated texture discrimination. PLoS Biology 6(8), e220 (2008)

[12] Pearson, M.J., Mitchinson, B., Sullivan, J.C., Pipe, A.G., Prescott, T.J.: Biomimetic vibrissal sensing for robots.. Philosophical transactions of the Royal Society of London. Series B, Biological Sciences 366(1581), 3085–3096 (2011)

[13] Gomez, G., Pfeifer, R.: Haptic discrimination of material properties by a robotic hand. In: ICDL, pp. 1–6 (July 2007)

[14] Mukaibo, Y., Shirado, H., Konyo, M., Maeno, T.: Development of a Texture Sensor Emulating the Tissue Structure and Perceptual Mechanism of Human Fingers. In: ICRA. Number, 2576–2581 (April 2005)

[15] Torres-Jara, E.M., Vasilescu, I.M., Coral, R.M., Asilescu, I.: A soft touch: Compliant Tactile Sensors for Sensitive Manipulation (2006)

[16] Lang, P.: Optical Tactile Sensors for Medical Palpation Pencilla Lang. (2004)

[17] Ferrier, N.J., Brockett, R.W., Hristu, D., Cdcss, T.R.: The performance of a deformable-membrane tactile sensor: basic results on geometrically-defined tasks. In: ICRA, vol. 0114, pp. 508–513 (2000)

[18] Dandekar, K., Raju, B.I., Srinivasan, M.A.: 3-D Finite-Element Models of Human and Monkey Fingertips to Investigate the Mechanics of Tactile Sense. Journal of Biomechanical Engineering 125(5), 682 (2003)

[19] Salehi, S., Cabibihan, J.J., Ge, S.S.: Artificial skin ridges enhance local tactile shape discrimination. Sensors 11(9), 8626–8642 (2011)

[20] Scheibert, J., Debregeas, G., Prevost, A.: A MEMS-based tactile sensor to study human digital touch: mechanical transduction of the tactile information and role of fingerprints. In: EPJ Web of Conferences, vol. 6, p. 21006 (June 2010)

Sensory Feedback of a Fish Robot with Tunable Elastic Tail Fin

Marc Ziegler and Rolf Pfeifer

University of Zurich, Department of Informatics,
Artificial Intelligence Laboratory, Switzerland
mziegler@ifi.uzh.ch
http://ailab.ifi.uzh.ch/

Abstract. Many designs that evolved in fish actuation have inspired technical solutions for propulsion and maneuvering in underwater robotics. However, the rich behavioral repertoire and the high adaptivity to a constantly changing environment are still hard targets to reach for artificial systems. In this work, we truly follow the bottom up approach of building intelligent systems capable of exploring their behavioral possibilities when interacting with the environment. The free swimming fish robot Wanda2.0 has just one degree of freedom for actuation, a tail fin with varying elasticity, and various on board sensors. In the data analysis we isolate the minimal set of sensory feedback to distinguish between different swimming patterns and elasticity of the tail fin.

Keywords: biomimetics, locomotion, sensory feedback, elastic tail fin.

1 Introduction

Biologists and engineers are investigating how fish swim and their approaches are manyfold: precise observation in testing environment of a flow tank Digital Particle Image Velocimetry, modeling (Computational Fluid Dynamic), or the synthetic approach, i.e. building artifacts that mimic a particular behavior or motion in the water. Many of the insights from various scientific fields (e.g. [1], [2], [3], [4], [5]) underline the importance of passive elements that improve the hydrodynamic performance. This is the skin property of a shark that influences the boundary layer to reduce friction between water and the surface, the riblets and tubercles on the leading edge of a whale flipper creating small turbulences, and of course the shape and elasticity dispersion of a fish tail fin, just to name a few biological optimizations.

A recent development in underwater vehicles are compliant actuator in fish robot. From experimental biology we know, that flexural stiffness of a fish is altered actively to achieve different task. For example in [6], it has been shown that a live sunfish increases its body stiffness by a factor of two relative to its passive state and that the active stiffening has a positive effect on swimming speed. In [7] and [8] a test platform was developed with an elastic two segment fin capable of changing the stiffness dynamically. It is shown that the optimal

N.F. Lepora et al. (Eds.): Living Machines 2013, LNAI 8064, pp. 335–346, 2013.

stiffness for thrust efficiency varies for different pitching angles and frequencies. Another approach is not only to have a compliant tail fin, but a fully compliant body and tail with a single actuator [9]. Despite the simplicity of the robot, high maneuverability and swimming speed is realized, reinforcing the insight, that simple constructions do perform complex task.

The 4 degrees of freedom fish robot platform developed by [10] and [11] is controlled by a three layer architecture (cognitive, behavior and swim pattern layer). Here, the behaviors and the swimming patterns are manually designed and are not emerging through the interaction with the environment. This is in contrast to our own approach where we showed that rich behavioral diversity can be rooted in the morphology and material properties of one actuator alone and does not have to be explicitly coded [12].

The fish robot Wanda2.0 is not mimicking a particular fish species nor competing against other underwater robots in terms of swimming performance. It was built to help to understand the importance of the morphology in interaction with the environment and how exploring new behaviors are represented in its sensory data. The underlying concept for the design of Wanda2.0 are two connected plates of the same size, and actuated on the axis of symmetry. This idea was already studied in [12]. A clear and simple construction helps to isolate and analyze the passive mechanism which again, has an influence on the behavioral repertoire. In order to break the symmetry, and to get propulsion in the first place, one of the two plates is partly made of flexible foils, whereas the rest of the body is rigid. A more in depth analysis of a tail fin with actively varying elasticity can be found in [13]. All sensors are placed on board to collect data from different swimming patterns.

The collected data was then analyzed in a second step. Here, the goal was to find a minimal set of sensor combinations that can be used to distinguish between different behaviors.

After a technical description of the robot fish Wanda2.0, the 10 different emerging behaviors are explained, followed by the experimental setup for collecting the raw data, the processing and correlation. Finally, we show how a change in behavior is represented in a minimal set of sensory input.

2   Experimental Setup

2.1   Fish Robot

The overall size is 60 cm in length, 23 cm in height and 7 cm at the thickest part, weight in air is 870 grams. The main driving motor is a Hitec HS-5945MG with 128 Ncm of torque and its axis is aligned with the axis of symmetry of the robot fish. A schematic overview with the naming of the main parts used throughout this text is provided in Fig. 1. There are further two small servo motors for varying the elasticity of the tail fin. The main frame and most of the mechanical parts were designed in SolidWorks CAD and made of ABS plastic on a Dimension BST 768 3D-printer. Apart from the three main micro controllers and connectors for charging, each component (sensors, motor, batteries etc.) is

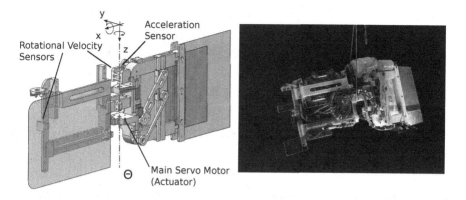

Fig. 1. A CAD drawing of Wanda2.0 and labeling of the most important parts and axis used throughout the text. The platform is symmetric around its vertical middle axis in size and weight distribution, i.e. the center of mass lies also on the middle axis. The right side shows a photography of the assembled robot fish Wanda2.0.

sealed individually, which allows easy reallocation all over the body. This is useful for later experiments or might be necessary when changing weight distribution or orientation of sensors. The sealed main box contains the three micro controllers and is placed in the upper front part of the body. A key factor to achieve rich behavioral diversity is the precise placement of weight and buoyancy. Based on the findings in [12] the best position of the two battery packs (weight) on the body is below the horizontal axis and far away from the middle axis, whereas the position of the two air cylinders is close to the middle axis and above the horizontal axis.

2.2   Tunable Elastic Tail Fin

The flexible tail fin consists of two main square foils of 13 cm length and 20 cm hight. In between these main foils, additional foils of various thickness and material can be inserted through a ellipse linkage. The positioning of the mechanism is actuated by two servo motors, one for each insertion foil. These foils can be inserted individually and to any length, from no insertion, for a soft tail fin, to full insertion resulting in a stiff tail fin. The enveloping main foils are not affected in terms of shape when additional foils are inserted and therefore the influence on the swimming performance due deformation (thickening) or size (length and hight) can be neglected. The mechanism for changing the elasticity of the tail fin is shown in fig. 2.

2.3   Data Acquisition and Controller

The 3-axis linear accelerometer (LIS3LV02DQ, STMicroelectronics) works in a range of ± 2g with 12 bit resolution and is mounted in alignment with the axis

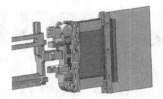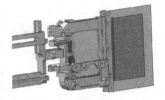

Fig. 2. Mechanism for stiffness change in tail fin. Please note that the inserted foils do not change the geometry of the tail fin.

of symmetry of the robot. On the same axis, an angular rate sensor ($\pm$ 150°s^{-1} and 11 bit AD resolution), MLX90609 from Melexis, is attached. There is also a second one collecting the angular rate of the roll motion on the x-axis of the body. Nevertheless, not all built-in sensors are measuring the swimming kinematics of the robot fish. In addition a digital compass module (HMC6352, Honeywell) is monitoring the orientation of the head plate at a resolution of 0.5° and a pressure sensor (MS5535C) from Intersema, with a pressure range of 0-14 bar and 16 bit AD resolution gives feedback about current swimming depth. Data collection, motor control and communication to a PC for data recording are orchestrated by three Atmel ATmega168 micro controllers built on Arduino pro mini boards. Communication at higher bandwidth, such as WLAN or Bluetooth does not work in water since the higher frequencies are absorbed too much. Here, we used a serial to radio solution from easy radio, ER400TRS-02, sending around 434 MHz with 10mW power output, which worked fine for the 10 m range of the experimental test pool.

3   Experiments and Results

3.1   Observed Behavior

A first step before starting the experiments was to categorize sets of different combinations of the values for the four main parameters, frequency, amplitude, offset, and stiffness into different behaviors. In this context, we use an observer-based notion of behavior, i.e. the behaviors are defined by a human observer looking at the running fish robot. The most obvious observed behavior is swimming straight forward; followed by turning left or right based on two criteria: (a) the radius for a full turn does not exceed 2 meters and (b) the swimming depth is constant. Due to the the shape of the robot and its weight distribution, diving left or right as a forth and fifth behavior is possible. How diving is achieved with just one degree of freedom (DOF) is explained in the next paragraph (or for a more detailed explanation, see [12]). Finally, when the experiments are performed with a hard tail fin rather than a soft one, the behaviors look very different (again, from a human perspective), which implies that there are another five distinct swimming patterns.

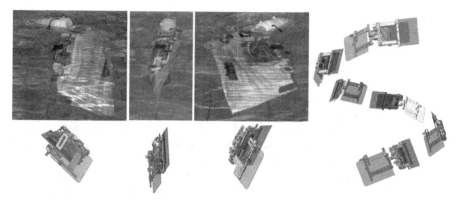

Fig. 3. 3 pictures on top and the corresponding CAD drawings below: The robot fish rolls to the right side when the main motor turns to the right, stays upright when main motor is centered and rolls to the left side when the main motor turns to the left. The right side shows the corkscrew like path when the robot fish is diving in open water as a sequence of CAD drawings. In this sequence the main motor runs with an offset to the right side.

Although the robot fish has only one DOF for actuation (one servo motor for moving the tail fin) it is able to swim in all three dimensions in the tank without any additional active control mechanism (e.g. artificial air bladder, weight shifting, or external rudder). The balance and distribution of the weights and the buoyant part is such, that while it is not moving or just swimming straight forward it floats and stays upright. On the other hand, if the main motor runs with an offset, the robot fish will roll to the right side when the offset is to the right and vice versa. This effect is caused by the weights in the front and rear part which are then not aligned anymore with the buoyant parts near the center. The robot fish rolls until the weights are back under the buoyant parts. All three positions are shown in Fig. 3. Due to this rolling effect the front plate points downwards and together with a minimum swimming speed will push the robot fish in a corkscrew like path deeper and deeper. The drag of the front plate works against the buoyant force as long as the swimming speed is high enough. The picture sequence in Fig. 3 illustrates the working principle. Summing up, the diving behavior depends on (a) the offset of the main motor and (b) swimming speed.

3.2 Parameter Space

The main motor placed in the middle of the robot fish follows a sine wave pattern

$$\theta(t) = a\sin(2\pi f t) + \rho \tag{1}$$

where $\theta(t)$ is the motor position (working range $\pm 90°$) as a function of time t, amplitude a, and frequency f. The offset parameter o has three states such as swimming straight, right or left. The corresponding offset value ρ for swimming straight is zero but for turning right or left the value depends on the current amplitude a of the main motor, following the rule: $\max(a) + o = 90°$. Preliminary experimental runs were performed in order to find a set of good parameter combinations for different observed behaviors. The three chosen amplitudes are $40°$, $60°$ and $80°$ with the corresponding offset values ρ for left and right of $\pm 50°$, $\pm 30°$ and $\pm 10°$. Each amplitude and offset parameter combination was tested on three frequencies, 0.5Hz, 0.67Hz and 1Hz. Altogether, with two possible states for the tail fin stiffness k (soft or hard), 54 possible combinations were examined in the pool and the data recorded for later analysis.

Each of the 54 combinations of $f/a/o/k$ has been categorized into one of the 10 different behaviors mentioned in section 3.1. The values for $f/a/o/k$ are such, that the effect of turning or diving etc. is clear and differentiable.

3.3   Results

Sensor Data Collection and Processing. All free swimming experiments were performed untethered, neither for power nor for communication, and in open water to minimize any external disturbance. The wireless recording of the data started after a steady swimming pattern was reached in order not to record any data from the starting maneuver. The sampling rate is 10Hz, which is relatively low for a precise kinematic analysis but sufficient for the goal of this experiment.

Kinematic data (acceleration (x, y, z-axis) and gyroscope (x and z-axis)) is compared in double strokes. A double stroke (left to right and back again) of the tail fin is the mechanical output of one full cycle of the main motor sine wave control signal. The sequence of each $f/a/o/k$ combination was chunked in individual double strokes, summed up and averaged to one double stroke. Pressure sensor and compass data is correlated to a longer sequence of strokes and not reduced to a single average double stroke. One typical output is shown in Fig. 4 with the phase plot of the kinematic data and the recording of the pressure and orientation for one full run.

Behavior and Kinematic Data Correlation. We introduce the notion of "kinematic intensity", which is the difference between the maximum and minimum value within an average double stroke. This means that each of the 54 $f/a/o/k$ combinations has five kinematic intensities (acceleration (x, y, z-axis) and rotational velocity (x and z-axis)). The motivation to reduce the information content of the sensor data even more is to find a minimalistic concept that shows the switch from one behavior to another.

The results are structured to answer the question of how the change from one behavior to another is represented in the kinematic intensity. A change in behavior results from changing one of the input parameter (f, a, o or k) as explained in 3.2. For some behavior transitions, the kinematic intensity alone does not work to differentiate between two behaviors and additional sensor information is needed, e.g. pressure or compass.

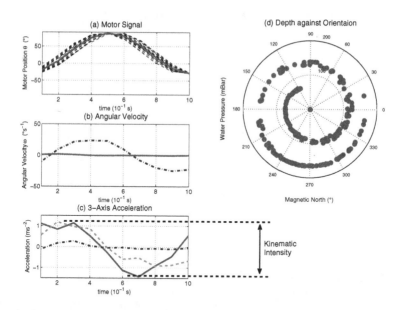

Fig. 4. Kinematic data of a full run. This is an overview of one kinematic data set out of 54 $f/a/o/k$ combinations: frequency 1Hz, amplitude 60°, offset 30°, and stiffness setting soft. In (a) the dashed lines represent the overlay of each double stroke of the main motor position and the solid line is the average double stroke for analysis. For (b), where the rotational velocity of the x-axis (roll) is a solid line and for the z-axis (yaw) a dashed line, only the final averaged values are plotted for better readability. Acceleration is plotted in (c) with x-axis (surge) in solid, y-axis (sway) in dashed and the z-axis (heave) in dash dot. For the orientation of the axes please refer to Fig. 1. Additionally the kinematic intensity (highest and lowest value withing a double stroke) for the acceleration of the x-axis is shown. The polar coordinate system (d) represents the collected data of one diving run with the radius is water pressure and the angle is the deviation from magnetic north.

First, the input parameter k is compared when changing from soft to hard. Looking at the difference of the kinematic intensity, that is the kinematic intensity of the soft configuration minus kinematic intensity of the hard configuration, the result shows a significant increase of the acceleration on the y-axis (sway) when swimming at a stiffer tail fin configuration. Fig. 5 provides the mean values for forward swimming, turning, and diving including the standard deviation. Only matching $f/a/o$ settings are compared to avoid their influence on the result.

The second input parameter that leads to a behavior change when varied is the offset o. Fig. 6 shows the change from zero offset, that is swimming straight forward to turning or diving (whether the robot starts to turn or to dive depends on the swimming speed, which again depends on the parameter settings of f and a). The difference in kinematic intensity is therefore zero offset minus offset

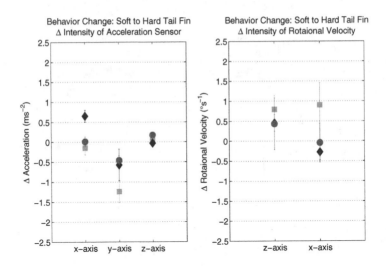

Fig. 5. Change of the kinematic intensity when switching from stiff to hard tail fin. The difference when switching from k soft to k hard is visible in the change of the kinematic intensity, which is only significant in the acceleration in the y-axis (pitch). For forward swimming the marker is a dot, turning (left and right) is diamond and for diving the marker is a square. All markers are plotted including the standard deviation from the experimental results.

right. Turning left does not differ from turning right in this context, which will be explained later. The kinematic intensity shifts when the robot fish starts to turn/dive with an increase in the acceleration in the x-axis (surge) and a decrease of kinematic intensity in the y-axis (pitch).

Another difference in the kinematic intensity is the change from turning to diving which is the case when either the frequency parameter changes or the amplitude. The difference between diving and turning is, that diving needs a minimum swimming speed, as explained in section 3.1. That is, in this experiment, at a frequency of 0.67Hz and above and a maximum amplitude of 60°. However, changing frequency or amplitude is clearly visible in the kinematic intensity, but does not necessarily change the behavior. In Fig. 7 all combinations between turning and diving behavior are summed up and averaged, including the standard deviation. The clearly most significant indication when switching from turning to diving in the kinematic intensity is in the rotational velocity of the x-axis (roll).

Finally, using the kinematic intensity alone one cannot distinguish between going left or going right. This is due to the symmetric motion of the robot fish, that the difference of the kinematic intensity for acceleration and rotational velocity is zero in all axes.

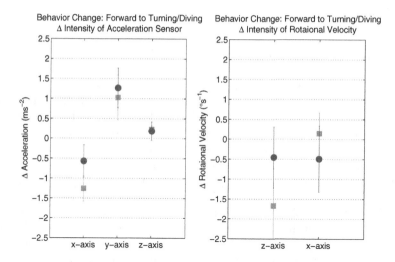

Fig. 6. Change of the kinematic intensity when switching the offset. The difference when switching from offset o zero to o right is visible in the change of the kinematic intensity, which is significant in the acceleration in the x and y-axis (surge and sway). For switching from forward swimming to turning behavior the marker is a dot and for switching from forward swimming behavior to diving the marker is a square. All markers are plotted including the standard deviation from the experimental results.

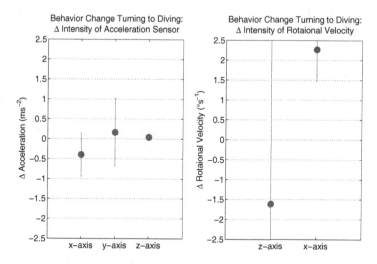

Fig. 7. Change of the kinematic intensity when switching from turning to diving. The difference when switching from turning to diving behavior is visible in the change of the kinematic intensity, which is significant in the rotational velocity of the x-axis (roll). All markers are plotted including the standard deviation from the experimental results.

4   Analysis

4.1   Minimum Kinematic Feedback

An increase in stiffness of the tail fin (k from soft to hard) goes along with a higher intensity of the acceleration on the y-axis (sway) of the sensor. The sensor is placed exactly along the line of symmetry in the middle of the robot fish, which means there is more motion in the middle section when the tail fin is stiffer, less motion when the tail fin is softer. In other words the head plate is moving less when the tail fin is soft, almost an oscillating tail fin, whereas for a stiff tail fin the head plate moves more against the tail fin, almost an undulatory motion.

The kinematic intensity indicates also the change of behavior from swimming straight forward to turning and diving. Due to the special morphology, the robot fish starts to roll and point downwards when it's bending onto one side. The rolling motion is more intense and therefore the kinematic intensity higher within a double stroke when turning/diving.

Finding a difference in the kinematic intensity when the behavior alters from turning to diving is not clearly possible with data from the linear acceleration sensor alone. Despite that the change in amplitude is represented in the linear acceleration, it is not clear where diving behavior starts and turning ends. But clearly the rolling of the robot fish is significantly higher while diving, which leads to the much higher kinematic intensity of the rotational velocity of the x-axis (roll) and therefore allows a separation between turning and diving behavior.

4.2   Adding Information Sources

Since the scope of this paper is to investigate the minimum sensory feedback required to differentiate between various behaviors, the variety of sensors in the analysis is kept small. Interestingly, in order to distinguish some very obvious behaviors, like turning left and turning right, additional sensors are needed. In order to do this, the data from the compass and the pressure sensor are added to the analysis.

In a turning/diving left behavior the compass value increases over time (except it goes back to zero when heading north again). In contrast the compass value decreases for turning/diving right. In other words, the feedback of the kinematic intensity can not tell the difference between left and right, but with an additional reference system, in this case the magnetic field of the earth, and the recordings of several strokes back in time, it becomes distinguishable.

At this point, all 10 initially introduced behaviors can be distinguished when switching from one behavior to another based on a minimum of sensory information. Nevertheless, by combining the information of the pressure sensor with the one from the compass, turning and diving left or right, can easily be separated. In Fig. 8 these four behaviors are plotted for comparison. Finally a complete overview is given in Tab. 1

Table 1. An overview of 4 changes from one observed behavior to another as discussed in section 4 where the up-arrow means an increase of the kinematic intensity (ki) and the down-arrow a decrease. Fields with - mean, that the data show no significant change in the ki. Of course all changes from one behavior to anther work also in the other way (e.g. turning ← diving), just with opposite values for the ki.

	lin. acc.	rot. velocity	compass/pressure
k soft → hard tail fin	↑ ki sway	-	-
o frwd → turn/dive	↑ ki surge, ↓ ki sway	-	(↕ compass)
$f\,a$ turning → diving	-	↑ ki roll	(↑ pressure)
o left → right	-	-	↑ compass

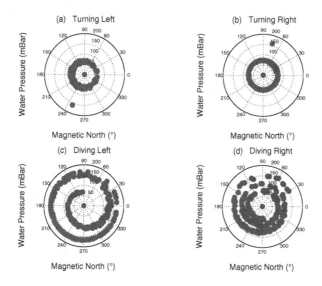

Fig. 8. These four plots show the correlation between the heading (referring to magnetic north) and the change in pressure while running at one of the following $f/a/o/k$ combination: (a) 0.67Hz, 80°, offset right, tail fin hard; (b) 1Hz, 80°, offset left, tail fin soft; (c) 1Hz, 60°, offset right, tail fin soft; (d) 1Hz, 40°, offset left, tail fin hard. While the behavior in (a) and (b) shows several full turns without any change in water depth, the behavior in (c) and (d) change to diving represented by the increasing distance to the origin of the polar coordinate system. Diving and turning left is plotted counter clock wise, turning and diving right clockwise.

5 Conclusions

We have shown in this experiment that various behaviors do not necessarily originate from a complex mechanism of the robot fish or a sophisticated feedback control system. With just one DOF, the right morphology and by exploiting the interaction with the environment some very fundamental swimming maneuvers (turning and diving) can be performed by the robot fish. Then, minimal sensory feedback is enough to distinguish between 10 different behaviors. The difference in the feedback signal, mainly in the kinematic intensity per double stroke or the correlation between compass and pressure values over several strokes, is clear and sufficient for a categorization. However, a future step would be to let the system find this correlation on its own using a suitable learning architecture which combines input parameters with sensory feedback while exploring the body dynamics in water (e.g. [14] and [15]). As we show here, data recorded from the sensors while swimming seems to be applicable.

References

1. Sfakiotakis, M., Bruce, L.D.M.J., Davis, C.: Review of fish swimming modes for aquatic locomotion. Journal of Oceanic Engineering 24, 237–252 (1999)
2. Colgate, J.E., Lynch, K.M.: Mechanics and control of swimming: A review. Journal of Oceanic Engineering 29, 660–673 (2004)
3. Blake, R.W.: Fish functional design and swimming performance. Journal of Fish Biology 65(5), 1193 (2004)
4. Lauder, G.V., Tytell, E.D.: Hydrodynamics of undulatory propulsion. Fish Physiology 23, 425–468 (2006)
5. Fish, F., Lauder, G.: Passive and active flow control by swimming fishes and mammals. Annual Review of Fluid Mechanics 38(1), 193 (2006)
6. McHenry, M.J., Pell, C., Long, J.H.: Mechanical control of swimming speed: Stiffness adn axial wave form in undulating fish models. Journal of Experimental Biology 198, 2293–2305 (1995)
7. Kobayashi, S., Ozaki, T., Nakabayashi, M., Morikawa, H., Itoh, A.: Bioinspired aquatic propulsion mechnisms with real-time variable apparent stiffness fins. In: IEEE Robotics and Biomimetics (2006)
8. Nakabayashi, M., Kobayashi, K.R.S., Morikawa, H.: A novel propulsion mechanism using a fin with a variable-effective-length spring. In: IEEE Robotics and Biomimetics (2008)
9. Mazumdar, A., Valdivia, P., Alvarado, Y., Youcef-Toumi, K.: Maneuverability of a robotic tuna with compliant body. In: IEEE Robotics and Automation (2008)
10. Liu, J., Hu, H.: Biological inspiration: From carangiform fish to multi-joint robotic fish. Journal of Bionic Engineering 7, 35–48 (2010)
11. Liu, J., Hu, H.: Biological inspired behaviour design for autonomous robotic fish. Journal of Automation and Computing 4, 336–347 (2006)
12. Ziegler, M., Iida, F., Pfeifer, R.: "cheap" underwater locomotion: Roles of morphological properties and behavioural diversity. In: CLAWAR (2006)
13. Ziegler, M., Hoffmann, M., Carbajal, J., Pfeifer, R.: Varying body stiffness for aquatic locomotion. In: 2011 IEEE International Conference on Robotics and Automation (ICRA), pp. 2705–2712 (May 2011)
14. Pfeifer, R., Gómez, G.: Creating Brain-like Intelligence: Challenges and Achievements. In: Sendhoff, B., Körner, E., Sporns, O., Ritter, H., Doya, K. (eds.) Creating Brain-Like Intelligence. LNCS, vol. 5436, pp. 66–83. Springer, Heidelberg (2009)
15. Iida, F., Gomez, G., Pfeifer, R.: Exploiting body dynamics for controlling a running quadruped robot. In: Proceedings of the 12th Int. Conf. on Advanced Robotics (ICAR 2005), Seattle, U.S.A, pp. 229–235 (2005)

Leech Heartbeat Neural Network on FPGA

Matthieu Ambroise, Timothée Levi[*], and Sylvain Saïghi

University of Bordeaux, IMS Laboratory, Talence, France
{matthieu.ambroise,timothee.levi,sylvain.saighi}@ims-bordeaux.fr

Abstract. Most of rhythmic movements are programmed by central pattern-generating networks that comprise neural oscillators. In this article, we implement a real-time biorealistic central pattern generator (CPG) into digital hardware (FPGA) for future hybrid experiments with biological neurons. This CPG mimics the Leech heartbeat neural network system. This system is composed of a neuron core from Izhikevich model, a biorealistic synaptic core and a topology to configure the table of connectivity of the different neurons. Our implementation needs few resources and few memories. Thanks to that, we could implement network of these CPG for instance to mimic the behavior of a salamander. Our system is validated by comparing our results to biological data.

Keywords: Biorealistic neural network, Central Patter Generator, FPGA.

1 Introduction

CPGs underlie the production of most rhythmic motor patterns and have been extensively studied as models of neural network function [1]. We can find half-center oscillators in Leech heartbeat [2]. In this paper, we implement a CPG that mimics the leech heartbeat neural network using digital neurons [3], [4].

2 Digital Topology

Izhikevich Model
The neuron model used is based on the Izhikevich model (IZH) [5]. This model depends on four parameters, which allow reproducing the spiking and bursting behavior of specific types of cortical neurons. So the good compromise between a simple model (Leak Integrate and Fire) and a complex one (Hodgkin-Huxley), is the IZH one.

Synapse Model
In order to simulate a CPG, we need a biorealistic synapse that can reproduce inhibitory behavior (GABA$_a$) and synaptic efficiency. GABA$_a$ is an inhibitory neurotransmitter that provides a hyperpolarization of a neuron's membrane.

---

[*] Corresponding author.

N.F. Lepora et al. (Eds.): Living Machines 2013, LNAI 8064, pp. 347–349, 2013.

A hyperpolarization means a restraint for K^+ and Na^+ ions, which will be represented by a negative contribution on the synaptic current. $GABA_a$ has an exponential decay propriety that will be implemented by the following equation:

$$I_{exp}(t+T) = (1-T/\tau).I_{exp} \qquad (1)$$

The synaptic efficiency is another biological phenomenon that consists to reduce a synaptic weight after a spike. In biology, each contribution provided by synapses are done thanks to synaptic vesicle.

Network Topology
In this design, there are 3 blocks: RAM, a neural computation core and a synaptic computation core (Fig. 1). The main idea is each data needed by a core are provided by the RAM and each new values computed by a core are stocked in the RAM.

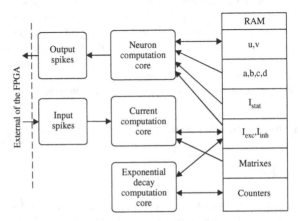

Fig. 1. Architecture of the network

3   Results

We implement a network of eight regular spiking neurons (Fig. 2) linked by inhibitory synapses. Like in biological leech heartbeat system, we observe in our digital board that the CPG activity is an alternation of bursts between the different pairs of neurons and the mean duration of the burst is 5 seconds [2].

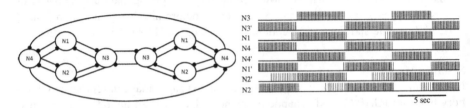

Fig. 2. Topology of CPG and measurement of the activities of our 8-neurons CPG into FPGA. The spike activity is important for keeping the dynamic of the network and to allow the hybrid experiments.

4   Conclusion

We implement into Xilinx Virtex 4 SX55 FPGA board our real-time biorealistic CPG. Our implementation uses around 5% of the resources (just one multiplier, 977 slices and 1598 4-LUT's) and 2% of the RAM (80 Kb).

Acknowledgement. This work is supported by the generous contribution of the European Commission ICT-FET FP7 (FET Young Explorers scheme) BRAINBOW (www.brainbowproject.eu), granted to the authors of this study.

References

1. Hooper, S.: Central Pattern Generators. Current Biology (2000)
2. Hill, A.A., Lu, J., Masino, M.A., Olsen, O.H., Calabrese, R.L.: A model of a segmental oscillator in the leech heartbeat neuronal network. Journal of Computational Neuroscience 10, 281–302 (2001)
3. Cassidy, A., Andreou, A.G.: Dynamical digital silicon neurons. In: IEEE Biomedical Circuits and Systems Conference, BioCAS 2008, November 20-22, pp. 289–292 (2008)
4. Linares-Barranco, A., Paz, R., Gómez-Rodríguez, F., Jiménez, A., Rivas, M., Jiménez, G., Civit, A.: FPGA Implementations Comparison of Neuro-cortical Inspired Convolution Processors for Spiking Systems. In: Cabestany, J., Sandoval, F., Prieto, A., Corchado, J.M. (eds.) IWANN 2009, Part I. LNCS, vol. 5517, pp. 97–105. Springer, Heidelberg (2009)
5. Izhikevich, E.M.: Simple model of spiking neurons. IEEE Transactions on Neural Networks 14(6), 1569–1572 (2003)

Artificial Muscle Actuators for a Robotic Fish

Iain A. Anderson[1,2], Milan Kelch[1], Shumeng Sun[1], Casey Jowers[1],
Daniel Xu[1], and Mark M. Murray[3]

[1] Biomimetics Laboratory, Auckland Bioengineering Institute, Level 6,
70 Symonds Street Auckland, NZ
i.anderson@auckland.ac.nz
[2] Engineering Science, Faculty of Engineering, University of Auckland, NZ
[3] Dept. of Mechanical Engineering, US Naval Academy, Annapolis, Maryland

Keywords: Soft robots, Dielectric elastomer.

1 Abstract

Biology is a source of inspiration for many functional aspects of engineered systems. Fish can provide guidance for the design of animal-like robots, which have soft elastic bodies that are a continuum of actuator, sensor, and information processor. Fish respond to minute pressure changes in water, generating thrust and gaining lift from obstacles in the current, altering the shape of body and fins and using sensory nerves in their muscles to control them. Dielectric Elastomer (DE) artificial muscles offer a mechanism for a fish muscle actuator. DE devices have already been shown to outperform natural muscle in terms of active stress, strain, and speed[1-3]. DE's also have multi-functional capabilities that include actuation, sensing, logic and even energy harvesting, all achievable through appropriate control of charge[4, 5]. But DE actuators must be designed so that they provide enough torque to drive the tail and develop useful forward thrust.

In this study bench-top measurements of maximum torque and deflection data for DE actuators have been collected and compared with active torques measured using an instrumented stepper motor driven robotic fish based on the New Zealand Snapper (*Pagrus auratus*). The rear half of the robot was driven at the mid-section by a stepper motor and a torque sensor interposed between motor and robot body measured swimming torque for a range of speeds and tail amplitudes. The candidate DE actuators were based on a double cone device first described by Choi and co workers[6], consisting of two convex conical membrane actuators held apart by a stiff central pin. Actuation on one side resulted in rotation of the robotic segment as depicted in figure 1.

Two actuator designs were evaluated: Model 1 supported 6 actuators, three each side (figure 1). For the second design we substituted a long tent-shaped actuator in the place of the trio of conical actuators. Instead of pins, rectangular pieces of acrylic (Model 2) held the membrane tents on opposing faces apart. A double thickness muscle for the Model 2 actuator was also tested. 3M 4905 VHB acrylic tape

N.F. Lepora et al. (Eds.): Living Machines 2013, LNAI 8064, pp. 350–352, 2013.
© Springer-Verlag Berlin Heidelberg 2013

(St. Paul, USA) was used for the dielectric membrane. From its original thickness of 500 µm the tape was stretched 3.4 by 3.4 to a final thickness of 43 µm. A conductive carbon powder (XC72 GP-3882, Cabot Vulcan, Boston, USA) was used for the surface electrodes. Each actuator consisted of 3 active DEA layers covered by an inactive and protective layer on each side. A soft conductive tape was used to connect the electrodes to a power supply (EAP Controller, Biomimetics Laboratory, Auckland).

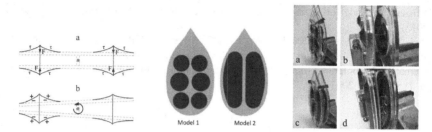

Fig. 1. On the left we show the basic principle of the conical actuator (Model 1). In the middle we show a schematic for Model 1 and the tent-shaped actuator of Model 2. On the far right we show photos of Model 1 (a and b), and Model 2 (c and d).

Of the two designs, Model 2 performed better and doubling the actuator thickness also doubled the blocked torque that reached 0.047 Nm at 3 kV. The Model 2 design also had a much greater range of active free motion at an angle of 3.7° swing between segments.

The robotic fish tail was driven at 0.5 and 1 Hz for angles of twist up to ±22° with realistic tail amplitudes of up to ±50mm (~0.1× body length). The peak measured torque per cycle was 0.08 Nm for 0.5 Hz at ±5.4° of angle and ±0.3 Nm for 1 Hz at ±22° of angle. Out of the water torques, measured to estimate the effects of friction, showed that friction could account for up to 0.05 Nm of torque in our system.

These results suggest that the torque provided by our current actuators is of the magnitude required to drive a robotic fish. Other actuator configurations are possible and under evaluation. In parallel with this we are developing electronics for DE self-sensing: an important step on the path to an animal-like robot.

Acknowledgements. The authors gratefully acknowledge the support provided by the Department of the Navy, Office of Naval Research Global (Grant number: N62909-12-1-7096). Any opinions, findings, and conclusions or recommendations expressed in this material are those of the authors and do not necessarily reflect the views of the Office of Naval Research Global.

References

1. Pelrine, R., Kornbluh, R., Joseph, J., Chiba, S.: Electrostriction of polymer films for micro-actuators. In: Tenth Annual International Workshop on Micro Electro Mechanical Systems, pp. 238–243 (1997)

2. Pelrine, R., Kornbluh, R., Joseph, H.J.R., Pei, Q., Chiba, S.: High-Speed Electrically Actuated Elastomers with Strain Greater Than 100%. Science 287, 836–839 (2000)
3. Madden, J.D.W., Vandesteeg, N.A., Anquetil, P.A., Madden, P.G.A., Takshi, A., Pytel, R.Z., Lafontaine, S.R., Wieringa, P.A., Hunter, I.W.: Artificial muscle technology: physical principles and naval prospects. IEEE Journal of Oceanic Engineering 29(3), 706–728 (2004)
4. Anderson, I.A., Gisby, T.A., McKay, T.G., O'Brien, B.M., Calius, E.P.: Multi-functional dielectric elastomer artificial muscles for soft and smart machines. Journal of Applied Physics 112(4), 041101 (2012)
5. O'Brien, B., Anderson, I.: An artificial muscle computer. Applied Physics Letters 102(10) (2013)
6. Choi, H.R., Jung, K.M., Kwak, J.W., Leea, S.W., Kim, H.M., Jeonb, J.W., Nam, J.D.: Digital Polymer Motor for Robotic Applications. In: Proceedings of the 2003 IEEE International Conference on Robotics & Automation, Taipei, Tairxo, September 14-19, pp. 1857-1862 (2003)

Soft, Stretchable and Conductive Biointerfaces for Bio-hybrid Tactile Sensing Investigation

Irene Bernardeschi[1,2,*], Francesco Greco[1], Gianni Ciofani[1], Virgilio Mattoli[1], Barbara Mazzolai[1], and Lucia Beccai[1]

[1] Center for Micro-BioRobotics @SSSA, Istituto Italiano di Tecnologia, Pontedera, Italy
[2] The BioRobotics Institute, Scuola Superiore Sant'Anna, Pontedera, Italy
{irene.bernardeschi,francesco.greco,gianni.ciofani,
virgilio.mattoli,barbara.mazzolai,lucia.beccai}@iit.it

Abstract. In this work we present the design, fabrication and characterization of polymer-based soft, stretchable and conductive biointerfaces for achieving both mechanical stimulation of cells and recording of their response. These biointerfaces represent the very first step for investigating potential bio-hybrid tactile sensing approaches. Cells transform mechanical stimuli into biological and biochemical signals (mechanotransduction): modifications on cell membrane structure lead to a final change in membrane electric potential. Therefore, an efficient way to quantify cellular response to mechanical stimuli is to evaluate the impedance variation of cells due to externally applied forces.

Keywords: Biointerfaces, bio-hybrid sensors, stretchable conductor, mechanotransduction, membrane potential, PDMS, PEDOT:PSS.

1 Introduction

Polymer-based soft, stretchable, and conductive biointerfaces can be built to embed biological components, obtaining bio-hybrid constructs that can be investigated to: *i*) increase knowledge on mechanical stimuli and cell response correlation; and *ii*) study possible bio-hybrid tactile sensing approaches. In nature tactile sensing is based on mechanotransduction, the mechanism by which cells convert mechanical stimuli into biological and biochemical signals, involving specialized mechanoreceptors [1, 2]. Mechanical stimulation alters cellular membrane structure inducing conformational changes in stress-sensitive ion channels, thus leading to a variation in their opening/closing rates. This causes an increase of ionic current fluxes into the cell (*e.g.*, Ca^{2+} entry), with a subsequent change of membrane potential that can be recorded as a variation of impedance by using proper conductive substrates and techniques. Electric cell-substrate impedance sensing (ECIS) is a static recording technique by which morphological changes and behaviour of cells (cultured on electrodes surfaces in different conditions) are assessed by monitoring impedance variation in time [3]. In our case indeed, the final aim is to quantify cellular response to a specified mechanical stimulus

---

*Corresponding author.

N.F. Lepora et al. (Eds.): Living Machines 2013, LNAI 8064, pp. 353–355, 2013.
© Springer-Verlag Berlin Heidelberg 2013

by recording impedance variation caused by cellular membrane potential change; thus dynamic stimulation and recording are needed. Dynamic stimulation has been previously implemented, but only calcium fluxes variation during stimulation was assessed [4]. Otherwise zebrafish embryo impedance variation was recorded as consequence of indentation force stimuli exerted using a micro-indenter, but no dynamic stimulation was performed [5]. Here we present a biointerface designed to deliver a uniaxial strain while allowing impedance recording. Moreover, human SH-SY5Y neuroblastoma cells are used to assess the biointerface biocompatibility.

2 Design and Characterization of the Soft, Stretchable and Conductive Biointerfaces

The biointerface design consists of two layers: *i*) an elastic substrate of poly(dimethyl siloxane) (PDMS) elastomer, that can be elastically deformed to apply the uniaxial strain for cells stimulation; *ii*) a more rigid and conductive film of the conjugated polymer poly(3,4-ethylenedioxythiophene):poly(styrene sulfonate) (PEDOT:PSS) on which the cells are cultured. The latter allows for recording the expected impedance variation. In order to enable the stretchability of the all-polymer conductive biointerface, we exploited the mechanism of spontaneous formation of wrinkles [6]. Surface wrinkling occurs when a stiff uncompressible film, attached to a soft, elastomeric substrate, is compressed. Different wrinkling morphologies are observed depending on materials and fabrication parameters (i.e. thickness and Young's modulus of the film and substrate). A PDMS (Sylgard 184, Dow Corning, 10:1 monomer/curing agent ratio) substrate of 130 μm thickness was pre-stretched (15% of the total length). A conducting layer with thickness $t = 80$ nm was deposited on top of PDMS by spin coating a PEDOT:PSS dispersion in water (Clevios PH1000, Heraeus Gmbh) doped with a 5% w/w of dimethylsulfoxide (DMSO). When sample is relaxed micro-wrinkles (Fig. 1.a) were formed on the surface, aligned along perpendicular direction to the pre-stretching one.. Wrinkles periodicity was estimated by Fast Fourier Transform (FFT) analysis of SEM images of samples (Fig. 1.b). Two characteristic wavelength types were observed: $\lambda_1 = 1.73$ μm and $\lambda_2 = 3.62$ μm. Strain cycles with 10 μm strain, corresponding to 0.05% of the total sample length, were imposed to five samples. Resistance variation was simultaneously recorded: $\Delta R = R - R_0$ and $\Delta R/R_0$ (R strained state and R_0 relaxed state resistances) were calculated for every strain cycle. The obtained average values were $\Delta R = 4.3 \pm 3.8$ Ω and $\Delta R/R_0 = 0.29 \pm 0.26$ %. These results confirmed that the conductive biointerfaces allow for recording of impedance variation due to just cell response to mechanical stimuli with strain ≤ 10 μm. Cell viability and proliferation on the substrates were assessed by means of Live/Dead® test (an assay that stains live cells in green and dead cells in red) and WST-1 assay (highlighting cell metabolic activity) respectively, after 3 and 6 days of culture: polystyrene dishes (control), smooth and micro-wrinkled samples were compared. Proliferation and viability were both satisfactory (Fig. 2.a,b,c). In particular, after 6 days of culture, proliferation was higher on both kinds of samples with respect to the control (Fig. 2.d). The proposed biointerface is completely polymeric, soft, stretchable, transparent and conductive, and its resistance is not affected by application of a 10 μm uniaxial strain. In future work, impedance variation recording will be performed by applying a 1 μm uniaxial strain, avoiding the possible influence of impedance variation due to cell deformability [7]. In parallel, calcium fluxes will be monitored to validate our approach.

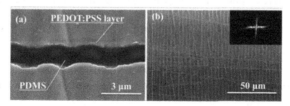

Fig. 1. (a) Biointerface surface and cross-sectional view in correspondence to a vertical cut along the stretching direction. (b) SEM image of surface morphology and relative FFT (inset).

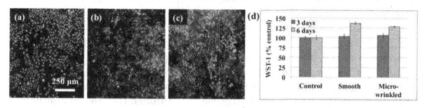

Fig. 2. Live-Dead® test results at 6 days culture: (a) control, (b) smooth samples, (c) micro-wrinkled (for colour pictures please refer to the online version); (d) WST-1 assay results

References

1. Ingber, D.E.: Cellular mechanotransduction: putting all the pieces together again. Faseb J. 20, 811–827 (2006)
2. Boulais, N., Misery, L.: The epidermis: a sensory tissue. Eur. J. Dermatol. 18, 119–127 (2008)
3. Wegener, J., Keese, C.R., Giaever, I.: Electric Cell–Substrate Impedance Sensing (ECIS) as a noninvasive means to monitor the kinetics of cell spreading to artificial surfaces. Exp. Cell Res. 259, 158–166 (2000)
4. Heo, Y.J., Iwase, E., Matsumoto, K., Shimoyama, I.: Stretchable substrates for the measurements of intracellular calcium ion concentration responding to mechanical stress. In: 20th Annual International Conference on Micro Electro Mechanical Systems, pp. 68–71. IEEE Press, Sorrento (2009)
5. Nam, J.H., Chen, P.C.Y., Lu, Z., Luo, H., Ge, R., Lin, W.: Force control for mechanoinduction of impedance variation in cellular organisms. J. Micromech. Microeng. 20, 025003 (2010)
6. Chung, J.Y., Nolte, A.J., Stafford, C.M.: Surface wrinkling: a versatile platform for measuring thin-film properties. Adv. Mater. 23, 349–368 (2010)
7. Kim, D., Choi, E., Choi, S.S., Lee, S., Park, J., Yun, K.: Measurement of single-cell deformability using impedance analysis on microfluidic chip. Jpn. J. Appl. Phys. 49, 127002 (2010)

Learning of Motor Sequences Based on a Computational Model of the Cerebellum

Santiago Brandi[1], Ivan Herreros[1],
Martí Sánchez-Fibla[1], and Paul F.M.J. Verschure[1,2]

[1] SPECS, Technology Department, Universitat Pompeu Fabra,
Carrer de Roc Boronat 138, 08018 Barcelona, Spain
[2] ICREA, Institució Catalana de Recerca i Estudis Avançats,
Passeig Lluís Companys 23, 08010 Barcelona
{santiago.brandi,ivan.herreros,marti.sanchez,paul.verschure}@upf.edu

Abstract. In classical conditioning, the repeated presentation of a Conditioning Stimulus (CS) followed by an Unconditioned Stimulus (US) establishes a basic form of associative memory. After several paired CS-US presentations, a Conditioned Response (CR) is elicited by the solely presence of the CS. It is widely agreed that this associative memory is stored in the cerebellum. However, no studies have link this basic form of cerebellar associative learning with the acquisition of sequences of motor actions. The present work suggests that through the Nucleo Pontine Projections (NPPs), a CR elicited by a first CS may be fed-back to the cerebellum, and that this CR can act as the CS for a subsequent CR. This process would allow a single CS to trigger a sequence of learned responses, having a total duration above the timespan of the cerebellar memory trace. We demonstrate this principle with a robotic experiment, where a computational model of the cerebellum that includes the NPPs controls a robot navigating a track with two turns. A predictive cue, the CS, precedes the first turn, but the second one can only be acquired if the previous turn is also used as a CS. After repeated training trials, the robot associates a sequence of two turns to the single CS. This result confirms that the positive feedback established via the NPPs allows the cerberllar model to control an action sequence, and that the duration of the whole sequence can exceed the timespan of a cerebellar memory trace.

It is known that the cerebellum plays an important role in the coordination of muscle activity to perform smooth movements and that it is involved in sensory prediction [1]. These two qualities allow the cerebellum to control the execution of precise anticipatory actions. Such is the case in the conditioning learning paradigm, in which after several presentations of a Conditioning Stimulus (CS) followed by an Uncoditioned Stimulus (US), the presentation of the CS alone elicits a Conditioned Response (CR) that anticipates the US. However, the time span of the cerebellar memory is limited to a CS-US Inter-Stimulus Interval (ISI) below one second. Indeed when the ISI is larger, the acquisition of CRs requires

N.F. Lepora et al. (Eds.): Living Machines 2013, LNAI 8064, pp. 356–358, 2013.
© Springer-Verlag Berlin Heidelberg 2013

the recruitment of additional brain areas to the cerebellum, such as the hippocampus of prefrontal cortex. This would limit the capacity of the cerebellum of performing sequences of motor actions when the interval between the CS and the last US exceeds that maximum ISI. However, here we propose that the cerebellum can overcome this limitation chaining several CS-US associations of short ISIs, using one CR as the CS for the learning of the next action, thus eliciting a sequence of actions with a single initial CS. We hypothesize that in nature, this recurrence is established through the Nucleo-Pontine Projections (NPPs), a set of connections that link the cerebellar output structures (the cerebellar nuclei) with a structure providing input to the cerebellum, i.e., the pontine nuclei.

We test the functionality of this hypothesis with a robotic collision avoidance task. The task is designed to be analogous to a classical conditioning experiment: the robot has to learn to perform a predictive turn, the CR, to avoid the collision with the wall, the US, following the perception of a predictive cue, the CS. We depart from a previous work where a cerebellar model was applied in the task of navigating, avoiding collisions, a track with a single turn [2]. We extend this work by adding a second turn to the track, and thus requiring the robot to acquire a series of two coordinated turns. We configure the track so that the second turn occurs too distant to the initial cue for the cerebellum to be able to directly link both signals. For this, instead of the CS, the cerebellum has to use the first turn signal, CR_1, as the predictive cue for the second turn, CR_2.

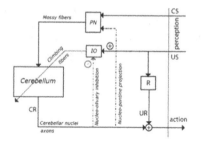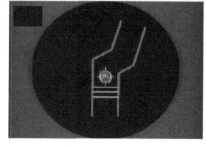

Fig. 1. (*Left*) Cerebellum as an adaptive filter, the CS is relayed by the mossy fibers and the US is relayed by the climbing fiber. The output of the cerebellum inhibits the Inferior Olive but also reaches the Pontine Nuclei, from where, through the mossy fibers, enters back into the cerebellum. (*Right*) Experimental setup: Virtual environment with the robot avatar.

We use an algorithmic model of the cerebellum along the lines of the adaptive filter theory [3][4], including the nucleo-olivary inhibition and the NPPs (Fig. 1 (*Left*)). Learning is performed by a de-correlation learning rule that updates the synaptic weights of the cerebellar model; see [2] for details on the algorithm. The computational model is then applied to a setup consisting of an epuck robot placed on a Mixed Reality Robot Arena (MRRA) [5]. The physical robot navigates a track back-projected onto a table (Fig. 1 (*Right*)). This setup mixes the physical constraints implied in the control of a real robot with the flexibility of using virtual scenarios.

The results show that when the ISI between the CS and the US_2 is too large for the cerebellum to learn a direct CS-CR_2 association, the sequence CS-CR_1-CR_2 is acquired chaining two stimulus-response associations, namely CS-CR_1 and CR_1-CR_2 (Fig. 2 (*Left*)) . This type of learning requires the CR_1 to be fed back to the cerebellum via the NPPs, for it to serve as a CS for the CR_2. Indeed, the disconnection of the NPPs prevents the acquisition of the second anticipatory turn (Fig. 2 (*Right*)). In conclusion, we demonstrate in a real robot experiment that an autonomous controller based on the cerebellum is capable of acquiring a sequence of anticipatory motor responses.

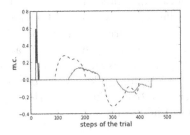 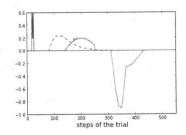

Fig. 2. Results: (*Left*) After several trials the robot performs two predictive turns [dashed line] (The USs are plotted in thick lines and the initial spikes in blue are the CS signal). (*Right*) When the NPPs are cut the ISI between the first CS and the second US is too large for the cerebellum model to associate them and the second predictive turn disappears.

References

1. Mial, R.: The cerebellum, predictive control and motor coordination (1998)
2. Herreros, I., Verschure, P.: Nucleo-olivary inhibition balances the interaction between the reactive and adaptive layers in motor control. Neural Networks (2013)
3. Fujita, M.: Adaptive filter model of the cerebellum. Biological Cybernetics 45(3), 195–206 (1982)
4. Dean, P., Porrill, J., Ekerot, C., Jörntell, H.: The cerebellar microcircuit as an adaptive filter: experimental and computational evidence. Nature Reviews Neuroscience 11(1), 30–43 (2010)
5. Fibla, M.S., Bernardet, U., Verschure, P.F.: Allostatic control for robot behaviour regulation: An extension to path planning. In: 2010 IEEE/RSJ International Conference on Intelligent Robots and Systems (IROS), pp. 1935–1942. IEEE (2010)

Bio-inspired Caterpillar-Like Climbing Robot

Jian Chen, Eugen Richter, and Jianwei Zhang

Institute of Technical Aspects of Multimodal Systems
Department of Informatics, University of Hamburg
Vogt-Koelln-Strasse 30, 22527, Hamburg, Germany
{jchen,erichter,zhang}@informatik.uni-hamburg.de
http://tams-www.informatik.uni-hamburg.de

Abstract. Caterpillars are extraordinary climbers in nature. The caterpillar-like climbing is in fact slow, but a caterpillar's multi-segment body trunk strongly enhances the climbing versatility and dexterity. In this study, we present a three-stage locomotion method to imitate the caterpillar-like climbing strategy. For practical implementation, a multi-segment modular robot is constructed, and the three-stage climbing method is successfully implemented on the robot, which confirms the effectiveness of the proposed climbing method.

Keywords: Climbing robot, bio-inspired robot, caterpillar-like robot.

1 Introduction

In the climbing robotics area, bio-inspired robots have attracted the most significant research attention due to the effectiveness of natural animal climbing patterns. Several successful imitative climbing robots have been developed in the existing literature [1], e.g., the gecko inspired climbing robots using synthetic adhesives [2][3], the cockroach and spider inspired climbing robots using micro-spine arrays [4][5], etc. In nature, caterpillars are extraordinarily successful climbers that can maneuver in complex three-dimensional environments, and can hold on the substrate using effective grasping system [6]. Due to kinematic redundancy of the body trunk, natural caterpillars demonstrate remarkable versatility and dexterity in performing various tasks [7], which has further inspired us in designing climbing robots.

In this paper, we present a three-stage locomotion method to imitate the caterpillar-like climbing strategy. A multi-segment modular robot is constructed to realize the caterpillar-like climbing pattern, and successful robot climbing experiment results confirm the effectiveness of the proposed climbing strategy.

2 Caterpillar-Like Climbing

Similar to the natural caterpillar wall climbing pattern, we decompose the wall climbing process of the caterpillar-like robot into three stages: lift the tail (Fig. 1a), propagate the body hump (Fig. 1b), and release the head (Fig. 1c).

N.F. Lepora et al. (Eds.): Living Machines 2013, LNAI 8064, pp. 359–361, 2013.

During the first stage of the caterpillar-like climbing strategy (Fig. 1a), by jointly actuating a group of motive joints, the robot gradually lifts the tail and forms the body hump. At the end of this process, the tail of the robot attaches to the wall again. Fig. 1b shows an instance of the second climbing stage. During this stage, a new segment is activated, and one segment is deactivated. The motive segments drive the end-effector towards the wall, which finally travels the body hump one segment forward. By iteratively perform the second stage, the robot travels the body hump from tail to head. Fig. 1c shows the third stage of the caterpillar-like climbing strategy, which is a reversal procedure of the first stage. During this stage, the robot gradually stretches its head and finally attaches all the segments on the wall.

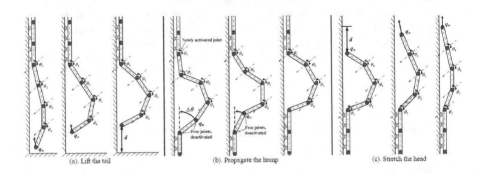

(a). Lift the tail   (b). Propagate the hump   (c). Stretch the head

Fig. 1. Three stages in caterpillar-like robot locomotion

3   Robot Component and Control System Design

To validate the feasibility of the proposed three-stage caterpillar-like climbing strategy, we designed and implemented a 7-segment modular robot, and designed a real-time locomotor for realizing the climbing strategy. The 7-segment caterpillar-like robot is shown in Fig. 2, where each segment has a mass of $242g$, a length of $7.7cm$, and the center of mass of each segment is of $2.6cm$ deviated from the revolute joint. By calculation, the moment of inertia with respect to the revolute joint is $3 * 10^3 g \cdot cm^2$.

(a)   (b)   (c)   (d)

Fig. 2. The caterpillar-like robot climbing the wall

The robot climbing experiment results are also shown by Fig. 2, where the images are collected from a video of 35s length. Fig. 2a shows the robot performing the first stage of the caterpillar-like climbing locomotion, namely, lifting the tail, where the body hump is gradually formed. The second stage of the caterpillar-like climbing locomotion is illustrated in Figs. 2b-c, where each image shows the robot forwards the body hump one segment ahead. Fig. 2d shows the final stage of the caterpillar-like climbing locomotion, where the robot gradually stretches the first three segments and finally attaches all the segments on the wall.

4 Conclusion

This paper presents our recent bio-inspired researches to study the natural caterpillar climbing patterns, which is imitated by a three-stage climbing method. We present a multi-segment modular robot to test the feasibility of the climbing method. Robot climbing experiment results successfully confirm the effectiveness of the proposed three-stage caterpillar-like climbing strategy.

Acknowledgments. This work was supported by Deutsche Forschungsgemeinschaft (DFG), Germany (Reference No. U-4604-DFG-1001).

References

1. Silva, M.F., Tenreiro Machado, J.A.: A Survey of Technologies and Applications for Climbing Robots Locomotion and Adhesion. In: Miripour, B. (ed.) Climbing and Walking Robots 2010, pp. 1–22. InTech (2010)
2. Kim, S., Spenko, M., Trujillo, S., Heyneman, B., Santos, D., Cutkosky, M.R.: Smooth Vertical Surface Climbing with Directional Adhesion. IEEE Trans. Robot. 24(1), 65–74 (2008)
3. Murphy, M.P., Kute, C., Menguc, Y., Sitti, M.: Waalbot II: Adhesion Recovery and Improved Performance of a Climbing Robot Using Fibrillar Adhesives. Int. J. Robot. Res. 30(1), 118–133 (2011)
4. Asbeck, A.T., Kim, S., Cutkosky, M.R., Provancher, W.R., Lanzetta, M.: Scaling Hard Vertical Surfaces with Compliant Microspine Arrays. Int. J. Robot. Res. 25(12), 1165–1179 (2006)
5. Spenko, M.J., Haynes, G.C., Saunders, J.A., Cutkosky, M.R., Rizzi, A.A.: Biologically Inspired Climbing with a Hexapedal Robot. J. Field Robot. 25(4-5), 223–242 (2008)
6. Mezoff, S., Papastathis, N., Takesian, A., Trimmer, B.A.: The Biomechanical and Neural Control of Hydrostatic Limb Movements in Manduca Sexta. J. Exp. Biol. 207, 3043–3053 (2004)
7. Van Griethuijsen, L.I., Trimmer, B.: Kinematics of Horizontal and Vertical Caterpillar Crawling. J. Exp. Biol. 212, 1455–1462 (2009)

The Green Brain Project – Developing a Neuromimetic Robotic Honeybee

Alex Cope[1], Chelsea Sabo[1], Esin Yavuz[2], Kevin Gurney[1], James Marshall[1], Thomas Nowotny[2], and Eleni Vasilaki[1]

[1] University of Sheffield, Sheffield S10 2TN, UK
a.cope@sheffield.ac.uk
[2] University of Sussex, Brighton BN1 9RH, UK

The development of an 'artificial brain' is one of the greatest challenges in artificial intelligence, and its success will have innumerable benefits in many and diverse fields, including the creation of autonomous robotic agents. Most research effort is spent on modelling vertebrate brains. Yet smaller brains can display comparable cognitive sophistication, while being more experimentally accessible and amenable to modelling.

It has been well established that the honeybee *Apis mellifera* has surprisingly advanced cognitive behaviours despite the relative simplicity of its brain when compared to vertebrates (e.g. [1]). These cognitively sophisticated behaviours are achieved despite the very limited size of the honeybee brain, on the order of 10^6 neurons. In comparison even rats or mice have brains on the order of 10^8 neurons. The greatly reduced scale and the experimental accessibility of the honeybee brain makes thorough neurobiological understanding and subsequent biomimetic exploitation much more practical than with even the simplest vertebrate brain.

Modelling the honeybee brain will focus on three brain regions. These are; the system for olfactory sensing, the system for visual sensing, and the mushroom bodies for multi-modal sensory integration. These systems are chosen as they are implicated in implementing complex cognitive behaviours essential to autonomous agents, which are not currently understood.

The models will integrate existing published neuroanatomical data with neural and behavioural experimental data. For the olfactory sensory system we will focus on *associative learning*. Honeybees are capable of both positive and aversive associative learning. They can learn complex associative rules, such as individual rewarded odours which are punished when combined. Aside from optic flow, little is known about the honeybee visual system. Therefore, we seek to develop novel mechanisms that demonstrate learning behaviour such as labyrinth navigation and development of *complex associations between objects and abstract rules*. The mushroom bodies are vital for of *multi-sensory integration* of learning and cross sensory transference of learning [2], and our model will perform these roles between the olfactory and visual models. These models will be able to account for existing experimental data and make testable predictions.

The neural models will be integrated into a 'Green Brain', which will be deployed for the real-time control of a flying robot able to sense and act autonomously.

N.F. Lepora et al. (Eds.): Living Machines 2013, LNAI 8064, pp. 362–363, 2013.

This robot testbed will be used to demonstrate the development of new biomimetic control algorithms for artificial intelligence and robotics applications. Further, by modelling complete sensorimotor loops endowed with behaviour, we will be able to begin examining the nature of embodied cognition in biological brains rather than abstract agents.

The 'Green Brain' will communicate with the robotic platform via wireless networking (Figure 1). Despite the small scale of the honeybee brain in comparison to vertebrate brains, computing a simulation in real-time presents a significant challenge. Traditionally, expensive supercomputers have been used to overcome this problem. The 'Green Brain' will instead run on the affordable massively-parallel GPU architecture of nVidia's Tesla, and will build on existing work simulating the honeybee mushroom body on GPUs (see Figure 1).

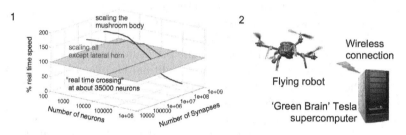

Fig. 1. (1) Benchmark figures for real-time implementation of honeybee brain model on nVidia hardware. Good scale-up to the full model should be possible with multiple devices. (2) The robotic hardware is controlled remotely.

In summary: the 'Green Brain' project will combine computational neuroscience modelling, learning and decision theory, modern parallel computing methods and robotics with data from state-of-the-art neurobiological experiments on cognition in the honeybee, to build and deploy a modular model of the honeybee brain describing detection, classification and learning in the olfactory and optic pathways as well as multi-sensory integration across these sensory modalities.

References

1. Reinhard, J., Srinivasan, M.V., Guez, D., Zhang, S.W.: Floral scents induce recall of navigational and visual memories in honeybees. The Journal of Experimental Biology 207(pt. 25), 4371–4381 (2004)
2. Devaud, J.M., Blunk, A., Podufall, J., Giurfa, M., Grünewald, B.: Using local anaesthetics to block neuronal activity and map specific learning tasks to the mushroom bodies of an insect brain. The European Journal of Neuroscience 26(11), 3193–3206 (2007)
3. Srinivasan, M.V.: Visual control of navigation in insects and its relevance for robotics. Current Opinion in Neurobiology 21(4), 535–543 (2011)

Efficient Coding in the Whisker System: Biomimetic Pre-processing for Robots?

Mathew H. Evans

Sheffield Centre for Robotics (SCentRo), University of Sheffield,
Sheffield, S10 2TN, U.K
mat.evans@shef.ac.uk

Introduction. The Efficient coding hypothesis [1, 2] proposes that biological sensory processing has evolved to maximize the information transmitted to the brain from the environment, and should therefore be tuned to the statistics of the world. Metabolic and wiring considerations impose additional sparsity on these representations, such that the activity of individual neurons are as decorrelated as possible [3]. Efficient coding has provided a framework for understanding early sensory processing in both vision and audition, for example in explaining the receptive field properties of simple and complex cells in primary visual cortex (V1) and the tuning properties of auditory nerve fibres [4].

Whisker sensing has been the subject of growing interest both in neuroscience [5] and robotics [6]. Rodent whiskers are an excellent model mammalian sensorimotor system: they are highly amenable to study with a range of genetic and optical tools, and the sensory and motor neural pathways are delineated to a greater degree than in other systems [7]. For robotics, whiskers are low powered, cheap to assemble and certain perceptual decisions, such as contact localisation along the length of the whisker, can be made with very simple computations [8]. Tactile discrimination in more complex conditions, however, remains a challenge for robotics. A given whisker-surface contact can have a number of unknown parameters (whisker speed, contact angle, surface texture *et cetera*), making surface identification difficult [9]. Whisker movement in exploring rodents and robots is highly variable, and trial to trial changes in whisker-object contact geometry complicates surface classification [10].

Can such an efficient coding approach be applied to understanding the rodent whisker system? In addition, does a biomimetic efficient coding scheme provide insights for improving tactile discrimination in robots? Here we present preliminary results from a meta-analysis of 'natural' whisker deflection data (analogous to natural images in computer vision) collected from seven published experiments with robotic whisker systems [11–14, 9, 10, 8]. We show that certain features from efficient and sparse coding of whisker deflections appear to be aligned with the tuning curves of rodent primary afferent and thalamic relay neurons, suggesting that this system is performing efficient coding.

Results. Whisker follicle primary afferent (PA) neurons have been broadly classified as 'contact', 'pressure' or 'detach' cells [15], as shown in Figure 1A, or as slowly and rapidly adapting units [16]. An common method of efficient encoding

N.F. Lepora et al. (Eds.): Living Machines 2013, LNAI 8064, pp. 364–367, 2013.

is principal components analysis (PCA). The first 3 eigenvectors from PCA of artificial whisker deflections from [14] is shown in Figure 1B. The gross shape of these PCs resemble the PA response properties in Figure 1A. This early result indicates that PA neurons may be performing an efficient transformation of natural whisker deflection signals. PCA is often used as an encoding step in its own right, but is also used as a 'whitening' pre-processing step. Flattening or whitening a signal is the operation of removing redundant correlations, ensuring that input signal features (such as spatial frequency in an image) have the same variance, and outputs are sparse.sparse coding algorithms tune basis function populations to represent input distributions evenly, is related to Independent Component Analysis, and has been used to explain end-stopping and non-classical receptive field surround suppression in V1 neurons [17]. Rat ventral posterior medial thalamic nucleus (VPm) neurons have been described as a population of diverse, precise kinetic feature detectors [18], the most common of these (from a larger distribution) are shown in Figure 1C. We tested a non-linear sparse coding approach [17] on artificial whisker data from [11] to determine whether VPm-like kinetic features would emerge. Some resultant basis functions are shown in Figure 1D (from a larger distribution). The shapes of these basis functions resemble VPm neuron responses (Figure 1C), suggesting that this structure may be performing sparse coding. Work is ongoing to rigorously compare the distribution of features generated through efficient and sparse coding to population neural tuning properties. For robotics we propose that the two coding schemes presented here could be applied in succession and considered a biomimetic filter cascade for pre-processing signals on artificial whisker systems. Such an approach may endow robots with the capabilities of biological

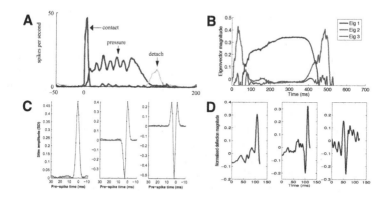

Fig. 1. Neural responses to whisker deflection resemble whisker deflection statistics. **A** Primary afferent responses from [15], adapted from [19]. **B** First 3 eigenvectors from PCA on artificial data resemble neural responses in **A**. **C** VPm thalamus responses from [18]. **D** Subset of sparsely generated basis functions (of a complete set of 64) that mimic the neural response properties shown in **C**. These results suggest rodent tactile sensory processing follows efficient coding principles, an approach which may improve robot sensing.

whisker systems, such as high representational capacity, sensory robustness to motor noise, and computational efficiency.

Acknowledgements. Funded by a University of Sheffield/EPSRC Doctoral Prize Fellowship.

References

1. Barlow, H.B.: Possible principles underlying the transformation of sensory messages. Sensory Communication, 217–234 (1961)
2. Atick, J.J.: Could information theory provide an ecological theory of sensory processing? Network: Computation in Neural Systems 3(2), 213–251 (1992)
3. Simoncelli, E.P.: Vision and the statistics of the visual environment. Current Opinion in Neurobiology 13(2), 144–149 (2003)
4. Olshausen, B.A., Field, D.J., et al.: Sparse coding of sensory inputs. Current Opinion in Neurobiology 14(4), 481–487 (2004)
5. Diamond, M., von Heimendahl, M., Knutsen, P., Kleinfeld, D., Ahissar, E.: 'Where' and 'What' in the whisker sensorimotor system. Nat. Rev. Neurosci. 9(8) (2008)
6. Prescott, T., Pearson, M., Mitchinson, B., Sullivan, J., Pipe, A.: Whisking with robots from rat vibrissae to biomimetic technology for active touch. IEEE Robotics and Automation Magazine 16(3), 42–50 (2009)
7. O'Connor, D.H., Huber, D., Svoboda, K.: Reverse engineering the mouse brain. Nature 461(7266), 923–929 (2009)
8. Evans, M.H., Fox, C.W., Lepora, N., Pearson, M.J., Sullivan, J.C., Prescott, T.J.: The effect of whisker movement on radial distance estimation: a case study in comparative robotics. Frontiers in Neurorobotics 6(12) (2012)
9. Fox, C.W., Evans, M.H., Pearson, M.J., Prescott, T.J.: Towards hierarchical blackboard mapping on a whiskered robot. Robotics and Autonomous Systems 60(11), 1356–1366 (2012)
10. Evans, M.H., Pearson, M.J., Lepora, N.F., Prescott, T.J., Fox, C.W.: Whiskered texture classification with uncertain contact pose geometry. In: 2012 IEEE/RSJ International Conference on Intelligent Robots and Systems (IROS), pp. 7–13 (2012)
11. Evans, M.H., Fox, C.W., Pearson, M.J., Prescott, T.J.: Spectral template based classification of robotic whisker sensor signals in a floor texture discrimination task. In: Proceedings of Towards Autonomous Robotic Systems (TAROS), pp. 19–24 (2009)
12. Evans, M.H., Fox, C.W., Pearson, M.J., Lepora, N.F., Prescott, T.J.: Whisker-object contact speed affects radial distance estimation. In: 2010 IEEE International Conference on Robotics and Biomimetics (ROBIO), pp. 720–725 (2010)
13. Fox, C.W., Evans, M.H., Lepora, N.F., Pearson, M., Ham, A., Prescott, T.J.: CrunchBot: A mobile whiskered robot platform. In: Groß, R., Alboul, L., Melhuish, C., Witkowski, M., Prescott, T.J., Penders, J. (eds.) TAROS 2011. LNCS (LNAI), vol. 6856, pp. 102–113. Springer, Heidelberg (2011)
14. Sullivan, J.C., Mitchinson, B., Pearson, M.J., Evans, M.H., Lepora, N.F., Fox, C.W., Melhuish, C., Prescott, T.J.: Tactile discrimination using active whisker sensors. IEEE Sensors Journal 12(2), 350–362 (2012)
15. Szwed, M., Bagdasarian, K., Ahissar, E.: Encoding of vibrissal active touch. Neuron 40(3), 621–630 (2003)
16. Lottem, E., Azouz, R.: A unifying framework underlying mechanotransduction in the somatosensory system. The Journal of Neuroscience 31(23), 8520–8532 (2011)

17. Lee, H., Battle, A., Raina, R., Ng, A.Y.: Efficient sparse coding algorithms. In: Advances in Neural Information Processing Systems, vol. 19, p. 801 (2007)
18. Petersen, R., Brambilla, M., Bale, M., Alenda, A., Panzeri, S., Montemurro, M., Maravall, M.: Diverse and temporally precise kinetic feature selectivity in the VPm thalamic nucleus. Neuron 60(5), 890–903 (2008)
19. Mitchinson, B., Gurney, K.N., Redgrave, P., Melhuish, C., Pipe, A.G., Pearson, M., Gilhespy, I., Prescott, T.J.: Empirically inspired simulated electro-mechanical model of the rat mystacial follicle-sinus complex. Proc. Biol. Sci. 271(1556), 2509–2516 (2004)

Octopus-Inspired Innovative Suction Cups

Maurizio Follador[1,2], Francesca Tramacere[1,2], Lucie Viry[1], Matteo Cianchetti[2],
Lucia Beccai[1], Cecila Laschi[2], and Barbara Mazzolai[1]

[1] Center for Micro-BioRobotics@SSSA, Istituto Italiano di Tecnologia,
Pontedera, Italy
[2] The BioRobotics Institute, Scuola Superiore Sant'Anna, Pisa, Italy
Maurizio.Follador@iit.it

Abstract. Octopus show great adhesion capabilities thanks to their
suckers covering their ventral side of their arms. Starting from biological investigation, we identified preliminary specifications for the design
of innovative artificial suction cups, which could be used in the field
of soft robotics. The main features of the biological sucker are maintained as leading criteria for the choice of the actuation technology and
mechanism. In this preliminary work, we focused on the imitation of the
functionality of the specific muscle bundles which generate suction to
obtain adhesion. Dielectric Elastomers Actuators (DEA) were identified
as a suitable solution. A study on materials and manufacturing techniques was made. Different possible solutions in the use of DEA are also
described.

Keywords: bioinspired wet adhesion, dielectric elastomers, soft robotics.

1   Functional Features of Octopus Suckers

Octopus sucker is a muscular hydrostat which consists of two main portions, an
upper hollow cup, the *acetabulum*, and a lower disk-like portion, the *infundibulum*. The two portions are connected by means of an orifice. The musculature
arrangement of octopus sucker consists of three kinds of fibers (Fig. 1(a)): radial
fibers (r) which extend throughout the thickness of the sucker; circular fibers
(ci) which are oriented parallel to the opening surface of the *infundibulum*; and
meridional fibers (m) which extend from the apex to the bottom of the whole
sucker. In order to obtain adhesion, a seal is formed thanks to the rim which
encircles the *infundibulum*, and then a the pressure differential is created in
between the internal cavity of the cup and the outside, by the contraction of
the radial muscles. Circular and meridional muscles are instead involved in the
detachment mechanism [1].

2   Specifications for the Artificial Sucker

The study of the biological sucker gives inputs for the definition of the design
specifications for the performances expected in the suction cup inspired by the

N.F. Lepora et al. (Eds.): Living Machines 2013, LNAI 8064, pp. 368–370, 2013.

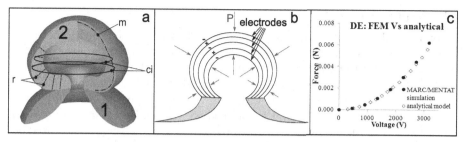

Fig. 1. (a) 3D CAD model of the octopus sucker 1- *infundibulum* 2-*acetabulum* r- radial muscles ci- circular mmuscles m- meridional muscles. (b) concept for the use of DEA, longitudinal section P- radial pressure generated by DEA. (c) Force-Voltage graph of actuator in planar stretching, analytical and FEM model comparison, ref. Sect. 3.1

natural counterpart. Octopus suckers are able to generate a differential pressure up to 0.27 MPa, in the time scale of milliseconds. This specifications was used to identify the most suitable technology for the realization of the bioinspired artefact. At this preliminary stage, we focused on the role of the radial muscles, which produce the differential pressure for the adhesion. The characteristics needed for this particular application are: good compliance, for the integration in a soft structure, capability to work in wet conditions and possibility of minia-turization. We identified dielectric elastomers as actuators able to mimic the contraction effect performed by radial muscles. The actuators developed, de-scribed in Sect. 3.1, can generate an internal stress up to 0.4 MPa with a speed of hundreds of milliseconds.

3 Design of the Artificial Sucker

Dielectric elastomers represent a good candidate for the characteristics men-tioned in Sect. 2. Moreover the fabrication procedure makes them suitable for the design of actuators with different shapes and geometries. In presence of an electric field, the dielectric membrane is subjected to a stress field described by the Maxwell stress tensor:

$$\sigma_M = \varepsilon E \otimes E - \frac{\varepsilon (E \cdot E) I}{2} . \tag{1}$$

where E is the electric field, I is the identity tensor and ε is the dielectric constant of the material. In the proposed solution, the actuator is composed by many dielectric elastomer layers, as represented in Fig. 1(b), which compose the acetabular portion of the suction cup. When the voltage is applied, the electrodes are attracted to each other, inducing a reduction of the thickness of the acetabular wall and an increase of the volume enclosed. If the rim of the *infundibulum* is sealed on the substrate, the volume of water inside the sucker remains constant. In this configuration, the isometric contraction of the actuator reduces the pressure of the internal fluid and adhesion force is generated.

3.1 Experimental Results

A commercial Fluorosilicone (Dow Corning 730) was selected for its dielectric properties [2]. This dielectric material is spin-coated to obtain a layer of 70 − 100 μm. The electrode, made of carbon black, is sprayed on the silicone surface after suspension in a solvent [3]. A procedure for the fabrication of multilayer actuators, in a fast and repeatable way, was defined. It consists of the following main steps: (i) the dielectric layer is spin-coated; (ii) the electrode is sprayed on it just before complete curing of the silicone; (iii) a second layer is spin-coated on the previous one. This steps are repeated to obtain the desired number of layers. The design of the actuator, as shown in Fig. 1(b), could present some issues in sustaining the pressure difference, due to the compliance of the actuator. In order to overcome this limit, we consider to embed an elastic element, with the aim to support the silicone active membrane and accumulate elastic energy. This passive elements work as antagonist respect to the membrane. When the actuator is in rest position it constrains the elastic element, which stores energy, that is released when the actuator is activated. A FEM model was developed in Marc/Mentat (Msc Software, CA, USA) to evaluate the performances of the actuators, the different geometries and the interaction with the rest of the soft structure. A good correspondence between the analytical model (1) and the simulation was found and shown in Fig. 1(c). The force produced by a rectangular actuator undergoing planar stretching in an isometric actuation, as illustrated in [4], was calculated with both methods and compared.

4   Discussion and Future Work

This work is the initial step through the design of an active suction cup bioinspired by the octopus sucker. This preliminary study on the capabilities of dielectric elastomers, demonstrates the possibility to fabricate an actuator that mimics the functionality of octopus sucker radial muscles. The use of FEM models will be used in the design process of the actuators in relation with the other part of the structure, which, for design specifications, are totally soft. The artificial sucker prototypes will be designed starting from the reconstruction of the biological sucker (Fig. 1(a)), embedding the actuated parts in the *acetabulum*.

References

1. Kier, W.M., Smith, A.M.: The Structure and Adhesive Mechanism of Octopus Suckers, 1153, 1146–1153 (2002)
2. Pelrine, R., Kornbluh, R., Joseph, J., Heydt, R., Pei, Q., Chiba, S.: High-field deformation of elastomeric dielectrics for actuators. Materials Science and Engineering: C 11(2), 89–100 (2000)
3. Carpi, F., Migliore, A., Serra, G., De Rossi, D.: Helical dielectric elastomer actuators. Smart Materials and Structures 14(6), 1210 (2005)
4. Kofod, G., Sommer-Larsen, P.: Silicone dielectric elastomer actuators: Finite-elasticity model of actuation. Sensors and Actuators A: Physical 122(2), 273–283 (2005)

A Cognitive Neural Architecture as a Robot Controller

Zafeirios Fountas and Murray Shanahan

Department of Computing, Imperial College London,
South Kensington Campus, London SW7 2AZ
{zfountas,m.shanahan}@imperial.ac.uk

Abstract. This work proposes a biologically plausible cognitive architecture implemented in spiking neurons, which is based on well-established models of neuronal global workspace, action selection in the basal ganglia and corticothalamic circuits and can be used to control agents in virtual or physical environments. The aim of this system is the investigation of a number of aspects of cognition using real embodied systems, such as the ability of the brain to globally access and process information concurrently, as well as the ability to simulate potential future scenarios and use these predictions to drive action selection.

Keywords: Spiking neural networks, action selection, cognitive architectures, global neuronal workspace.

1 Introduction

Computational models of cortical structures and the basal ganglia usually fall into two categories. On the one hand some high-detail bottom-up models aim at the validation of particular neurophysiological theories [1–3] while on the other hand, faster but less realistic top-down models focus on the robust cognitive control of robotic systems [4]. Hence, the majority of the current robotic implementations are likely to neglect features that are vital for the cognitive function of the nervous system. The model we propose here is an attempt to reduce this gap and to provide a working robot controller that could be used in neurobehavioural studies of action selection, in the same fashion as in [5, 6].

An early version of this neural architecture has been used as the main controller of NeuroBot [7, 8], an autonomous agent (bot) designed to control avatars within the environment of the computer game Unreal Tournament 2004, displaying human-like behaviour. The performance of NeuroBot was tested by competing in the 2011 BotPrize competition, a Turing test for AI bots, where it came a close second, achieving a humanness rating of 36%, while the most 'human' human reached 67%. Despite its success, this system was subjected to some limitations, such as its purely reactive nature and the lack of biological plausibility in the mechanism for action selection.

In this work we re-engineered the cognitive architecture of NeuroBot in order to overcome the above limitations and to introduce new cognitive features such as

N.F. Lepora et al. (Eds.): Living Machines 2013, LNAI 8064, pp. 371–373, 2013.

approximations to the concepts of imagination and emotion. The new improved neural system is based on the cognitive architecture proposed in [6] and it is able to simulate potential future scenarios, given the currently selected motor output. These predictions are used by the system, when the outcome of the selected motor output is not yet clear, and they modulate action selection through the affective responses they evoke.

2   Architecture and Methods

The Architecture is depicted in figure 1 and comprises four main areas that are functionally analogous to corresponding brain structures. These include the basal ganglia (BG), thalamus, amygdala (Am) and the cortex.

Inputs to the system are encoded in spiking spatio-temporal patterns which travel through the thalamus to the cortex and cause motor cortical responses. The BG then handle the selection of the most salient motor action, though in case that this salience is weak, they also impose a veto to its execution until the higher-order loops conclude whether it leads to a rewarding of punishing state.

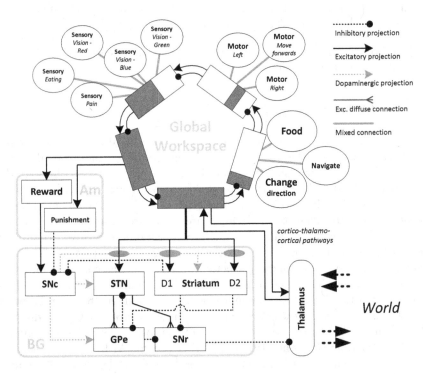

Fig. 1. The neural architecture. Excitation of SNc leads to the increase of the dopamine levels in the connected areas. Reward neurons excite SNc while punishment neurons inhibit it. Finally, the presence of internal noise in the same neurons keeps the level of dopamine constant, in the absence of an input signal.

The cortex analogue is based on a model of information flow described in [9] and its extended version used in NeuroBot, while the circuitry of the BG is based on the model of Humphries et al [1].

Methods. All neurons of this system are governed by the Izhikevich simple model [10] with a different set of parameters for every type of neuron. The majority of the parameters are taken from [11]. An exception is the case of the medium spiny projection neurons of neostriatum, where an extension of the simple model is taken from [2]. Also, to model the synaptic dynamics, we used a standard conductance-based approach and dopamine-modulated spike-timing-dependent plasticity as described in [3].

Acknowledgments. This work was supported by an EPSRC doctoral training grand.

References

1. Humphries, M.D., Stewart, R.D., Gurney, K.N.: A physiologically plausible model of action selection and oscillatory activity in the basal ganglia. The Journal of Neuroscience 26(50), 12921–12942 (2006)
2. Humphries, M.D., Lepora, N., Wood, R., Gurney, K.: Capturing dopaminergic modulation and bimodal membrane behaviour of striatal medium spiny neurons in accurate, reduced models. Frontiers in Computational Neuroscience 3(26) (2009)
3. Izhikevich, E.M., Edelman, G.M.: Large-scale model of mammalian thalamocortical systems. Proc. of the National Academy of Sciences 105(9), 3593–3598 (2008)
4. Eliasmith, C., Stewart, T.C., Choo, X., Bekolay, T., DeWolf, T., Tang, C., Rasmussen, D.: A large-scale model of the functioning brain. Science 338(6111), 1202–1205 (2012)
5. Prescott, T.J., Montes González, F.M., Gurney, K., Humphries, M.D., Redgrave, P.: A robot model of the basal ganglia: behavior and intrinsic processing. Neural Networks 19(1), 31–61 (2006)
6. Shanahan, M.: A cognitive architecture that combines internal simulation with a global workspace. Con&Cog 15, 433–449 (2006)
7. Fountas, Z., Gamez, D., Fidjeland, A.K.: A neuronal global workspace for human-like control of a computer game character. In: IEEE Conference on Computational Intelligence and Games (CIG), pp. 350–357 (2011)
8. Gamez, D., Fountas, Z., Fidjeland, A.K.: A Neurally Controlled Computer Game Avatar With Humanlike Behavior. IEEE Transactions on Computational Intelligence and AI in Games 5(1), 1–14 (2013)
9. Shanahan, M.: A Spiking Neuron Model of Cortical Broadcast and Competition. Con&Cog 17, 228–303 (2008)
10. Izhikevich, E.M.: Simple model of spiking neurons. IEEE Transactions on Neural Networks 14(6), 1569–1572 (2003)
11. Izhikevich, E.M.: Dynamical systems in neuroscience: the geometry of excitability and bursting. MIT Press, Cambridge (2006)

A Small-Sized Underactuated Biologically Inspired Aquatic Robot

Max Fremerey, Steven Weyrich, Danja Voges, and Hartmut Witte[*]

Technische Universität Ilmenau, Chair of Biomechatronics, Ilmenau, Germany
{maximilian-otto.fremerey,steven.weyrich,danja.voges,
hartmut.witte}@tu-ilmenau.de

Abstract. This extended abstract introduces the biologically inspired swimming robot URMELE *light*. The robot embodies the strategy of massive underactuation. Propulsion is generated using a central module with one single actuator and coupled passive, compliant tail modules. Therewith this robot differs from other modular swimming robots like e.g. AmphiBot [3] or ACM-R5 [6] which feature motor-gear combinations in each module without defined in- and inter-module compliance. It also differs from fish-like robots like e.g. MT1, a robot which generates propulsion by a single drive and a "C-bending" tail structure [8] due to the explicit use of spring elements for locomotion.

The swimming behavior of URMELE *light* is changeable: we aim at a shift between as well anguilliform as thunniform swimming modes due to tunable compliant elements. Therewith URMELE *light* could be used as reproducibly displaceable experimental platform for investigation of swimming abilities of fishes, e.g. energy efficiency of different swimming modes. Aside it may serve as a mobile sensing unit for control of water quality (e.g. detection and tracking of oil or chemicals) with minimized environmental disturbances. Currently URMELE light performs a thunniform swimming mode.

Keywords: tunable compliance, biologically inspired swimming robots, modular robotics, autonomous systems.

1 Introduction and State of the Art

Focusing on compliant elements Cavagna et al. 1977 postulated the spring-mass interaction as one of the 'two basic mechanisms for minimizing energy expenditure' [2] in terrestrial locomotion. In 1986/1989 Blickhan introduced the first calculus for spring-mass models, making the effects of compliance mathematically manageable [1]. Furthermore he and his co-workers (Wagner, Seyfarth, Geyer) showed the importance of compliant elements for self-stabilization of pedal locomotion. Therewith robustness widely replaces neural control, 'intelligent mechanics' (Witte & Fischer, SAB '98) allows to avoid (or at least to reduce) the amount of neural resp. technical calculation.

---

[*] Corresponding author.

N.F. Lepora et al. (Eds.): Living Machines 2013, LNAI 8064, pp. 374–377, 2013.
© Springer-Verlag Berlin Heidelberg 2013

Aligning these results with the requirements for mobile robots, directed tunable compliance in the drive chain of the robot would offer several benefits: Robust locomotion and an adaption to different substrates due to a real-time shift between known gaits could be achieved as well as an optimization of the overall power requirement – with reduces control effort.

Transfer of the biological principles of tuneable compliant elements to techniques is done by several groups (van Ham et al., Tonietti et al., Sugar et al.). Mainly in the research field 'compliant actuators' several prototypes were introduced, e.g. MACCEPA [10] or the 'jack spring' actuator [7]. Using these scientific findings, the authors want to provoke concerted changes in locomotion patterns of mobile robots by the usage of tuneable spring elements. Therewith an approach detailed in [5] is followed, which currently is also focused by [11]. To allow neglection of the effects of gravitation during investigation (the dominating dynamic component during terrestrial locomotion - even in fast gaits – and arboreal locomotion) successive approach compensates this influence by lift in a fluid.

2   URMELE *light*

2.1   Design

During this study, URMELE *light was* composed of a base structure and two tail modules. The base structure's dimension is about 150 mm · 75 mm · 33 mm, the size of a tail module amounts 30 mm · 47 mm · 0.3 mm. The overall weight of the robot is about 190 g. Among each other's, the tail modules are coupled by a pair of passive pull-springs (spring stiffness during study: 0.03 N/mm). A multi-body model for exploring the region of other suited values of spring stiffness is introduced in [4].

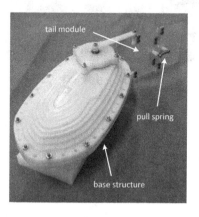
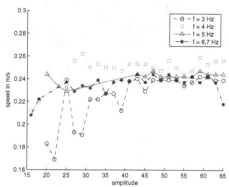

Fig. 1. left: Robot URMELE *light* **right**: Experiments with URMELE *light*, here variation of amplitude and frequency of servo driving sinusoidal function and tracking resulting speed

The actuator for generating thrust by elongating the tail modules is a single common RC servo drive (Modelcraft inc., type VSD-5E-HS, torque 100 Nmm). Core element of electronics is an ARDUINO™ Nano microcontroller. Control signals and status are transferred wireless between URMELE *light* and a host PC using XBEE modules. In order to obtain an oscillating locomotion pattern a sinusoidal function drives the servo:

$$x(t) = A \cdot \sin(2 \cdot \pi \cdot f \cdot t) + b \tag{1}$$

with x(t) = position of servo drive, A = amplitude, t = time, f = frequency, b = bias (offset)

2.2 Experiments

The successively described setup investigates the range of suited control parameters for URMELE *light*; in detail influence of frequency f and amplitude A in terms of resulting speed are point of interest. Therefore a systematic parameter search was performed. Featuring two tail modules, URMELE *light* performs a thunniform locomotion type during successive experiments [9].

During all runs, URMELE *light* had to pass a defined distance of two meters. Tracking time speed was calculated subject to frequency and amplitude adjusted. To ensure comparable conditions during all runs bias was adapted to guarantee straightforward swimming.

Analyzing previous experiments best results are achieved at a frequency of 4 Hz and an amplitude of 28 degrees. Therewith URMELE *light* attains a speed of $\approx$ 0.27 m/s (two times body length per second). Further experiments, especially with more tail modules and different spring stiffness's, are in progress.

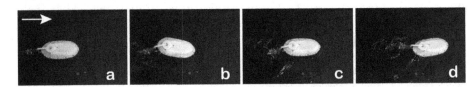

Fig. 2. Swimming sequence of URMELE *light*, speed is about 0.25 m/s

References

[1] Blickhan, R.: The spring-mass model for running and hopping. J. Biomech. 22, 1217–1227 (1989)
[2] Cavagna, G.A., Heglund, N.C., Taylor, C.R.: Mechanical work in terrestrial locomotion: two basic mechanisms for minimizing energy expenditure. Am. J. Physiol. 233, 243–261 (1977)
[3] Crespi, A., Badertscher, A., Guinard, A., Ijspeert, A.: AmphiBot 1: an undulant snake-like robot. Robotics and Autonomous Systems 50, 163–175 (2005)
[4] Fremerey, M., Fischheiter, F., Mämpel, J., Witte, H.: Reducing complexness of control by intelligent mechanics in undulant swimming robots. Int. J. of Design & Nature and Ecodynamics 7(1), 1–13 (2012)

[5] Fremerey, M., Fischheiter, L., Mämpel, J., Witte, H.: Design of a single actuated, undulant swimming robot. In: Proceedings of 3rd International Symposium on Mobiligence, Awaji, Japan, pp. 174–178 (2009)

[6] Hirose, S., Yamada, H.: Snake-Like Robots - Machine Design of Biologically Inspired Robots. IEEE Robotics & Automation Magazine, 88–98 (2009)

[7] Hollander, K., Sugar, T., Herring, D.: Adjustable robotic tendon using a 'jack spring'. In: Proc. 9th Int. Conf. Rehabilitation Robotics (ICORR 2005), pp. 113–118 (June-July 2005)

[8] Liu, J., Dukes, I., Hu, H.: Novel Mechatronics Design for a Robotic Fish. In: Intelligent Robots and Systems (IROS 2005), pp. 807–812 (2005)

[9] Sfakiotakis, M., Lane, D.M., Davies, J.B.: Review of fish swimming modes for aquatic locomotion. IEEE Journal of Oceanic Engineering 24, 237–252 (1999)

[10] Van Ham, R., Van Damme, M., Verrelst, B., Vanderborght, B., Lefeber, D.: 'MACCEPA', the mechanically adjustable compliance and controllable equilibrium position actuator: A 3DOF joint with 2 independent compliances. Int. Appl. Mech. 4, 130–142 (2007)

[11] Ziegler, M., Hoffmann, M., Carbajal, J.P., Pfeifer, R.: Varying body stiffness for Aquatic Locomotion. In: 2011 IEEE International Conference on Robotics and Automation, Shanghai, China, pp. 2705–2712 (2011)

Neural Networks Learning the Inverse Kinetics of an Octopus-Inspired Manipulator in Three-Dimensional Space

Michele Giorelli*, Federico Renda, Gabriele Ferri, and Cecilia Laschi

The BioRobotics Institute, Scuola Superiore SantAnna, Pisa, Italy
{michele.giorelli,federico.renda,cecilia.laschi}@sssup.it,
ferri@cmre.nato.int

Abstract. The control of octopus-like robots with a biomimetic design is especially arduous. Here, a manipulator characterized by the distinctive features of an octopus arm is considered. In particular a soft and continuous structure with a conical shape actuated by three cables is adopted. Despite of the simple design the arm kinetics model is infinite dimensional, which makes exact analysis and solution difficult. In this case the inverse kinetics model (IK-M) cannot be implemented by using mathematical methods based on Jacobian matrix, because the differential equations of the direct kinetics model (DK-M) are non-linear. Different solutions can be evaluated to solve the IK problem. In this work, a neural network approach is employed to overcome the non-linearity problem of the DK-M. The results show that a desired tip position can be achieved with a degree of accuracy of 1.36% relative average error with respect to the total length of the arm.

The solution to the *inverse kinetics* problem of soft manipulators is essential to generate paths in the task space in order to perform grasping or other tasks. To address this issue, researchers have proposed different iterative methods based on the Jacobian matrix. Although these methods have been traditionally applied with success to piecewise constant curvature manipulators, they can also be employed with non-constant curvature manipulators. However, in this case, particular attention should be taken on the feasibility aspect. This issue is investigated in our previous work, where the Jacobian method has been successfully applied to non-constant curvature manipulators driven by only two cables [1], because, in this case, a *linear* differential equation is able to represent the direct kinetics model.

In this work the inverse kinetics problem is addressed for a non-constant curvature manipulator driven by three cables (Fig. 1). The cables are arranged at an angle of $2\pi/3$ rad. An exact geometrical model of this manipulator has been employed [2]. The differential equations of the mechanical model are *non-linear*,

---

* This work was supported by the European Commission in the ICT-FET OCTOPUS Integrating Project, under contract no. 231608.

N.F. Lepora et al. (Eds.): Living Machines 2013, LNAI 8064, pp. 378–380, 2013.

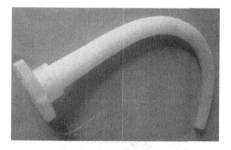

Fig. 1. Three-cable conical shape manipulator made of silcone

therefore the analytical solutions are difficult to calculate. Since the exact solutions of the mechanical model are not available, the elements of the Jacobian
matrix cannot be calculated. To overcome intrinsic problems of the methods
based on Jacobian matrix, a neural network learning the inverse kinetics of a
soft manipulator is proposed. In particular, a feed-forward neural (FNN) network has been chosen for its effectiveness at approximating functions. To authors
knowledge this is the first case of a model learning such as a neural network is
employed to solve the inverse kinematics problem of a soft robot. A complete
review of supervised learning methods used for rigid robots is provided by [3] [4].
The neural network implemented here is a fully-connected FNN with one hidden layer and using the hyperbolic tangent as activation function. The training
algorithm used is a back-propagation algorithm, which implements the gradient
descent method including a momentum term. The FNN takes the tip position as
input and provides the cable tensions as output. The data set has been generated
by means of the direct kinetics model [2]. The size of data set is 500 samples. The
data set has been divided first of all in training set (80%) and test set (20%).
The training set is also split in estimation set (80%) and validation set (20%).
The training set is used during the learning phase, whereas the test set is only
employed to evaluate the performance of the FNN. The validation set is used for
model selection and for avoiding the overfitting. The model selection provides
the right number of neurons in the hidden layer (N_H^*). In our case 34 neurons
are enough to solve the IK problem. After the training, the FNN approximate
the function which maps the manipulator tip position onto the forces applied
to the cables. The results on test set (Fig. 2) show that a desired tip position
can be achieved with an average absolute error of 4.2mm, i.e with a degree of
accuracy of 1.36% relative average error with respect to the total length of the
arm. The computational cost of the FNN is $7N_H^* + 3 = \mathcal{O}(N_H^*)$, and it takes
0.162ms on a 2.2GHz microprocessor.

The feasibility study accomplished in this work represents a preliminary step
for further investigation on a real prototype. A set-up will be implemented with
video-cameras and force sensors to collect the data. An aspect particularly interesting to be investigated is how the NNs are able to adapt to robot variation.
The soft structure makes soft robots particularly sensitive to the environment

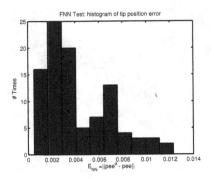

Fig. 2. Histogram of the FNN tip errors (in m) calculated using the test set

variation. In the future, a sensitivity analysis with respect to the variation of the NN parameters will be conducted and adaptive methods will be considered to overcome the lack of robustness of the soft structure.

References

1. Giorelli, M., et al.: A Two Dimensional Inverse Kinetics Model of a Cable Driven Manipulator Inspired by the Octopus Arm. In: Proc. IEEE Int. Conf. on Robot. and Automat., St. Paul, USA, pp. 3819–3824 (2012)
2. Renda, F., Laschi, C.: A general mechanical model for tendon-driven continuum manipulators. In: Proc. IEEE Int. Conf. on Robot. and Automat., St. Paul, USA, pp. 3813–3818 (2012)
3. Sigaud, O., Salaün, C., Padois, V.: On-line regression algorithms for learning mechanical models of robots: a survey. Robotics and Autonomous Systems 59(12), 1115–1129 (2011)
4. Nguyen-Tuong, D., Peters, J.: Model learning for robot control: a survey. Cognitive Processing 12(4), 319–340 (2011)

A Minimal Model of the Phase Transition into Thermoregulatory Huddling

Jonathan Glancy*, Roderich Groß, and Stuart P. Wilson

The University of Sheffield, Sheffield, UK
{j.glancy,r.gross,s.p.wilson}@sheffield.ac.uk

Abstract. Huddling by endotherms is an important model through which to study the emergence of complexity. Canals et al. (2011) have recently described the emergence of huddling in rodents as a phase transition mediated by the ambient environmental temperature [1]. We present an agent-based model as a minimal account of the reported transition to huddling at low temperatures. Simulation results suggest that the huddle self-organises as ambient temperature changes drive individuals from 'orient-from-contacts' to 'orient-to-contact' behaviours.

Keywords: self-organisation, agent-based model, thermoregulation.

1 Introduction

Aggregate behaviours in groups of animals are often the result of self-organisation. Without requiring explicit blueprints, templates, recipes, or leadership to construct patterned behaviour, self-organising systems emerge from individual responses to local environmental cues [2]. The thermoregulatory huddling behaviours of endotherms, such as rodents, has been proposed as a model self-organising system [3].

Juvenile rodents huddle at low ambient temperatures. During early postnatal development, the huddle allows individuals to efficiently regulate their body temperature and conserve energy. Whilst huddling, individuals are able to consume less oxygen, operating at lower metabolic rates [4]. The efficiency of the huddle is mostly attributable to a reduction in the exposed surface area of each pup, and the energy saved in the micro-climate of the huddle can instead be focused towards development and growth.

Canals et al. [1] studied the huddling behaviour of white mice whilst controlling the ambient temperature of the environment. They found that huddling exhibits a second order critical phase transition, driven by changes in the ambient temperature, such that huddling behaviour is only active below a critical ambient temperature. Here we describe the development of a minimalistic model of huddling that reproduces the critical phase transition found by ref. [1].

---

* Corresponding author.

N.F. Lepora et al. (Eds.): Living Machines 2013, LNAI 8064, pp. 381–383, 2013.

2   Methods

We developed an agent-based model of huddling, where each pup is an agent based upon the Braitenberg type IIb vehicle [5]. Accordingly, each pup is modeled as a circle (of radius r_{pup}), with each half of the body acting as one of two sensors controlling the drive speeds of a motor on the opposite side. The circular 'skin' of each is discretised into 1000 'taxels', which act as temperature sensors. If a taxel is exposed, it detects the ambient temperature (T_A), else if it intersects with another pup it detects body temperature ($T_B = 37°$). For each sensor, ϵ is the proportion of taxels exposed to the ambient temperature, and at each time step, the sensor value is:

$$s = T_A \epsilon + T_B (1 - \epsilon). \tag{1}$$

Each pup is allocated a finite (arbitrarily chosen) energy to convert into forward velocity, $M = 0.5$. This energy is divided between two 'legs' (i.e., wheels) based on the difference between the two sensor values (in crossed-wires configuration):

$$m_{\text{left}} = M \frac{s_{\text{right}}}{s_{\text{left}} + s_{\text{right}}}, \quad m_{\text{right}} = M \frac{s_{\text{left}}}{s_{\text{left}} + s_{\text{right}}}. \tag{2}$$

The motor drives are used to estimate the rates of translation and rotation:

$$\frac{\Delta \mathbf{x}}{\Delta t} = \frac{1}{2} M \begin{bmatrix} \cos(\theta) \\ \sin(\theta) \end{bmatrix}, \quad \frac{\Delta \theta}{\Delta t} = \tan^{-1} \left(\frac{2 r_{\text{pup}}}{m_{\text{left}} - m_{\text{right}}} \right), \tag{3}$$

which are used to update the position and orientation as follows:

$$\mathbf{x}_{t+1} = \mathbf{x}_t + \frac{\Delta \mathbf{x}}{\Delta t} h, \quad \theta_{t+1} = \theta_t + \frac{\Delta \theta}{\Delta t} h, \tag{4}$$

where $h = 0.01$ is the integration time step. Collisions between two pups, or between a pup and wall of the arena (where the ratio $r_{\text{arena}}/r_{\text{pup}} = 10$ sets the size of the circular arena), is handled using simple spring equations:

$$\mathbf{x}_{\text{post}} = \mathbf{x}_{\text{pre}} - kX\hat{\mathbf{v}}h, \tag{5}$$

where $\hat{\mathbf{v}}$ is the unit vector between the pup and the point of collision, X is the extent of the overlap between them, and $k_{\text{arena}} = 50.0$ and $k_{\text{pup}} = 75.0$ are spring constants.

3   Results

We ran simulations consisting of 15 agents with random initial positions and orientations for 1000 time steps. To quantify huddling we measured the average proportion of taxels in contact across all agents and time steps. A higher proportion shows that the pups have successfully reduced their exposed surface area, and are thus able to maintain a lower metabolic rate.

We investigated the robustness of the phase transition in two ways, firstly by introducing noise to each of the taxels (each taxel detects $T \cdot \mathcal{N}(1, \sigma^2)$).

And secondly, by changing the way that sensors measure ambient and body temperatures:

$$s = T_A \epsilon + \alpha T_B (1 - \epsilon). \tag{6}$$

The phase transition occurs at the critical temperature (T_C) when $T_A = \alpha T_B$. For ambient temperatures $T_A < \alpha T_B$, the simulated pups orient towards contacts, and when $T_A > \alpha T_B$ they orient away. We can then manipulate the parameter α to tune the model to have a desired point of transition. The results of these experiments can be seen in Figure 1.

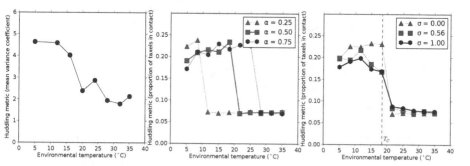

Fig. 1. *Left:* Data from mice showing a second-order phase transition into huddling at low ambient temperatures (height variance indicates huddling; modified from [1], with error bars removed). *Center:* Results from our minimal model also reveal a switch to huddling at low ambient temperatures (the free parameter $\alpha = 0.5$ was used to set the switching temperature $T_C = \alpha T_B = 17.5°C$). *Right:* Adding noise to each pup's sensors smooths the change in group behaviour, revealing a second-order phase-transition that provides a good qualitative match to the experimental data. At around T_C, individual pup behaviours transition between 'orient-towards-contact' and 'orient-from-contact' control, suggesting that simple mechanisms may also mediate the self-organised huddling of endotherms.

References

1. Canals, M., Bozinovic, F.: Huddling Behavior as Critical Phase Transition Triggered by Low Temperatures. Complexity 17(1), 35–43 (2011)
2. Camazine, S., Deneubourg, J.L., Franks, N.R., Sneyd, J., Theraulaz, G., Bonabeau, E.: Self-Organization in Biological Systems. Princeton University Press (2003)
3. Schank, J.C., Alberts, J.R.: Self-organized huddles of rat pups modeled by simple rules of individual behavior. Journal of Theoretical Biology 189(1), 11–25 (1997)
4. Alberts, J.R.: Huddling by rat pups: group behavioral mechanisms of temperature regulation and energy conservation. Journal of Comparative and Physiological Psychology 92(2), 231–245 (1978)
5. Braitenberg, V.: Vehicles: Experiments in Synthetic Psychology. MIT Press (1984)

Towards Bio-hybrid Systems Made of Social Animals and Robots

José Halloy[1], Francesco Mondada[2], Serge Kernbach[3], and Thomas Schmickl[4]

[1] LIED, Université Paris Diderot, Paris, France
jose.halloy@univ-paris-diderot.fr
[2] Laboratoire de Systèmes Robotiques, Ecole Polytechnique Fédérale de Lausanne, Switzerland
francesco.mondada@epfl.ch
[3] Cybertronica Research, Research Center of Advanced Robotics and Environmental Science,
Stuttgart, Germany
serge.kernbach@cybertronica.co
[4] Artificial Life Lab of the Department for Zoology, Karl-Franzens University Graz, Austria
thomas.schmickl@uni-graz.at

Abstract. For making artificial systems collaborate with group-living animals, the scientific challenge is to build artificial systems that can perceive, communicate to, interact with and adapt to animals. When such capabilities are available then it should be possible to built cooperative relationships between artificial systems and animals. Machines In this framework, machines do not replace the living agents but collaborate and bring new capabilities into the resulting mixed group. On the one hand, such artificial systems offer new types of sensors, actuators and communication opportunities for living systems; on the other hand the animals bring their cognitive and biological capabilities into the artificial systems. Novel bio-hybrid modeling frameworks should be developed to streamline the implementation issues and allow for major time saving in the design and building processes of artificial agents. We expect strong impacts on the design of new intelligent systems by merging the best of the living systems with the best of ICT systems.

Keywords: Mixed society, bio-hybrid systems, collective intelligence, social emergence, collective robotics, behavioral biology.

1 Introduction

The scientific field of animal-machine interaction at the collective level has been barely explored. We envision the design of intelligent artificial systems capable of closing the loop of interaction between animals and robots. Our envisioned methodology allows numerous social interactions among individuals of a mixed society composed of animals and robots, finally showing novel system properties [4]. Robots that interact with animals and can participate in their social activity can form a mixed robot-animal society, which is coherent at the collective level. Mixed societies are dynamical systems where animals and artificial agents interact and cooperate to

N.F. Lepora et al. (Eds.): Living Machines 2013, LNAI 8064, pp. 384–386, 2013.

produce shared collective intelligence. Artificial agents do not replace the animals but collaborate and bring new capabilities into the mixed society that are inaccessible to the pure groups of animals or artificial agents. Each category of agents, living or artificial, may react to signals or perform tasks that the other category does not detect or perform.

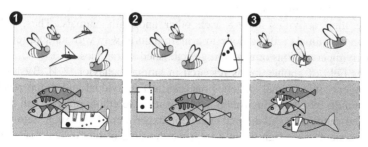

Fig. 1. Concepts of artificial (1) mobile, (2) static and (3) mounted nodes interacting with different types of animal societies

Artificial agents interacting with living animals may be designed in different forms (see Fig.1): (1) Mobile nodes: autonomous mobile robots that mix with living animals. (2) Static nodes: distributed immobile sensor-actuator units. (3) Mounted nodes: sensor-actuator units mounted on the animals themselves and conferring them new capabilities. The problems that have to be solved and a large part of the solution for these problems are unconventional compared to today's state of the art in ICT. The challenge for robotics and ICT is to design novel systems that are capable of handling natural collective intelligence in real-time heterogeneous environments [4].

2   Proposed Methodology

Mixed societies of animals and robots are at the crossroad of animal social behavior and collective robotics, as well as on the intersection of natural and artificial collective intelligence [3,4]. The development of such bio-hybrid systems requires parallel and coordinated research tracks between behavioral biology and robotics. The first technological challenge is to make the robots accepted by the animal societies by finding appropriate channels of communication corresponding to specific animal traits such as motion patterns, visual, olfactory, sound cues and cognitive processes They require the development of specific, novel and safe sensors and actuators. For example, good results with cockroaches were obtained by using olfactory cues [2]. These cues were calibrated olfactory signals that were simply deposited on the robot bodies. To establish interactions between chicken and autonomous mobile robots filial imprinting, an innate learning window can be used to create social attachments with robots [1]. Quantitative studies of animal social behavior have to be undertaken to build mathematical models, necessary to design the behavioral modules and programs used in robots [4]. The challenge is to put together interdisciplinary research teams involving behavioral biology and ICT engineering and sharing common scientific frameworks.

3   Expected Impact of Bio-hybrid Social Systems

One potential field of application is management of domestic animal stocks that all are social animals. This may lead to various agricultural applications such as low-stress management of livestock. Optimal management of such systems could be achieved by integrating artificial adaptive system elements into animal groups to coordinate their activities among each other and with animals. Thus, behavioral modulation of animals will be done in a more natural way without suppressing their instincts, living conditions or neglecting established hierarchies.

Another field of application is research in animal behavior [2,4]. By introducing artificial agents into animal societies one can test individual and group reactions to various stimuli; by combining robots with quantitative ethograms one can achieve an unparalleled automation of animal behavior experimentation. Such automated and robotized systems may significantly improve the field of biomedical research that is using model animals.

Social bio-hybrid systems could also be used to manage wildlife animal pests or resources in particular group living species. We envision artificial intelligent systems capable of interacting and modulating the behavior of wild species treated as pests or as valuable resources.

Acknowledgements. This work was supported by the following grants: EU-FP7 project 'ASSISI|bf', no. 601074; EU-FP7 project "CoCoRo", no. 270382; FWF (Austrian Science Fund) "REBODIMENT", no. P23943-N1.

References

1. Gribovskiy, A., Halloy, J., Deneubourg, J.L., Bleuler, H., Mondada, F.: Towards mixed societies of chickens and robots. In: Proceedings of the IEEE/RSJ International Conference on Intelligent Robots and Systems (IROS), pp. 4722–4728 (2010)
2. Halloy, J., et al.: Social integration of robots into groups of cockroaches to control self-organized choices. Science 318(5853), 1155–1158 (2007)
3. Martinoli, A., Mondada, F., Correll, N., Mermoud, G., Egerstedt, M., Hsieh, M.A., Parker, L.E., Støy, K. (eds.): Distributed Autonomous Robotic Systems. Springer Tracts in Advanced Robotics, vol. 83 (2013)
4. Mondada, F., Halloy, J., Martinoli, A., et al.: A General Methodology for the Control of Mixed Natural-Artificial Societies. In: Kernbach, S. (ed.) Handbooks of Collective Robotics. Pan Stanford Publishing (2013)

Biomimetic Spatial and Temporal (4D) Design and Fabrication

Veronika Kapsali, Anne Toomey, Raymond Oliver, and Lynn Tandler

P3i Research Group, Northumbria University, London, United Kingdom
{veronika.kapsali,anne.toomey,raymond.oliver,
lynn.tandler}@northumbria.ac.uk

Abstract. We imagine the built environment of the future as a 'bio-hybrid machine for living in' that will sense and react to activities within the space in order to provide experiences and services that will elevate quality of life while
coexisting seamlessly with humans and the natural environment. The study of
Hierarchical design in biological materials has the potential to alter the way
designers/ engineers/ craftsmen of the future engage with materials in order to
realise such visions. We are exploring this design approach using digital manufacturing technologies such as jacquard weaving and 3D printing.

Keywords: Biomimetics, 3D Printing, 4D Design, Digital Jacquard.

1 Introduction

The study of structural hierarchy in biological materials can alter the role of material
selection in the design or engineering process. Nature shows us that you can achieve
advanced, complex behaviours combining simple materials in clever structural com-
posites. The driver behind this is survival; organisms in nature rarely exist in environments with surplus resources, therefore those that fail to optimise the use of
raw materials simply do not survive. There is much we can learn as designers from
Nature's 'lean' operation.

We rely on the properties of materials to deliver a system, so we currently operate
in a space where the needs of the *system* inform the selection of *material*. We rely on
the material to deliver properties such as strength, toughness etc. When we need a
structure to demonstrate a specific property and we do not have a material that delivers the performance, we synthesize one that does. As a result there are over 300 man-
made polymers used commercially.

In Nature protein and polysaccharide are the two main polymers that form the basis
of all biological materials and structures [1]. Variations in the assembly of these materials deliver the vast range of properties demonstrated in biological materials. Insect
cuticle, for instance, is made from protein yet can be stiff or flexible, opaque or trans-
lucent, depending on the way the raw materials are put together [2]. In Nature, *material* forms the *system*.

N.F. Lepora et al. (Eds.): Living Machines 2013, LNAI 8064, pp. 387–389, 2013.
© Springer-Verlag Berlin Heidelberg 2013

Organisms have multiple levels of hierarchy. The organisation of raw materials within and across each level is what enables the rich diversity in properties demonstrated by biological structures [3]. We tend to use less complex non-hierarchical design processes, generally because it is more cost effective to invest in material than skilled labour/ craftsmanship necessary to achieve more complex structures.

In this hierarchical classification, the Eiffel tower is a third order design while metal frameworks forming the skeleton of conventional skyscraper buildings are classified as first order structures. The structure of the Eiffel tower is an iron lattice work made from relatively short bars of metal bolted into a shape (1st order), these configurations are assembled into greater structures (2nd order) these in turn are joined to compose the tower (3rd order). The metal framework of conventional buildings is composed of long lengths of structural steel that are bolted together (usually at right angles) to form a 1st order structure.

Iron is a relatively weak material especially when compared to the qualities of structural steel used in construction today. Many believed, at the time of its erection, that the Eifel tower would collapse because the quality of the material used to make it was not strong enough to support the weight of the structure. In fact Lakes (1993) estimated that the relative density ρ/ρ_0 (density ρ as mass per unit volume of structure divided by density ρ_0 of material of which it is made) of the Eifel tower is 1.2×10^{-3} times that of iron, while the metal skeleton of a skyscraper has relative density 5.7×10^{-3} of structural steel [4].

Designing with hierarchy can deliver strong structures from weak materials by managing strength and stiffness of composite systems. If this approach can deliver material systems with counterintuitive properties, can it aid us in the transition from static design to spatial and temporal engineering?

2   Biomimetic Realisation

Although Biomimetics has long reaching applications in both the digital and virtual worlds, conventional making processes do not generally lend themselves to mimicking complex architectures. Layered manufacturing methodology is an effective way of exploring artificial muscles and smart soft composite prototypes cheaply and efficiently [5,6] but designs are limited to linear structures.

Researchers at Reading University were able to mimic the microfibril orientation of cellulose in wood fibre cell walls, which is responsible for the stiff and ductile properties of wood, into a macro scale prototype. In order to build the prototype, that relied on accurate controlling the orientation of the fibres within a matrix, the team had to design and build a custom machine [7]. 'Technical plant stem' is a commercially scalable textile composite that combines high strength and impact resistance with minimal use of material [8]. The design combined knowledge of fibre orientation in wood with the anatomy of the giant reed stem. The prototype was made using advanced braid protrusion machinery at the Institute of Textile Technology and Process Engineering (ITV) Germany.

Recent advances in 3D printing technology in terms or resolution (micron versus previous millimetre scale) and range of useable materials have created a new platform for the exploration and experimentation of biologically inspired stimuli responsive 4D systems. This state of the art equipment is currently used in the biomedical sector for the creation of tissue scaffolds, which draw on the fine resolution capacity and the ability to print using high spec biomaterials [9].

We have identified the design principles behind hygroscopic seed dispersal mechanisms primarily in dehiscent legume pods as an ideal paradigm for technology transfer. Study of the hierarchical system reveals that the seedpod valves are simple bi-layers systems composed primarily of cellulose. Depending on the degree of difference in orientation of the cellulose microfibrils between these layers, the pods either twist or bend in dry conditions but always revert to their original shape when exposed to moisture. We are exploring shape change for the design of 4D composite systems using the orientation capabilities of advanced 3D Fibre deposition and digital weaving technologies. We wish to present an overview of this work in progress.

References

1. Vincent, J.: Structural biomaterials. Princeton University Press (2012)
2. Vincent, J.F., Wegst, U.G.: Design and mechanical properties of insect cuticle. Arthropod Structure & Development 33(3), 187–199 (2004)
3. Tirrell, J.G., Fournier, M.J.: Biomolecular materials. Chemical and Engineering News 72(51), 40–51 (1994)
4. Lakes, R.: Materials with structural hierarchy. Nature 361(6412), 511–515 (1993)
5. Weiss, L.E., Merz, R.: Shape deposition manufacturing of heterogeneous structures. Journal of Manufacturing Systems 16(4), 239–248 (1997)
6. Ahn, S.-H., Lee, K.-T.: Smart soft composite: An integrated 3D soft morphing structure using bend-twist coupling of anisotropic materials. International Journal of Precision Engineering and Manufacturing 13(4), 631–634 (2012)
7. Jeronimidis, G.: The fracture behaviour of wood and the relations between toughness and morphology. Proceedings of the Royal Society of London. Series B, Biological Sciences, 447–460 (1980)
8. Milwich, M., Speck, T.: Biomimetics and technical textiles: solving engineering problems with the help of nature's wisdom. American Journal of Botany 93(10), 1455–1465 (2006)
9. Moroni, L., de Wijn, J.R.: 3D fiber-deposited scaffolds for tissue engineering: Influence of pores geometry and architecture on dynamic mechanical properties. Biomaterials 27(7), 974–985 (2006)

A Swimming Machine Driven
by the Deformation of a Sheet-Like Body
Inspired by Polyclad Flatworms

Toshiya Kazama[1,2], Koki Kuroiwa[1], Takuya Umedachi[3,4], Yuichi Komatsu[1],
and Ryo Kobayashi[1,2]

[1] Hiroshima University, 1-3-1, Kagamiyama, Higashi-Hiroshima, 739-8526, Japan
[2] JST, CREST, 5, Sanbancho, Chiyoda-ku, Tokyo 102-0075, Japan
[3] JSPS, Kojimachi Business Center Building, 5-3-1 Kojimachi, Chiyoda-ku, Tokyo
102-0083, Japan
[4] Tufts University, 200 Boston Av. Rm 2613, Medford, MA. 02155, USA

Abstract. A swimming robot driven by the deformation of a sheet-like
body was developed, inspired by polyclad flatworms. The robot consists
of a soft oval rubber sheet and three motors. Two motors operate the
lateral flaps, and the third operates the body axis. The robot showed a
variety of speeds and swimming patterns depending on parameter values
such as the frequency and amplitude of flapping. We found that the robot
moved at essentially the same speed with different swimming patterns.

1 Introduction

Polyclad flatworms, an order in Turbellaria, show many types of locomotion, including swimming behaviours (Newman *et al.* 2003). The flat, oval body of these
animals is very soft and is used for swimming in an undulating or flapping manner (Fig.1a). However, the mechanism of locomotion, including the relationship
between the soft-body deformation and the speed, direction of motion, attitude,
and locomotion type (*e.g.*, hovering, quick turning, etc.), is not fully understood. To address this issue, we constructed a soft robot (Pfeifer *et al.* 2012;
Chu *et al.* 2012) and conducted experiments in the real world.

Recent advances in underwater robotics have provided ways of mimicking the
soft, flexible motion of living creatures using soft materials (Pfeifer *et al.* 2012)
and soft actuators (Chu *et al.* 2012), However, few studies have developed ways
to control the deformation of soft material or the propulsive mechanism, including the interaction between the soft material and fluid dynamics during underwater locomotion (Kazama et al. 2013a). This paper reports a preliminary study
designed to address these issues in soft robotics by studying polyclad flatworms.

2 Results

While swimming, polyclad flatworms undergo considerable deformation of the
body (Fig. 1a). In this study, we postulated that the deformation was generated

N.F. Lepora et al. (Eds.): Living Machines 2013, LNAI 8064, pp. 390–392, 2013.

by the frontal and lateral flaps and the body axis, as shown in Fig. 1b. Fig. 1c shows our robot. A rubber sheet was used for the soft, flat body. The flapping of the body axis and lateral flaps was controlled by motors, as shown in Fig. 1d. The angles of the motors, which were controlled by the equations in Fig. 1d, regulated the frequency f, amplitude A, and phase difference ϕ, of flapping. The robot could swim in water (Fig. 2c-d).

We investigated the dependence of the speed of the robot v on the different parameters. Fig. 2a shows the relationship between A and v; v increased with A. Fig. 2b shows the relationship between f and v; v increased with f. The trends of the graphs of $A \times v$ and $f \times v$ are within the acceptable parameter range of our robot. Differences in these parameters altered the behaviour of the robot. Snapshots of the swimming robot are shown in Fig. 2c ($A = 55$, $f = 1.0$ Hz) and d ($A = 40$, $f = 1.5$ Hz). The flapping amplitude of the robot in Fig. 2c is greater than that in Fig. 2d. Interestingly, despite such essentially different patterns of swimming, the speeds in both cases were similar, as shown in Fig. 2c and 2d. The use of different swimming patterns to achieve movement at the same speed might be used in different environments, $e.g.$, in an open area, a flatworm could flap with high amplitude and low frequency, whereas in a narrow area, a flatworm could flap with low amplitude and high frequency. This behavioural diversity might also be adaptive behaviour, reflecting the environmental adaptability of real flatworms.

In this robot, we also obtained locomotion diversity by changing the phase difference ϕ (Kazama $et\ al.$ 2013b). At the poster session, we will introduce our robot and report the body deformation analysis to explain the above experimental results in detail.

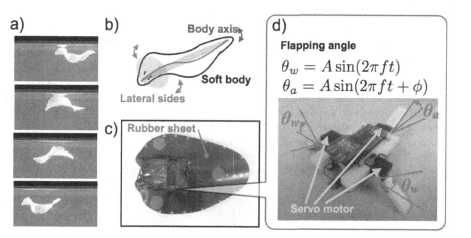

Fig. 1. a) Snapshots of the swimming behaviour of a polyclad flatworm (side view). **b)** Schematic of our robot. **c)** Our robot (top view). **d)** Control procedure for the robot.

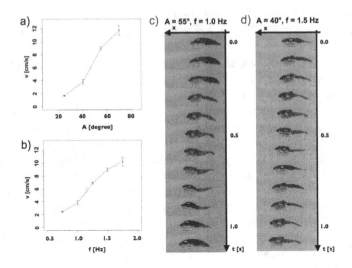

Fig. 2. a) The relationship between the amplitude A and speed v, with $f = 1.0$ Hz.
b) The relationship between the frequency f and speed v, with $A = 40$. The error
bar represents the standard deviation (n = 3). Snapshots of the robot swimming in
water at **c)** $A = 55$, $f = 1.0$ Hz and **d)** $A = 40$, $f = 1.5$ Hz. The robot is 27 cm long.
In the snapshots, the robot is swimming from right to left. The vertical width of the
space within which the robot was moving was greater in c) than in d). $\phi = 0.5\pi$ in all
experiments.

References

[Chu et al. 2012]Chu, W., et al.: Review of biomimetic underwater robots using smart
    actuators. International Journal of Precision Engineering and Manufacturing 13,
    1281–1292 (2012)
[Kazama et al. 2013a]Kazama, T., et al.: A development research for the swimming
    soft-robot with the mechanism of mutually feedback between fluid sensing and
    propulsive motion inspired by Polyclad flatworms. In: Proc. of 25th SICE Sym-
    posium on Decentralized Autonomous Systems, pp. 145–148 (2013)
[Kazama et al. 2013b]Kazama, T., et al.: Locomotion diversity emergence in the un-
    derwater soft-robot inspired by Polyclad flatworm (in review)
[Newman et al. 2003]Newman, L.J., et al.: Marine Flatworms:the world of polyclad.
    Csiro (2003)
[Pfeifer et al. 2012]Pfeifer, R., et al.: The Challenges Ahead for Bio-Inspired 'Softf'
    Robotics. Communications of the ACM 56, 76–87 (2012)

Towards a Believable Social Robot

Nicole Lazzeri, Daniele Mazzei,
Abolfazl Zaraki, and Danilo De Rossi

Research Center "E. Piaggio", Via Diotisalvi 2, 56126, Univ. of Pisa (Italy)
n.lazzeri@centropiaggio.unipi.it
http://www.faceteam.it

Abstract. Two perspectives define a human being in his social sphere: appearance and behaviour. The aesthetic aspect is the first significant element that impacts a communication while the behavioural aspect is a crucial factor in evaluating the ongoing interaction. In particular, we have more expectations when interacting with anthropomorphic robots and we tend to define them believable if they respect human social conventions. Therefore researchers are focused both on increasingly anthropomorphizing the embodiment of the robots and on giving the robots a realistic behaviour.

This paper describes our research on making a humanoid robot socially interacting with human beings in a believable way.

Keywords: Believability, social robots, human-robot interaction.

1   Introduction

Believable characters have always been a central issue since the first attempts in the entertainment industry for making attractive movies, cartoons and games. From 1970s to the present days, with the rapid advances in computer graphics, robotics technology and artificial intelligence, many fictional characters are becoming reality. The type of interaction is definitely based on the nature of the robot itself and on the aim which guided its development.

Social robots represent an emerging field of research focused on developing a "social intelligence" in order to maintain the illusion of dealing with a real human being. The term "social intelligence" implies the ability to interact with other people, to interpret and convey emotional signals and to perceive and react to people's intentions. On first encounter, the believability of a robot is communicated through its physical embodiment which strongly influences people's expectations about how it behaves. Later on the perception of believability of the robot is given by its expressiveness, behaviour and reactions to external stimuli which can make a human-robot interaction more or less natural and lifelike.

The purpose of this article is to show the architecture of a system for controlling a social robot in a believable human-like way. In particular, we focus the attention on integrating a reactive system to quickly respond to external signals in real time with a more expressive deliberative model to deal with complex situations and to plan actions compatible with the knowledge base of the robot.

N.F. Lepora et al. (Eds.): Living Machines 2013, LNAI 8064, pp. 393–395, 2013.

2   Towards a Hybrid Architecture

FACE (Facial Automaton for Conveying Emotions) is a humanoid female robot, developed in collaboration with Hanson Robotics, used as an emotion conveying system. It consists of an artificial skull covered by a porous elastomer material called Frubber™ and a passive mannequin body. Inside the skull, 32 servo motors are positioned as the major facial muscles making the robot capable to simulate realistic facial expressions. Fig. 1 shows the current architecture for controlling FACE and its future implementations.

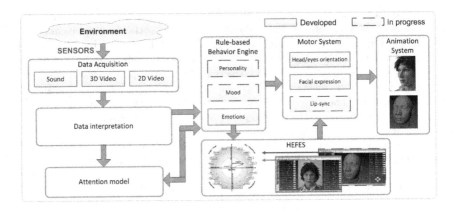

Fig. 1. The general architecture for controlling FACE

A set of sensors are used to capture and extract verbal/non-verbal cues of the people in the surrounding environment such as facial expressions, gestures, position, and speaker's position, and features of the scene such as colour and intensity. This information is processed by the data interpretation module to create a meta-scene, i.e. a description of the current scene which results from the integration of social relevant cues coming from the people analysis and the saliency map based on the pixel-based analysis [1]. Based on the meta-scene, the attention model determines what is the target point towards which the gaze attention of the social robot should be directed and how this target point should be looked at by the robot in terms of eyes and head movements.

The behaviour engine is based on CLIPS [2], a rule-based expert system which includes: a *working memory* which defines the affective status of the robot in terms of emotions and mood and describes the scene in terms of asserted facts; the *knowledge base* which is applied as if-then-else rules (deliberative behaviour) and functions (reactive behaviour) and represents the reasoning and planning capability of the robot; an *inference engine* which connects the working memory with the knowledge base to perform the actions coherently with the status of the robot and with what is happening in the surrounding environment.

If "behaviour" indicates a purely reflexive action in a reactive paradigm, the term is nearer to the concept of "skill" in a deliberative paradigm. The first paradigm is a world of reflexive behaviours which control the reactive aspects while the second paradigm is the world of symbolic reasoning which controls the deliberative aspects. For example, the information "I have detected a person in the scene" in the first paradigm could become "There is John, I have already seen him" in the second one. Indeed detecting a person means to only store the coordinates of the subject in the first paradigm or to predicate something more starting from the knowledge base and applying the rules in the second paradigm. Therefore, in order to make the robot socially believable, its behaviour has to emerge from the combination of the reaction to external stimuli in real time, through functions which transform sensor data into actuator commands, with the a priori knowledge of its world, through rules which operate on symbolic information.

At the same time the behaviour engine influences the attention model as our attitude affects the point and the way we look at [3]. The output of the behaviour engine is a configuration of different actions which include facial expressions, the orientation of the head and eyes and in the future lip-sync motions. All the possible expressions performable by FACE are generated by HEFES [4], an engine for synthesizing new expressions and interpolating them in an emotional space.

3   Conclusion and Discussion

Starting from a pure reactive system [5] we moved towards a hybrid structure in which the reactive behaviour is integrated with a deliberative reasoning taking into account the current status, the social context and the environmental conditions. The reactive level ensures that the robot can handle the real time challenges of its environment appropriately while the deliberative level endows the robot with the ability to perform more complex tasks that require reasoning.

References

1. Butko, N.J., Zhang, L., Cottrell, G.W., Movellan, J.R.: Visual saliency model for robot cameras. In: Proceedings of the 2008 IEEE International Conference on Robotics and Automation (ICRA), pp. 2398–2403 (2008)
2. Wygant, R.M.: CLIPS - A powerful development and delivery expert system tool. Computers & Industrial Engineering 17(1-4), 546–549 (1989)
3. Argyle, M., Cook, M.: Gaze and mutual gaze. Cambridge University Press, New York (1976)
4. Mazzei, D., Lazzeri, N., Hanson, D., De Rossi, D.: HEFES: a hybrid engine for facial expressions synthesis to control human-like androids and avatars. In: 4th IEEE RAS & EMBS International Conference on Biomedical Robotics and Biomechatronics (BioRob), pp. 195–200 (2012)
5. Mazzei, D., Lazzeri, N., Billeci, L., Igliozzi, R., Mancini, A., Ahluwalia, A., Muratori, F., De Rossi, D.: Development and evaluation of a social robot platform for therapy in autism. In: Proceedings of the 33rd Annual International Conference of the IEEE Engineering in Medicine and Biology Society (EMBS), pp. 4515–4518 (2011)

Towards a Roadmap for Living Machines

Nathan F. Lepora[1,2], Paul F.M.J. Verschure[3,4], and Tony J. Prescott[1,2]

[1] Sheffield Center for Robotics (SCentRo), University of Sheffield, UK
[2] Department of Psychology, University of Sheffield, UK
{n.lepora,t.j.prescott}@sheffield.ac.uk
[3] SPECS, Department of Technology, Universitat Pompeu Fabra, Barcelona, Spain
[4] ICREA, Barcelona, Spain Barcelona, Spain
paul.verschure@upf.edu

Abstract. A roadmap is a plan that identifies short-term and long-term goals of a research area and suggests potential ways in which those goals can be met. This roadmap is based on collating answers from interview with experts in the field of biomimetics, and covers a broad range of specialties. Interviews were carried out at events organized by the Convergent Science Network, including a workshop on biomimetics and Living Machines 2012. We identified a number of areas of strategic importance, from biomimetic air and underwater vehicles, to robot designs based on animal bodies, to biomimetic technologies for sensing and perception.

Keywords: Biomimetics, roadmap, interviews, living machines.

A roadmap is a plan that identifies short-term and long-term goals of a research area and suggests potential ways in which those goals can be met. Developing a roadmap has three major uses. It helps reach a consensus about a set of goals and risks for a research area; it provides a mechanism to help forecast research developments and it provides a framework to help plan and coordinate research. The aim of this roadmap is to identify current trends relating to living machines together with their implications for future research.

We have been constructing a roadmap on living machines, by collating answers from interviews with experts in the field of biomimetics across a broad range of specialties. Interviews lasted about an hour each and were carried out at events organized by the Convergent Science Network [1]: (a) A Biomimetics week for the 2011 Barcelona Cognition, Brain and Technology Summer School (BCBT); (b) The 2012 Living Machines conference in La Pedrera, Barcelona [2].

From these interviews, we identified several areas of strategic importance where the biomimetics of living machines can either further our understanding of biological systems or could lead to new technologies in robotics and engineering. A complete description of the strategic goals and risks of these areas will be given in a full article to complement our recent paper *The state of the art in biomimetics* [3] published in the journal Bioinspiration and Biomimetics. For the time being, we summarize briefly these various areas concerning living machines.

Self-assembly, microstructures and micro-machines. Bio-inspired robotic systems, with features at microscopic length scales are at the forefront of current

N.F. Lepora et al. (Eds.): Living Machines 2013, LNAI 8064, pp. 396–398, 2013.
© Springer-Verlag Berlin Heidelberg 2013

challenges and opportunities in robotic hardware. They have the potential to integrate nano-technology with macroscopic fabrication methods, and may lead to novel new methods of design and fabrication using bottom-up self-assembly rather than conventional top-down engineering approaches.

Biomimetic micro air vehicles (MAVs). Recently, a new class of MAVs are being developed that take inspiration from flying insects or birds to achieve unprecedented flight capabilities. Biological systems have inspired the application of unsteady aerodynamics to robots with flapping wings, while also motivating the use of other aspects of animal flight control such as distributed sensing.

Biomimetic autonomous underwater vehicles (AUVs). A new trend in the AUV community is to mimic designs found in nature. Although most are currently in their experimental stages, these biomimetic vehicles could be able to achieve higher degrees of efficiency in propulsion and maneuverability by copying successful designs in nature. A variety of sensors can be affixed to AUVs to measure the concentration of various elements or compounds, the absorption or reflection of light, and the presence of microscopic life.

Insect-based robotics. Insects are useful for biomimetic robotics because their morphologies and nervous systems are simpler than other animal species. Also, complex behaviors can be attributed to just a few neurons. Analysis of the walking behavior and neural architecture of insects can be used to improve robot locomotion. Alternatively, biologists can use insect-based robotics for testing biological hypotheses about the neurophysiological mechanisms involved in insects.

Soft and compliant robotics. The tentacles of squid, trunks of elephants, and tongues of mammals are examples of muscular hydrostats that inspire soft robots with the potential for dextrous manipulation in hazardous or unstructured environments. They are also a relatively safe technology to use around humans because of their compliant nature, making them ideal for the assistive or care applications, such as for the elderly.

Bipedal and quadrupedal robots. Types of bipedal movement include walking, running and hopping, while types of quadruped locomotion include trotting, pacing and bounding. Important principles include passive methods for un-powered walking, which has lead to novel designs for walking robots and furthered our understanding of the function of human and animal physiology. Legged robots have significant advantage over wheeled technologies in accessing areas currently unreachable.

Humanoid robotics. A humanoid design for a robot might be for functional purposes, such as interacting with human tools and environments, or for experimental purposes, such as the study of bipedal locomotion. Although the initial aim of humanoid research was to build better orthosis and prosthesis for human beings, knowledge has been transferred between both disciplines.

Social robotics and human-robot interaction. A social robot is an autonomous robot that interacts and communicates with humans or other social agents. A leading assumption is that social robots must develop a sense of self as to overcome the fundamental problem of social inference

Brain-based robotics. One goal of brain-based robotics is to use robots for embodying the principles underlying animal behavior. This can include using robots to study the neural mechanisms underlying movement control, perception and learning in animals, and in return to take inspiration from animals to design new control and sensing methods for robotics. Another important application is to test biological hypotheses that would be difficult or impossible otherwise.

Artificial olfaction and chemosensing. Devices that sense the existence of a particular chemical concentration in air or water is becoming an increasingly important requirement for modern robotics and automated systems. Applications include: quality control in food processing; detection and diagnosis in medicine; detection of drugs, explosives and dangerous or illegal substances; military and law enforcement; disaster response; and environmental monitoring.

Artificial audition and echolocation. Bats and cetaceans, such as whales and dolphins, use sound for perception and have far superior sensing capabilities than existing technologies. Applications include object localization in environments where vision is impaired, such as the dark, and even discriminating the material properties based on acoustic energy.

Artificial touch. Nature has provided examples of many different types of touch sensor to inspire artificial devices. Examples include the human fingertip, skin and tactile whiskers employed by animals such as rodents. Tactile sensors can be used to sense a diverse range of stimulus ranging from detecting whether an object has been grasped to a complete tactile image.

Grasping and manipulation with robot hands. An important application of tactile sensing is to help reproduce in some way the grasping and manipulation capabilities of humans. Even though many robotic devices have been developed, from very simple grippers to very complex anthropomorphic robotic hands, their usability and reliability still lags far behind human capabilities.

Acknowledgments. We thank Robert Allen, Joseph Ayers, Dieter Braun, Yoseph Bar-Cohen, Mark Cutkovsky, Yiannis Demiris, Frank Grasso, Mitra Hartmann, Auke Ijspeert, William Kier, Danica Kragic, Maarja Kruusma, David Lane, David Lentink, Tim Pearce, Giovanni Pezzulo, Andrew Phillipides, Barry Trimmer, Ian Walker and David Zipser for contributing to the roadmap. This work was supported by the EU coordination action 'Convergent Science Network (CSN)' (ICT-248986).

References

1. Convergent Science Network, http://www.csnetwork.eu
2. Prescott, T.J., Lepora, N.F., Mura, A., Verschure, P.F.M.J. (eds.): Living Machines 2012. LNCS, vol. 7375. Springer, Heidelberg (2012)
3. Lepora, N.F., Verschure, P., Prescott, T.J.: The state of the art in biomimetics. Bioinspiration & biomimetics 8(1), 013001 (2013)

Acquisition of Anticipatory Postural Adjustment through Cerebellar Learning in a Mobile Robot

Giovanni Maffei[1], Ivan Herreros[1], Martí Sánchez-Fibla[1],
and Paul F.M.J. Verschure[1,2]

[1] SPECS, Technology Department, Universitat Pompeu Fabra,
Carrer de Roc Boronat 138, 08018 Barcelona, Spain
[2] ICREA, Institució Catalana de Recerca i Estudis Avançats,
Passeig Lluís Companys 23, 08010 Barcelona
{giovanni.maffei,ivan.herreros,marti.sanchez,paul.verschure}@upf.edu

Abstract. Anticipatory Postural Adjustments (APAs) are motor responses which anticipate a perturbation on the current body position caused by a voluntary act. Here we propose that APAs can be decomposed into a compensatory and an anticipatory component and that the cerebellum might be involved in the acquisition of such responses. To test this hypothesis, we use a cerebellar model to control the acquisition of an APA in a robotic task: we devise a setup where a mobile robot is trained to acquire an APA which minimizes a perturbation in its speed after a collision with an obstacle. Our results show that the same cerebellar model can support the acquisition of an APA separately learning its two sub-components. Moreover, our solution suggests that the acquisition of an APA involves two stages: acquisition of a compensatory motor response and prediction of an incoming sensory signal useful to trigger the same response in an anticipatory manner.

Anticipatory Postural Adjustments (APA) are motor responses which anticipate a perturbation caused by a voluntary act with the goal of minimizing its effect on the current body position. Patients suffering cerebellar lesions lack this kind of anticipatory actions, making the cerebellum the ideal candidate for modulating APAs [1] [2].

Here, we propose that APAs can be decomposed into a compensatory and an anticipatory component. The former acts as a fast feed-forward controller that corrects the effect of a perturbation after it has been experienced. The latter is responsible for learning to anticipate the sensory signal that causes the perturbation in a way that the compensatory action can be initiated in advance. Moreover, we propose that both components can be acquired by separate cerebellar controllers.

To test this hypothesis, we use a cerebellar model to control the acquisition of an APA in a robotic task: we devise a setup where a mobile robot is trained to acquire an APA which minimizes a perturbation in its speed after a collision with an obstacle (Fig. 1 *right*).

N.F. Lepora et al. (Eds.): Living Machines 2013, LNAI 8064, pp. 399–401, 2013.
© Springer-Verlag Berlin Heidelberg 2013

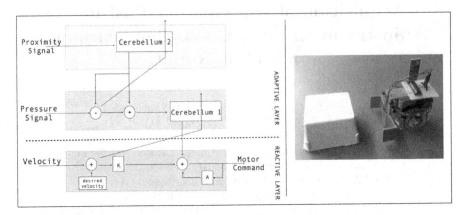

Fig. 1. *Left.* System design: the control architecture is organized into two different layers: the reactive layer adjusts the speed of the robot through slow feedback control. The adaptive layer includes two different cerebellar controllers respectively responsible for fast feed-forward control and sensory prediction. *Right* . Mobile robot approaching the obstacle.

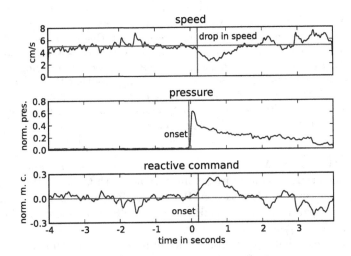

Fig. 2. Dynamics of the perturbation (average of 5 trials): *Top*: the robot navigates at constant speed (5 cm/s) experiencing a perturbation after the collision with the obstacle (red line). *Center*: activity of the pressure sensor during the collision. The red line marks the onset of the signal centered at second 0. *Bottom*: activity of the reactive controller during the collision. The red line marks the onset of the slow feedback-control signal compensating for the perturbation.

The robot has to traverse a straight track maintaining constant speed and controlling only one degree of freedom. A cubic object is placed on the track in a way that the robot will collide with it, experiencing a drop in velocity. The robot is equipped with sensors through which it detects the proximity to the object (infra-red sensors) and the magnitude of the impact with it (pressure capacitive sensor)(Fig. 2 *Center*). In addition the robot computes its own velocity through a tracking system implemented within the arena.(Fig. 2 *Top*)

The control architecture is organized into two different layers of behavior. The reactive layer adjusts the speed of the robot through slow feedback control (Fig. 2 *Bottom*). The adaptive layer includes two different cerebellar controllers: the first one maps a pressure signal into a fast compensatory motor command, minimizing the duration of the drop in speed after the collision. The second one predicts the pressure signal from information received via the proximity sensors, such that the corrective action can be initiated ahead of the collision. Both adaptive controllers implement the same abstract cerebellar algorithm based on decorrelation learning [3] .

Our results show that a cerebellar model can support the acquisition of an APA, learning its two separate components: the compensatory and the anticipatory one. Moreover, our solution suggests that the acquisition of an APA has two stages: first a compensatory action is acquired, and then the anticipatory controller learns to trigger the compensatory action in a predictive manner. To conclude, this work demonstrates how the same cerebellar learning algorithm can be applied to two different sub-problems within the same task.

References

1. Horak, F.B., Diener, H.C.: Cerebellar control of postural scaling and central set in stance. Journal of Neurophysiology (72) (1994)
2. Timmann, D., Horak, F.: Perturbed step initiation in cerebellar subjects: 2. Modification of anticipatory postural adjustments. Journal of Neurophysiology (141) (Exp. Brain Res.)
3. Herreros, I., Verschure, P.: Nucleo-olivary inhibition balances the interaction between the reactive and adaptive layers in motor control. Neural Networks (2013)

Using a Biological Material to Improve Locomotion of Hexapod Robots

Poramate Manoonpong[1], Dennis Goldschmidt[1], Florentin Wörgötter[1],
Alexander Kovalev[2], Lars Heepe[2], and Stanislav Gorb[2]

[1] Bernstein Center for Computational Neuroscience (BCCN)
University of Göttingen, D-37077 Göttingen, Germany
{poramate,goldschmidt,worgott}@physik3.gwdg.de
[2] Department of Functional Morphology and Biomechanics, Zoological Institute
Christian-Albrechts-Universität zu Kiel, D-24098 Kiel, Germany
{akovalev,lheepe,sgorb}@zoologie.uni-kiel.de

Abstract. Animals can move in not only elegant but also energy
efficient ways. Their skin is one of the key components for this achievement. It provides a proper friction for forward motion and can protect
them from slipping on a surface during locomotion. Inspired by this, we
applied real shark skin to the foot soles of our hexapod robot AMOS.
The material is formed to cover each foot of AMOS. Due to shark skin
texture which has asymmetric profile inducing frictional anisotropy, this
feature allows AMOS to grip specific surfaces and effectively locomote
without slipping. Using real-time walking experiments, this study shows
that implementing the biological material on the robot can reduce energy
consumption while walking up a steep slope covered by carpets or other
felt-like or rough substrates.

Keywords: Shark skin, Biomechanics, Walking robots, Frictional
anisotropy.

Animals show fascinating locomotor abilities. They are able to traverse a wide
range of surfaces in an energy efficient manner. During traversing, their locomotion can also adapt to a change of terrain. In addition, their movements are
elegant and versatile. Biological studies reveal that these capabilities are the
result of a combination of neural control and biomechanics including proper
material properties [1]. While neural control generates movement and allows
for adaptation, biomechanics provides shape, support, stability, and movement
to the body as well as enables energy efficient locomotion without high control
effort. Over the past decade, roboticists have tried to mimic such natural features
with their artificial systems [2,3] in order to approach animals in their levels of
performance and to understand the functions of their biomechanics and neural
mechanisms. To tackle this challenging problem towards animal-like locomotor
abilities, we have developed the AMOS series of biologically-inspired hexapod
robots, in a stepwise manner during the last years [4]. AMOS (Fig. 1a, left)
has now achieved a multitude of different walking patterns as well as adaptable

N.F. Lepora et al. (Eds.): Living Machines 2013, LNAI 8064, pp. 402–404, 2013.
© Springer-Verlag Berlin Heidelberg 2013

locomotion [4]. It is under real-time neural control by ways of a modular neural network allowing it to walk at different gaits and adapt its locomotion to traverse rough terrains, or to climb over obstacles [4]. Although AMOS has shown a certain degree of complex locomotor behavior under neural control, it still requires very high energy consumption to walk up a steep slope (e.g., a 17° slope covered by carpets). During walking up the slope, its legs slip since its rubber feet do not provide enough adhesive friction. Thus, a central question of this work is "Can we improve the locomotion of AMOS during walking on specific terrains (i.e., slope covered by carpets or other felt-like or rough substrates) without high control effort?"

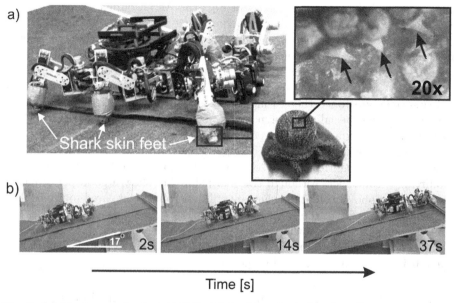

Fig. 1. (a) The hexapod robot AMOS with basking shark skin feet. Zoom panels show a shark skin foot formed to cover a rubber foot and a close up view (20X) of basking shark skin texture taken from a microscope. The basking shark skin has asymmetric profile like a sloped array of spines. (b) Snap shots of walking up a 17° slope covered by carpets where AMOS uses the shark skin feet.

To answer this question, in this study we have investigated different types of materials including biological ones (e.g., polishing papers, the skin of harbor seal (Phoca vitulina), and the skin of basking shark (Cetorhinus maximus) at head area) for using them as feet of AMOS. Among them, the shark skin provides important features which are appropriate for the task. It has texture having asymmetric profile (see Fig. 1a, right) which induces frictional anisotropy. This way, it acts as a locking mechanism which can prohibit slipping motion. In addition, it can be easily formed to have a cup-like shape (see Fig. 1a, middle) such that we can simply attach it to AMOS' foot. Shark skin feet were prepared from hydrated shark skin tightly pressed in a negative wooden form resembling the geometry of the robot feet. Shark skin was then dried for several days and remained stable in the robot feet geometry after removal from the form.

To evaluate the performance of the shark skin feet, we covered original rubber feet of AMOS by the shark skin feet (see Fig. 1a, left) and let AMOS walk with a wave gait up a $17°$ slope covered by carpets. Note that this angle is steep enough making AMOS difficult to walk up using its rubber feet. We performed five runs each and then compared walking efficiency using the shark skin feet with the one using the rubber feet. Here, the walking efficiency is measured by the specific resistance given by: $\epsilon = \frac{P}{mgV}$, where P is power consumption, mg is the weight of AMOS, i.e., 56.84 N, and V is walking speed. Low ϵ corresponds to highly efficient walking. An illustration of real-time walking experiments is shown as snap shots in Fig. 1b. We encourage readers to also see the video of AMOS walking behavior using the shark skin feet and the rubber feet at http://manoonpong.com/LM2013/S1.wmv. The average specific resistances with standard deviations of AMOS walking using the shark skin feet and the rubber feet from five runs each are 48.56±19.3 and 168.61±18.43, respectively. The experimental result shows that using the shark skin feet leads to low specific resistance, thereby highly efficient walking compared to the rubber feet. Due to the special shark skin profile (see Fig. 1a, right), it allows AMOS to grip specific surfaces (i.e., carpets, felt-like, and rough surfaces) and effectively locomote without slipping. This preliminary result reveals that utilizing material with strong frictional anisotropy can improve robot locomotion without modifying a controller. Although the shark skin feet show a good performance and improve robot locomotion, they are still less robust compared to the rubber feet since the profile of the shark skin feet was destroyed after a few successive runs. In our next step, we will investigate and create synthetic nano-structured surfaces that attempt to mimic aspects of the shark skin system for robot feet while being robust to prolonged usage.

Acknowledgments. This research was supported by the Emmy Noether Program (DFG, MA4464/3-1) and BCCN II Göttingen (01GQ1005A, project D1). We thank Anja Huss for producing a close-up view of the shark skin and Joachim Oesert for technical assistance.

References

1. Dickinson, M., Farley, C., Full, R., Koehl, M., Kram, R., Lehman, S.: How animals move: An integrative view. Science 288(5463), 100–106 (2000)
2. Hawkes, E.W., Eason, E.V., Asbeck, A.T., Cutkosky, M.R.: The Gecko's Toe: Scaling Directional Adhesives for Climbing Applications. IEEE/ASME Transactions on Mechatronics 18(2), 518–526 (2013)
3. Lewinger, W.A., Quinn, R.D.: Neurobiologically-based Control System for an Adaptively Walking Hexapod. Ind. Robot 38(3), 258–263 (2011)
4. Manoonpong, P., Parlitz, U., Wörgötter, F.: Neural Control and Adaptive Neural Forward Models for Insect-like, Energy-Efficient, and Adaptable Locomotion of Walking Machines. Front. Neural Circuits 7, 12 (2013), doi:10.3389/fncir.2013.00012

Angle and Position Perception
for Exploration with Active Touch

Uriel Martinez-Hernandez[1,3], Tony J. Dodd[1,3], Tony J. Prescott[2,3],
and Nathan F. Lepora[2,3]

[1] ACSE Department, University of Sheffield, U.K.
[2] Department of Psychology, University of Sheffield, U.K.
[3] Sheffield Centre for Robotics (SCentRo), University of Sheffield, U.K.
{uriel.martinez,t.j.dodd,t.j.prescott,n.lepora}@sheffield.ac.uk

Over the past few decades the design of robots has gradually improved, allowing them to perform complex tasks in interaction with the world. To behave appropriately, robots need to make perceptual decisions about their environment using their various sensory modalities. Even though robots are being equipped with progressively more accurate and advanced sensors, dealing with uncertainties from the world and their sensory processes remains an unavoidable necessity for autonomous robotics. The challenge is to develop robust methods that allow robots to perceive their environment while managing uncertainty and optimizing their decision making. These methods can be inspired by the way humans and animals actively direct their senses towards locations for reducing uncertainties from perception [1]. For instance, humans not only use their hands and fingers for exploration and feature extraction but also their movements are guided according to what it is being perceived [2]. This behaviour is also present in the animal kingdom, such as rats that actively explore the environment by appropriately moving their whiskers [3].

We are interested in the development of methods for perceiving the world through the sense of touch. Even though other sensory modalities such as vision have been widely studied, we are principally motivated by the way humans and animals rely on their sense of touch, such as in situations where vision is partial or occluded, as in a smoke-filled buildings or darkness. In this study we show how active Bayesian perception is important for decision making with a robotic fingertip. First, we present an approach based on maximum likelihood using one tactile contact for decision making. Then, we show an improvement using an active Bayesian perception approach where the tactile sensor performs contacts until the system has sufficient confidence to make a decision. To test the different approaches, we choose the simple but illustrative task of contour following. The aim of the task is to perceive the angle and position where the robotic fingertip is in contact with the edge of the object. For this task, the robot needs to make decisions about *what to do next* and *where to move next*, which results in an active exploration strategy that extracts the shape of the object.

First, we developed and implemented a tactile perception method based on maximum likelihood where only one tactile contact is required for making a

N.F. Lepora et al. (Eds.): Living Machines 2013, LNAI 8064, pp. 405–408, 2013.

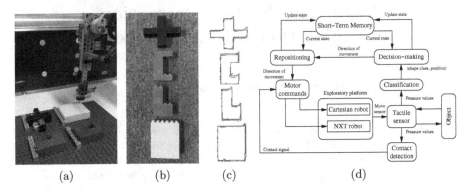

Fig. 1. Active perception based on maximum likelihood. (A) Test robot. (B) Test objects. (C) Results showing the traced contours. (D) Sensorimotor architecture for robot control.

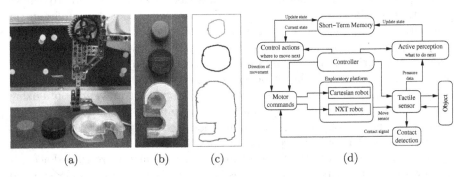

Fig. 2. Active Bayesian perception approach. Labels as in Fig. 1.

decision [4,5,6]. We used data collected from an edge of 4 different angles separated by 90 degs. For each angle we gathered 18 position classes, to give a total 72 perceptual classes for discrimination. Two data sets were collected to have one for training and one for testing. The training was performed off-line and the testing performed in real-time on-line using a sensorimotor architecture controlled by the tactile perception. The objects used for this experiment and the results are shown in Figure 1. The robotic fingertip was able to successfully trace the contour of the objects using this approach (Figure 1c). This was accomplished by actively moving the fingertip perpendicularly to the current edge while following the contour of the object. This active repositioning is responsible for ensuring that the tactile sensor will be placed in an optimal location for improving perception.

We developed a second method to improve the perception of complex object shapes, covering a wide range of angle classes. To achieve this improvement, we developed an active Bayesian perception method inspired by previous studies where different stimuli were successfully classified [7,8,9].

In active Bayesian perception, the robot accumulates perceptual evidence through successive contacts with an object. Once the evidence accumulated is sufficient for the perceptual beliefs to cross a threshold, a decision about the angle and position classes is triggered. This method is inspired by the way animals make decisions based on the experience accumulated through the interaction with the environment [5]. Here, we collected 72 angle classes in 5 deg steps from a circle covering 360 degs. For each angle, we gathered 18 position classes, giving a total of 1296 distinct perceptual classes for discrimination. The training was performed off-line, with results that showed the best location for perception is in the centre (9 mm position class) of the fingertip sensor, which corresponds with previous work [6,7,9]. Also, in the off-line analysis we estimate the mean angle and position errors against belief threshold and reaction time (number of taps) to make a decision. For passive perception we obtained mean angle and position errors of about 12.2 degs and 0.8 mm, whilst for active perception we obtained substantially improved results of 3.3 degs and 0.2 mm. The reaction time for making a decision was 4-6 taps.

To demonstrate the accuracy of the method with the robot, we implemented a sensorimotor architecture to trace the contour of different objects (Figure 2d). This architecture implements *what to do next* and *where to move next* decisions based on the tactile perception interacting with the object properties. We used two circles with 2 cm and 4 cm diameters and one asymmetric object (Figure 2b). The contours around the objects were successfully traced (Figure 2c) using a belief threshold of 0.8 for making a decision at each position of the contour. An average of 6 taps were needed for the robot to make a decision. From these results, we observe that the traced contours accurately represent the different object shapes. This shows that the robot is able to behave correctly according to the task specified.

The methods presented in this study showed that active perception provides an effective way of dealing with uncertainty. Also, we observed how active perception can be used for controlling a system according to the perceptual information received. We expect that this approach will lead to robust and accurate methods for autonomous robots to interact with, explore and perceive their environment.

Acknowledgments. This work was supported by FP7 grants EFAA (ICT-270490) and the Mexican National Council of Science and Technology (CONACYT).

References

1. Bajcsy, R.: Active perception. Proceedings of the IEEE 76(8), 966–1005 (1988)
2. Lederman, S.J., Klatzky, R.L.: Hand movements: A window into haptic object recognition. Cognitive Psychology 19(19), 342–368 (1987)
3. Prescott, T.J., Diamond, M.E., Wing, A.M.: Active touch sensing. Phil. Trans. R. Soc. B 366, 2989–2995 (2011)

4. Lepora, N.F., Evans, M., Fox, C.W., Diamond, M.E., Gurney, K., Prescott, T.J.: Naive Bayes texture classification applied to whisker data from a moving robot. In: The 2010 International Joint Conference on Neural Networks (IJCNN), pp. 1–8 (2010)

5. Lepora, N.F., Fox, C.W., Evans, M., Diamond, M.E., Gurney, K., Prescott, T.J.: Optimal decision-making in mammals: Insights from a robot study of rodent texture discrimination. Journal of The Royal Society Interface 9(72), 1517–1528 (2012)

6. Martinez-Hernandez, U., Dodd, T.J., Natale, L., Metta, G., Prescott, T.J., Lepora, N.F.: Active contour following to explore object shape with robot touch. In: 2013 IEEE International Conference on World Haptics (WHC) (2013)

7. Lepora, N.F., Martinez-Hernandez, U., Prescott, T.J.: Active touch for robust perception under position uncertainty. In: 2013 IEEE International Conference on Robotics and Automation (ICRA) (2013)

8. Lepora, N.F., Martinez-Hernandez, U., Prescott, T.J.: Active bayesian perception for simultaneous object localization and identification (under review)

9. Martinez-Hernandez, U., Dodd, T.J., Prescott, T.J., Lepora, N.F.: Active Bayesian perception for angle and position discrimination with a biomimetic fingertip (under review)

Toward Living Tactile Sensors

Kosuke Minzan[1], Masahiro Shimizu[1], Kota Miyasaka[2],
Toshihiko Ogura[2], Junichi Nakai[3], Masamichi Ohkura[3], and Koh Hosoda[1]

[1] Graduate School of Information Science and Technology, Osaka University,
2-1 Yamada-oka, Suita, 565-0871, Japan
[2] Institute of Development, Aging and Cancer (IDAC), Tohoku University,
4-1, Seiryo, Aoba, Sendai, 980-8575, Japan
[3] The Brain Science Institute, Saitama University,
2-1 Hirosawa, Wako-shi, Saitama 351-0198, Japan
minzan.kousuke@ise.eng.osaka-u.ac.jp,
{m-shimizu,koh.hosoda}@ist.osaka-u.ac.jp,
{k.miyasako,ogura}@idac.tohoku.ac.jp,
{jnakai,mohkura}@mail.saitama-u.ac.jp

Abstract. The authors aim to realize a "Cell Tactile Sensor" which
is integrated tactile receptors built by cultured cells. We regard the
cells cultured on the PDMS membrane as the force receptors. We here
report on interesting experimental results as follows: at first, we designed
the positions of cells by self-organization which is due to mechanical
stimulations; secondly, we visualized the tactile information of cell tactile
sensors by observing of the Ca^{2+} induced from mechanical stimulations.
Therefore, a new tactile sensor which has environment adaptability is
developed.

Keywords: Cell Tactile Sensor, self-organization, mechanical stimulation, G-CaMP.

1 Introduction

Living biological systems can adapt by changing not only control strategy but
also mechanical property. On the hand, to have environmental adaptability,
traditional robotics mainly advances control systems rather than mechanical
systems. Because online modification of mechanical system of the robot is difficult. Under this circumstance, by applying biological materials(*e.g.*, cultured
cells) to robots, its mechanical system is expected to achieve living organisms'
intrinsic prominent functions. Cells, that are units of biological constitution,
induces calcium ion signals due to mechanical stimulations[1]. And, applying
this mechanical stimulations also leads morphological and functional changes
of cells[2]. Furthermore, cells make change their orientations depending on the
applied mechanical stimulation during culturing[3]. These studies indicate that
we can use cells as mechanical receptors, capable of tactile sensing in robots(in
the following, we call this a **Cell Tactile Sensor**).

N.F. Lepora et al. (Eds.): Living Machines 2013, LNAI 8064, pp. 409–411, 2013.

This study intends to deal with development of the cell tactile sensor which has environmental adaptability. More specifically, we can expect that this cell tactile sensors intrinsically have "growing ability". That is this sensor becomes selective detector of a certain external force after culturing with mechanical stimulus. Compared with traditional engineering tactile sensors[4], the cell tactile sensor has significantly-small structure, because each cells generally range in physical size from 1[μm] to 100[μm]. For the realization of this, **1.**we design the positions and orientations of cells by exploiting self-organization due to mechanical stimulations. Which means the design of configuration structure of cell tactile sensors. **2.**We introduce the plasmid of gene of G-CaMP[5] into mouse myoblast cell line(C2C12 cells). The G-CaMP is calcium ion detective probe. Therefore, we can expect that the fluorescence strength increase at each cell, when we apply mechanical stimulations to cells. Which means that we can visualize the tactile information with cell tactile sensors.

2 Experimental Results

In this study, at first we culture the C2C12 cells on the PDMS silicon thin membrane as the substrate. Next, we introduce the G-CaMP into C2C12 cells. During culturing, we apply the outer force to the opposite side of the membrane, making a "Culture Stimulation" to induce self-organization. This process leads to adaptive changes of positions and orientations of the cells(first achievement of this study). After culturing, we made an "Evaluation Stimulation" to verify its function as a tactile sensor. Then, the C2C12 cells respond to mechanical stimulation by a change in their fluorescence to acquire tactile information.

The experimental results are indicated in Fig. 1. Here, cultured cells are also visualized by FFT image of their ROI(Region of Interest), which emphasizes

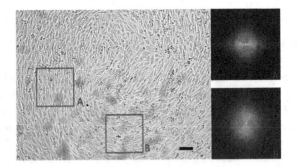

Fig. 1. The photo of C2C12 cells(left) and FFT images of their ROI(right) after culture stimulation. We applied culturing stimulation (5[mm] stroke of upthrust, stimulus area is a disc-like shape which is 2[mm] in diameter, time duration of 0.5[Hz]) at the center of the PDMS chamber for 3 hours. The scale bar indicates 100[μm]. High image frequency of the horizontal(image A) or the vertical(image B) directions appear in each FFT images. This means that the cells are widely oriented in a concentric-circle-like patterns.

Fig. 2. Fluorescence observations of C2C12 before(left) and after(right) applying evaluation stimulation(the stimulus protocol is same as Fig.1)

how the orientation of cells appears. In FFT image(A, B) about upthrust stimulation, the component of the radial direction strongly appears. Specifically, the components of the horizontal and the vertical directions appear in FFT image A and B respectively. This means that the cells are widely oriented in a concentric-circle-like patterns. Note here that the cells self-organize reflecting culture stimulations. After culturing in each condition, we made an evaluation stimulation to verify its function as a tactile sensor. We observed cell fluorescence before and after applying evaluation stimulation to cells. As evaluation stimulation, we applied the upthrust stimulation at once. The experimental result is shown in Fig. 2. As in the figure, the fluorescent results interestingly represent the significant increase of the emission of light after evaluation stimulation. The upthrust stimulation is selectively detectable. Note here that we achieved the visualization of tactile information by introducing G-CaMP to C2C12 cells. The verification of reliability and repeatability of this are ongoing works.

Acknowledgments. This work was supported partially by KAKENHI (24680023, 23220004 and 22127006), Japan. The authors would like to thank Prof. Shigeru Kondo and Prof. Masakatsu Watanabe, Osaka University, for stimulating, and helpful suggestions.

References

1. Yamamoto, K., Furuya, K., Nakamura, M., Kobatake, E., Sokabe, M., Ando, J.: Visualization of flow-induced ATP release and triggering of Ca2+ waves at caveolae in vascular endothelial cells. Journal of Cell Science 124, 3477–3483 (2011)
2. Engler, A.J., Sen, S., Lee Sweeney, H., Discher, D.E.: Matrix Elasticity Directs Stem Cell Lineage Specification. Cell 126, 677–689 (2006)
3. Shimizu, M., Yawata, S., Miyamoto, K., Miyasaka, K., Asano, T., Yoshinobu, T., Yawo, H., Ogura, T., Ishiguro, A.: Toward Biorobotic Systems with Muscle Cell Actuators. In: The Proc. of AMAM 2011, pp. 87–88 (2011)
4. Dahiya, R.S., Metta, G., Valle, M., Sandini, G.: Tactile Sensing—From Humans to Humanoids. IEEE Transactions on Robotics 26, 1–20 (2010)
5. Nakai, J., Ohkura, M., Imoto, K.: A high signal-to-noise Ca2+ probe composed of a single green fluorescent protein. Nature Biotechnology 19, 137–141 (2001)

Virtual Chameleon: Wearable Machine to Provide Independent Views to Both Eyes

Fumio Mizuno[1], Tomoaki Hayasaka[2], and Takami Yamaguchi[3]

[1] Department of Electronics and Intelligent Systems,
Tohoku Institute of Technology, Sendai, Japan
[2] Department of Biomengineering and Robotics, Tohoku University, Sendai, Japan
[3] Department of Biomedical Engineering, Tohoku University, Sendai, Japan

Abstract. We developed a Virtual Chameleon that is a wearable device that provides independent fields of view to the eyes. It consists of two independently controlled CCD cameras and a head-mounted display. The Virtual Chameleon artificially enables the user to vary the directions of the visual axes of both eyes to arbitrary directions independently and to perceive each field of view simultaneously. Humans move both eyes in various styles of eye movements. But an attitude control of camera platforms equipped with the system was only corresponding to smooth eye movements because a control method of the system depends on only arm movements. It was assumed that visual stimuli induced by various eye movements affect predominance occurred in binocular rivalry. Therefore, in this work, we focused on patterns of eye movements and control method, and implemented additional functions to make users control Virtual Chameleon in various way by mounting pointing devices.

Keywords: Wearable robotics, Binocular rivalry, Chameleon.

1   Introduction

Numerous species of reptiles, herbivores and fish are capable of conscious moving their eyes independently and perceiving fields of views of the surrounding environment for the purpose of watching prey or avoiding danger. Among these species, especially, chameleons can locate their prey with the large and independent saccades of their highly mobile eyes. The anatomical characteristics and patterns of monocular movements of chameleons are similar to higher primates, and obeys Listing's Law[1], but the oculomotor system and cognitive functions of humans differ greatly from those of chameleons.

Giving a normal person the same oculomotor ability to control both eyes independently as chameleons do is expected to induce binocular rivalry[2]. Humans have the capability to flexibly adapt to visual stimulation, such as spatial inversion in which a person wears glasses that display images upside down for long periods of time[3]. As with adaptation to spatial inversion, it is assumed that humans can also adapt when given the ability to control both eyes independently. Therefore, we developed a portable device that provides independent fields of view to the eyes for extended periods of time[4][5]. The device,

N.F. Lepora et al. (Eds.): Living Machines 2013, LNAI 8064, pp. 412–414, 2013.

Virtual Chameleon, consists of two independently controlled CCD cameras and a head-mounted display (HMD). The Virtual Chameleon enables the user to vary the directions of the visual axes of both eyes to arbitrary directions and to perceive each field of view simultaneously. The successful users of the system were able to actively control visual axes by manipulating 3D sensors held by their both hands, to watch independent fields of view presented to the left and right eyes, and to look around as chameleons do.

In previous work, although humans moves their both eyes in various styles of eye movements, an attitude control of camera platforms equipped with Virtual Chameleon was only corresponding to smooth eye movements because a control method of the system depends on only arm movements. It was assumed that visual stimuli induced by various eye movements affect predominance occurred in binocular rivalry. In this research, we focused on influenced of patterns of eye movements on visual perception with binocular rivalry, and implemented additional functions to make users control Virtual Chameleon in various styles of the attitude control of camera platforms by mounting pointing devices.

2 System

An exterior view of the Virtual Chameleon is shown in Fig. 1, and a system configuration is shown in Fig. 2. This device is composed of a 3D tracking sensor system, compact trackballs equipped with a momentary on-off switch, a camera positioning system and a display system. The camera positioning system independently controls the postures of the two CCD cameras to follow the positions of the 3D sensors held in both hands and each trackball input. Control timings between arm movements and trackball inputs are managed by states of corresponding momentary switches. Images taken by the cameras are processed and projected onto the HMD which project images on each eye independently in real time.

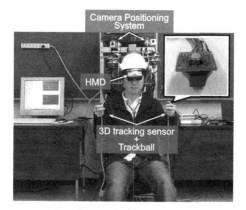

Fig. 1. Exterior view of Virtual Chameleon

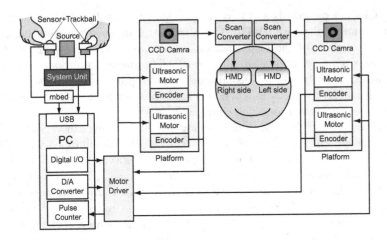

Fig. 2. System configuration of Virtual Chameleon

3   Conclusions

In this work, we implemented additional functions to make users control Virtual Chameleon in various styles attitude control of camera platforms such as saccade movements, smooth movements and pusuit movements.

In future work, performance assessments and psychophysical experiments with use of the Virtual Chameleon are necessary to investigate effects of various styles of attitude contorol of cameras on perception and performance of the user.

Acknowledgments. This work was supported by the Ministry of Education, Culture, Sports, Science and Technology(MEXT) in Japan, Grant-in-Aid for Young Scientists (B) 24700122.

References

1. Peteret, T.S., et al.: Chameleon eye position obeys Listing's Law. Vision Research 141, 2245–2251 (2001)
2. Blake, R., Logothetis, N.K.: Visual Competition. Nature Reviews Neuroscience 3, 1–11 (2002)
3. Stratton, G.M.: Vision without inversion of the retinal image -1. The Psychological Review 4, 341–360 (1887)
4. Mizuno, F., Hayasaka, T., Yamaguchi, T.: Virtual Chameleon - A System to Provide Different Views to Both Eyes. In: IFMBE Proceedings, vol. 25, pp. 169–172 (2009)
5. Mizuno, F., Hayasaka, T., Yamaguchi, T.: A fundamental evaluation of human performance with use of a device to present different two-eyesight both eyes. In: 5th European Conference of International Federation Medical and Biological Engineering 2011, Proc. of IFMBE, vol. 37, pp. 1176–1179 (2011)

Bioinspired Design and Energetic Feasibility of an Autonomous Swimming Microrobot

Stefano Palagi, Francesco Greco, Barbara Mazzolai, and Lucia Beccai

Center for Micro-BioRobotics@SSSA, Istituto Italiano di Tecnologia, Pontedera, Italy
{stefano.palagi,francesco.greco,barbara.mazzolai,
lucia.beccai}@iit.it

Abstract. A mobile microrobot is an untethered robotic device with typical size ranging from few micrometres to few millimetres. Endowing such a microrobot with autonomy-oriented capabilities, e.g. self-propulsion and self-powering, represents a scientific and technological challenge that requires innovative approaches. Bioinspiration provides fundamental cues for designing microrobots, enabling the development of working devices. Here we present the conceptual design of an autonomous swimming microrobot relying on biomimetic glucose-based powering, reporting a preliminary analysis on its energetic feasibility.

Keywords: swimming microrobot, bioinspiration, power, glucose, fuel cells.

1 Introduction

Mobile swimming microrobots are untethered devices with typical size ranging from few micrometres to few millimetres and able to move in liquids. Recently, they have drawn increasing attention due to their envisioned applications in medicine and to basic scientific and technological interests [1-2]. Current mobile microrobots either rely on wireless actuation by external purposely-provided sources of energy, e.g. magnetic fields [3], or consist of swimming microorganisms, possibly controlled by external fields, adopted as such, modified or attached to artificial microstructures [4].

Endowing an artificial swimming microrobot with autonomy-oriented capabilities, such as self-propulsion and self-powering, could be the first step toward a new generation of advanced microrobots [5]. In addition, dealing with this challenge could drive the demand for technological solutions at the microscale.

2 Conceptual Design

Two critical aspects in the design of artificial autonomous swimming microrobots are self-propulsion (propulsion by internal actions) and self-powering (internal generation of power), since adopting traditional components is not possible at these scales. Therefore, innovative design approaches are needed. Among them, bioinspiration provides essential design cues that enable the development of working devices [1].

N.F. Lepora et al. (Eds.): Living Machines 2013, LNAI 8064, pp. 415–417, 2013.

The issue of self-propulsion involves two different problems, namely: adopting propulsive body deformations; achieving these through embedded actuators. In previous works [6-7] we addressed the former by defining a propulsion mechanism inspired by metachronal waves in ciliates. This is based on a small-amplitude, longitudinally-travelling sinusoidal perpendicular deformation of an active surface enveloping the microrobot. The envisioned microrobot has a cylindrical shape with hemispherical front and rear (Fig. 1). In addition, we preliminary addressed the design of the active surface, emphasizing the limitations of current microscale actuation technologies. Since their suitability for autonomous microrobots is strictly constrained by power aspects, the estimation of the available power is crucial in microrobots design.

Microorganisms, as living beings, draw energy from their environment. This allows them to carry none or very limited energy on-board. Scavenging some form of energy from the surroundings could be an interesting approach for self-powered microrobots, as well [2]. A number of devices scavenging different forms of energy, e.g. chemical, thermal or mechanical, have been developed at the milli/microscale [8].

Considering their envisioned applications involving navigation in bodily fluids, the environment of swimming microrobots will be rich in chemicals to be exploited for generating power, such as the ubiquitous glucose. On the contrary, natural thermal gradients will be almost absent due to homeostasis. In addition, vibrations and other sources of mechanical power are not suitable to this particular mobile embodiment. Therefore, autonomous swimming microrobots powered by scavengers of chemical energy, such as fuel cells relying on the oxidation of glucose, can be envisioned [2].

3   Estimation of Available Power

Glucose Fuel Cells (GFCs) are 2D devices that generate electrical power by oxidising the glucose dissolved in the fluid they are in contact with. They rely either on noble-metals catalysts (non-enzymatic GFCs) [8] or on biocatalysts (enzymatic GFCs) [9]. In physiological conditions, enzymatic GFCs have higher power outputs (>100 $\mu W/cm^2$) but lower duration (days) than non-enzymatic ones (<10 $\mu W/cm^2$, months), which can also be microfabricated with standard lithographic techniques [8].

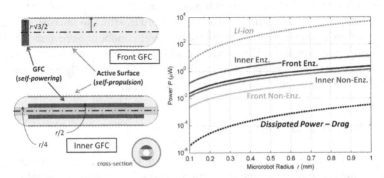

Fig. 1. Estimated available power for different microrobot designs: the two configurations of the microrobot (left); power from different GFC configurations (right): as reference we report the power from a hypothetic microscale Lithium-ion battery and the power dissipated by drag

Since in the proposed design the lateral surface of the microrobot is dedicated to propulsion, two viable ways for having the GFC in contact with the fluid are either to place a GFC on the front and rear surfaces or within an inner channel, with the fluid passing through the microrobot). Here we analytically estimated the available power from a GFC-based power supply considering two different designs of the microrobot, one with a front GFC (simpler but with smaller active surface) and one with a GFC placed in an inner channel (more complex but with larger active surface – see Fig. 1). For each design we considered both the enzymatic and non-enzymatic case (assumed output surface power densities $p_{Enz\text{-}GFC} = 100$ $\mu W/cm^2$ and $p_{NEnz\text{-}GFC} = 10$ $\mu W/cm^2$). We considered the radius r of the microrobot varying in the range $0.1 - 1$ mm, with the other dimensions varying accordingly (total length $l_{TOT} = 10r$, length of the active surface for propulsion $l_{ACT} = 8r$). This resulted in the area of the GFC electrodes A_{GFC} of both configurations varying, as well. The results reported in Fig. 1 show that even the worst configuration (front non-enzymatic GFC) would supply an amount of power ($P_{supply} = p_{GFC} \cdot A_{GFC}$) much larger than that dissipated by dragging the microrobot in the fluid at a reference relative speed of 1 bodylength/s ($P_{drag} = C_d \cdot \eta \cdot r \cdot v^2$). In addition, the results show that the available power could be further increased by adopting more advanced solutions, such as enzymatic catalysts and/or an inner channel design. Hence, this analysis returns fundamental power constraints for the design of the actuation elements for self-propelled and self-powered artificial swimming microrobots.

References

1. Abbott, J.J., Nagy, Z., Beyeler, F., Nelson, B.J.: Robotics in the Small, Part I: Microbotics. IEEE Robot Autom. Mag. 14(2), 92–103 (2007), doi:10.1109/MRA.2007.380641
2. Nelson, B.J., Kaliakatsos, I.K., Abbott, J.J.: Microrobots for Minimally Invasive Medicine. Annu Rev. Biomed. Eng. 12(1), 55–85 (2010), doi:10.1146/annurev-bioeng-010510-103409
3. Peyer, K.E., Zhang, L., Nelson, B.J.: Bio-inspired magnetic swimming microrobots for biomedical applications. Nanoscale 5, 1259–1272 (2013), doi:10.1039/C2NR32554C
4. Martel, S.: Microrobotics in the vascular network: present status and next challenges. J. Micro-Bio. Robot, 1–12 (2013), doi:10.1007/s12213-012-0054-0
5. Palagi, S., Pensabene, V., Mazzolai, B., Beccai, L.: Novel Smart Concepts for Designing Swimming Soft Microrobots. Procedia Computer Science 7, 264–265 (2011), doi:10.1016/j.procs.2011.09.074
6. Palagi, S., Mazzolai, B., Beccai, L.: Modeling of a propulsion mechanism for swimming microrobots inspired by ciliate metachronal waves. In: 2012 4th IEEE RAS & EMBS International Conference on Biomedical Robotics and Biomechatronics (BioRob), June 24-27, pp. 264–269 (2012), doi:10.1109/BioRob.2012.6290760
7. Palagi, S., Jager, E.W.H., Mazzolai, B., Beccai, L.: Propulsion of swimming microrobots inspired by metachronal waves in ciliates: from biology to material specifications. Bioinspiration & Biomimetics (submitted, 2013)
8. Rapoport, B.I., Kedzierski, J.T., Sarpeshkar, R.: A Glucose Fuel Cell for Implantable Brain–Machine Interfaces. PLoS One 7(6), e38436 (2012)
9. Ivanov, I., Vidaković-Koch, T., Sundmacher, K.: Recent Advances in Enzymatic Fuel Cells: Experiments and Modeling. Energies 3(4), 803–846 (2010)

Climbing Plants, a New Concept for Robotic Grasping

Camilla Pandolfi[1], Tanja Mimmo[2], and Renato Vidoni[2]

[1] Advanced Concepts Team, European Space Research and Technology Centre,
Noordwijk, The Netherlands
[2] Faculty of Science and Technology, Free University of Bolzano,
Piazza Università 5, 39100 Bolzano
camilla.pandolfi@unifi.it

Abstract. Climbing plants represent an outstanding example of a grasping strategy coming from the plant Kingdom. Tendrils are the filiform organs devoted to the task and are extremely flexible and sensitive to touch. In this preliminary contribution we present some of the observed key features of tendrils. Then a robotic approach is exploited to describe and simulate a bio-inspired robotic tendril from a kinematic point of view.

Keywords: Climbing plants, tendril, robotic grasping.

Over the last several decades the research on robotic manipulators has focused mainly on designs that resemble the human arm. But, if we examine the manipulators available in nature, we will see a plethora of other possibilities. Animals such as snakes, elephants, and octopuses can produce motions from their appendages or bodies that allow the effective manipulation of objects, even though they are quite different in structure compared to the human arm. Several serpentine and continuum robots have already been designed and developed, taking inspiration from animals world: some examples are elephant trunks, octopus and squid tentacles, snakes and caterpillars.

Moving from the animal to the vegetal world, we can also find some examples of grasping structures that have been overlooked so far. Climbing plants are capable to grasp and climb the surrounding environment with the goal to achieve maximum sun exposure while avoiding the energy expenditure of developing a supporting trunk [1]. The dedicated climbing organs are called tendrils and their strategy represents an evolutionary success. In fact, it allows them to succeed with the ability to ascend over would-be competitors at the cost of relatively small energetic investment.

Plants lack of a nervous system, and the exploration of the environment and the execution of mechanical actions rely efficiently on simple reflex-like behaviours [3]. Tendrils are highly touch-sensitive, and after the mechanical stimulus has been perceived by epidermal cells, plant hormones serve as mediators of the coiling response [2]. Using this basic sensory-motoric loop without centralized sensing and control, plants can blindly rely on organs that will eventually find and coil around supports, providing a successful grasping method.

N.F. Lepora et al. (Eds.): Living Machines 2013, LNAI 8064, pp. 418–420, 2013.
© Springer-Verlag Berlin Heidelberg 2013

In the present contribution we studied the basic rules that govern the tendrils of *Passiflora spp.* (Passion vine, Figure 1.a). In particular we addressed a set of experiments to understand the three different stages of the grasping: (i) the circumnutation is important to find the support; (ii) the contact coiling to grasp it; and (iii) the free-coiling to secure the hold and get closer to the support. For our biomimetic purposes the latter two are of major interest and to our knowledge, no biomimetic results or attempts to reproduce them can be found in literature. *Passiflora* tendrils are able to recognize supports and obstacles on the overall surface bending in different directions; the capability to produce multiple coils allows the plant to work as a winch. Thanks to the multiple coils, in fact, the tendril apex experiences a small and negligible tension, that increases from the apex to the inner touching point. Furthermore, the increased touching surface avoids the slippage due to the related increased friction. The signal transmission is yet not well known even if experimental observations show a sort of modular behavior. This means that the zone that senses a support induces a contraction phase to the near fibers; after that, if the contact increases, i.e. the touched nearest zones sense a contact, the bending signal is transmitted creating the overall tendril motion and grasping. Thanks to this, a distributed reflex control is made, allowing the activation of the motion locally and only when necessary. This is surely an advantage from an energetic point of view since the modules sleep in a normal phase and are activated only when necessary. Moreover, if the free-coiling phase is considered, other important features can be highlighted. The free-coiling phase allows to pull the stem towards the grasped support, by creating a zero-torsion helical spring. Indeed, by coming back to the original-intrinsic helical shape, the tendril shortens the distance between the fixed end-points. Finally, the helical-spring shape is perfectly tuned to resist to external loads and disturbances. Thus, all these desirable characteristics induced us to evaluate and investigate how to mimic the plant tendril system structure.

Here we propose a preliminary bio-robotic tendril that encapsules the sensing and actuating ability of colimbing plants. The overall structure has been considered from a kinematic point of view. The model is conceptualized and simplified dividing the tendril in two main parts (Figure 1.b): the first part (FC) mainly devoted to the free-coiling and pulling phase, which can be viewed as a single actuator that changes its shape from a linear wire to a helical spring; and the second part (GC) devoted to the coiling and grasping phase and considered as subdivided in n-sections. Kinematics of GC involves two main steps: first, the GC tendril kinematics problem is approached by means of a series of substitutions applied to a modifed homogeneous transformation based on the Denavit-Hartenberg (DH) notation; second, velocity kinematics could be solved by computing the Jacobian using standard techniques and then by chaining together the Jacobians (Figure 1.c). In order to simulate the kinematics of a bio-inspired tendril, the exploited formulas have been implemented in a Matlab simulator. The contact has been implemented by searching for each module if there is intersection between the segment that connects the two

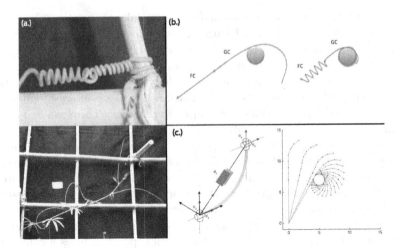

Fig. 1. The morphology of a tendril and its conceptualization. (a) A *Passiflora* tendril grasped to a support; (b-c) GC and FC modules; (d) a *Passiflora* plant; (e) equivalence wire - robot; (f) Kinematics of grasping.

universal/revolute joints. The FC behavior has been implemented as a pulling motion driven by the prismatic joint.

The sensory and actuation system may be less dynamic than our human senses and muscles but still have the advantage of greater autonomy. Although no work to mimic specifically climbing plants via coiling has been found within bio-inspired robotic literature. Some interesting results have been obtained in [4], where an hyperredundant serpentine robot able of pole climbing have been developed. Tactile sensing and new decentralized control law are indeed the main innovative points of this preliminary study. The possibility of relying on neuromorphic devices able to process information locally and at the same time actuate the single unit for producing a global coiling behaviour would be a desiderable feature that could advance the tecnology of grasping.

References

1. Isnard, S., Silk, W.K.: Moving with climbing plants from charles darwins time into the 21st century. American Journal of Botany 96(7), 1205–1221 (2009)
2. Jaffe, M., Galston, A.: The physiology of tendrils. Annual Review of Plant Physiology 19(1), 417–434 (1968)
3. Trewavas, A.: Green plants as intelligent organisms. Trends in Plant Science 10(9), 413–419 (2005)
4. Cowan, L.S., Walker, I.D.: Soft continuum robots: the interaction of continuous and discrete elements. Artificial Life 11, 126 (2008)

Biomimetic Lessons for Natural Ventilation of Buildings

A Collection of Biomimicry Templates Including Their Simulation and Application

David R.G. Parr

Welsh School of Architecture, Cardiff University, UK
DavidRGParr@Gmail.com, DavidRGParr@biomimicron.wordpress.com

Abstract. A variety of Biomimicry Templates were developed from peer-reviewed research into biological ventilation methods. These templates were then ranked with the templates that were the most supported and potentially beneficial being developed through CFD into concept designs. The objective of this was to incorporate them into more energy-efficient building designs. A range of concepts were tested, with some leading to significant improvement in ventilation rates, while others demonstrated issues in scaling up/adaption to human dwellings.

Keywords: Biomimetic, building, energy efficiency, ventilation, prairie dogs, ants, termites.

Biomimicry is a growing design technique in the environmental design of buildings[1]. A number of ventilation processes from the natural world including termite's mounds[2], prairie dog's burrows[3] and ant's nests[4] have been analysed and focused through a literature review into a set of Biomimicry Templates. These templates address a range of applications concerning the inducement of flow, increased gas transport velocity, the transfer of heat between solids and gases, the enhanced transport/diffusion of oxygen and the efficient utilisation of active processes that would otherwise constitute wasted energy.

A small selection of these Biomimicry Templates has then been examined in CFD tests. The goal of this process was threefold. Firstly, this was to attempt to verify the processes that had been reported from the previous work. Secondly, this tested the limitation of scale inherent in the templates selected. Thirdly, this investigated the possibility of successful adaptation of the templates previously modelled to contexts more applicable to the human built environment.

Of those tested, it was found that while the templates selected were verified according to the best available data, that flow behaviour deviated when these models were scaled up, making it preferable to implement these structures at the scale they are found in the natural world. The subsequent adaptation of the ants nest template to a context more useful to human habitation also demonstrated some aberrant flow

N.F. Lepora et al. (Eds.): Living Machines 2013, LNAI 8064, pp. 421–423, 2013.

behaviour; however, there was partial success in that the structure did induce flow. The adaptation of the prairie dog template was more successful, showing an enhancement of flow rate of between 1.5 to 5 times on the base case, dependent on domain wind speed.

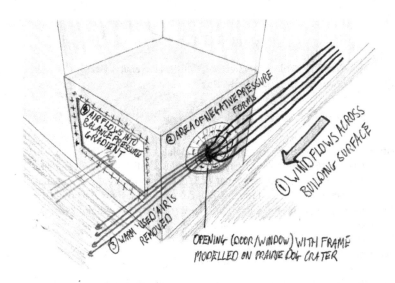

Fig. 1. Concept sketch using a Prairie Dog inspired structure to increase ventilation rate

Template: scale (lo) 1 - 5 (hi)	Termite mound			Ant Mound			Bird Lungs
	Wind Harnessing: Low frequency filter	Enhanced Diffusion: Structural resonance	Thermal Mass: Ground heat sink + chimney	Wind Harnessing: Wind velocity profile + Bernoulli	Thermal Mass: Ground heat sink	Efficient mechanical process	Efficient use of surface area
Clarity of template	2	2	2	4	3	4	3
Ease of development	1	1	2	3	4	2	3
Potential of benefit	4	4	3	4	3	3	2
Total	7	7	7	11	10	9	8

Template: scale (lo) 1 - 5 (hi)	Prairie Dog			Phragmites australis		
	Wind Harnessing: Wind velocity profile	Thermal Mass: Ground heat sink	Increased Humidity: Drawing air from porous soil	Wind Harnessing: Wind velocity profile	Enhanced Diffusion: Tissue pressurisation	Roots as Gas Transport Network
Clarity of template	5	3	1	5	4	4
Ease of development	4	4	2	4	1	4
Potential of benefit	2	3	4	2	4	1
Total	11	10	7	11	9	9

Fig. 2. Table ranking examined Biomimetic Templates in order to select those to be tested

Fig. 3. Model 'rooms' used in Autodesk Simulation CFD analysis. Interior volume is $2m^3$, ventilation hole diameter is 200mm.

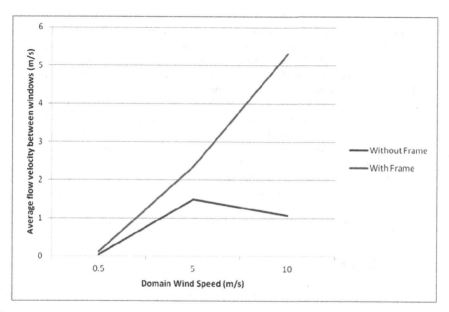

Fig. 4. Results of testing above structures, demonstrating the increase in flow velocity of including 'Prairie Dog Portholes' in the design

References

1. Pawlyn, M.: Biomimicry in Architecture, 1st edn. RIBA, London (2011)
2. Turner, J.: Ventilation and thermal constancy of a colony of a southernAfrican termite (Odontotermes transvaalensis: Macrotermitinae). Journal of Arid Environments 28, 231–248 (1994)
3. Vogel, S., Ellington, C., Kilgore, D.: Wind-Induced Ventilation of the Burrow of the Prairie-Dog, Cynomys ludovicianus. Journal of Comparative Physiology A 85, 1–14 (1973)
4. Kleineidam, C., Ernst, R., Roces, F.: Wind-induced ventilation of the giant nests of the leaf cutting ant Atta vellenweideri. Naturwissenshaften (88), 301–305 (2001)

Sub-millilitre Microbial Fuel Cell Power for Soft Robots

Hemma Philamore, Jonathan Rossiter, and Ioannis Ieropoulos

Bristol Robotics Laboratory
University of the West of England, Frenchay Campus,
Coldharbour Lane Bristol BS16 1QY, U.K.

Abstract. Conventional rigid-body robots operate using actuators which differ markedly from the compliant, muscular bodies of biological organisms that generate their energy through organic metabolism. We consider an 'artificial stomach' comprised of a single microbial fuel cell (MFC), converting organic detritus to electricity, used to drive an electroactive artificial muscle. This bridges the crucial gap between a bio-inspired energy source and a bio-inspired actuator. We demonstrate how a sub-mL MFC can charge two 1F capacitors, which are then controllably discharged into an ionic polymer metal composite (IPMC) artificial muscle, producing highly energetic oscillation over multiple actuation cycles. This combined bio-inspired power and actuation system demonstrates the potential to develop a soft, mobile, energetically autonomous robotic organism. In contrast to prior research, here we show energy autonomy without expensive voltage amplification.

Keywords: Microbial fuel cell, artificial muscle, energetic autonomy.

1 Introduction

Current robotics research is focused on combining self-fuelling mechanisms with robust mechanical designs, resulting in systems that can operate, unassisted, in terrain too hostile or inaccessible for humans [1]. The EcoBot robot series utilises rigid MFC technology, with raw organic fuel, to power conventional electromechanical actuator, sensor and communication systems, imitating the energy sustenance mechanisms of natural organisms. [2].

The biomimetic design of artificial muscles, including those comprised of electroactive polymers (EAPs), resembles the soft physical structure of muscular organisms. These materials can respond with greater compliance to varied and unpredictable environments and have excellent thermodynamic efficiencies and low mass to power ratios compared to conventional electromechanical actuators [3]. Among the EAP technologies, IPMCs have been implemented in a number of applications including propulsion [4], stirring and cilia-like motion [5]. IPMCs are capable of significant actuation at low voltages (ca. 1-3V) due to induced ionic migration within the polymer layer when a potential is applied to the two noble metal electrodes.

In this work we demonstrate how an MFC 'stomach' can directly drive IPMC actuator 'muscles' though a low resistance switching circuit.

N.F. Lepora et al. (Eds.): Living Machines 2013, LNAI 8064, pp. 424–426, 2013.

2 Method and Results

2.1 Energy Harvesting from Sub-millilitre Scale MFC

The performance of two replicated systems, each comprising an MFC, charging two 1F super-capacitors in parallel (Fig.1a), was measured using a Pico Technology ADC-24 data logger. The capacitors were charged to a mean voltage of 475.4mV. Charging was terminated on day 6, after which negligible voltage increase was observed. The MFC (Fig.1a) anode chamber held 0.2mL of anolyte, and was open on one side, where a cation selective membrane (VWR, UK) of 15mm diam. was attached. A moistened, open to air cathode, was fitted against the exterior side of this membrane. Anode and cathode electrodes were made from carbon fibre veil, with surface areas of 1800mm² and 4500mm², respectively. The anolyte was sewage sludge mixed with tryptone (10%) and yeast extract (5%).

2.2 Actuation of IPMC Artificial Muscle

Two capacitors, charged from the output of a single MFC, were connected in series and discharged to a hydrated 2 x 1cm IPMC rectangular cantilever strip, actuated in free air (Fig.1b). The polarity of the applied voltage was switched, using an externally powered relay stage, with a frequency of 1Hz. Voltage across the IPMC was measured using a National Instruments PCI-6229 board. Actuation displacement was recorded using a Keyence laser sensor. The IPMC sample was fabricated from Nafion 112 (DuPont), coated with gold electrodes using electroless plating.

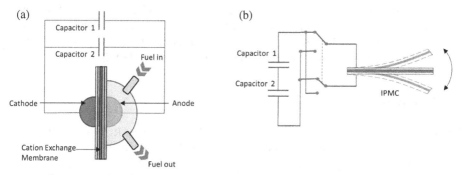

Fig. 1. Schematic of system configurations for **(a)** energy harvesting and **(b)** IPMC actuation

A clear actuation response to the applied voltage is shown by the IPMC artificial muscle (Fig.2). Voltage decay from 890mV to 282mV was accompanied by a decrease in amplitude of displacement from 0.2mm to 0.13mm per stroke over 60 actuations. Energy stored by the IPMC per actuation was calculated using Equation 1.

$$E = \frac{1}{2}CV_{final}^2 - \frac{1}{2}CV_{initial}^2 \qquad (1)$$

where V_{final} and $V_{initial}$ are the respective final and initial voltages across the IPMC per stroke and C is the capacitance of the IPMC. The average decrease in energy per actuation stroke was only 3%, indicating the feasibility of multiple oscillations from a single capacitive charge supply.

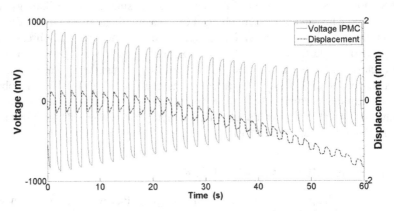

Fig. 2. IPMC displacement over 60s, driven directly from alternating supply charge, at 890mV

3   Summary

We have demonstrated the ability of a sub-mL MFC-powered system for sustained undulating soft-robotic actuation. The system achieved 60 actuations over a 1 minute interval from one capacitor charge cycle of 890mV. Potential uses for this low power generation and actuation mechanism include propulsion in a mobile robot and stirring in an MFC anode chamber. Further optimisation of the low-power switching mechanisms will lead to more effective power delivery to the IPMC actuator.

References

1. Tadesse, Y., Villanueva, A., Haines, C., Novitski, D., Baughman, R., Priya, S.: Hydrogen-fuel-powered bell segments of biomimetic jellyfish. Smart Mater. Struct. 21(4), 733–740 (2012)
2. Ieropoulos, I., Greenman, J., Melhuish, C., Horsfield, I.: EcoBot-III: a robot with guts. In: Artificial Life XII, E-Book, pp. 733–740 (2010) ISBN-10:0-262-29075-8
3. Shaninpoor, M., Kim, K.L., Mojarrad, M.: Artificial Muscles: Application of Advanced Polymeric Nanocomposites, pp. 20–47. Taylor & Francis (2007)
4. Kim, B., Kim, D.-H., Jung, J., Park, J.-O.: A biomimetic undulatory tadpole robot using ionic polymer–metal composite actuators. Smart Mater. Struct. 14(6), 1579–1585 (2005)
5. Ieropoulos, I., Anderson, I., Gisby, T., Wang, C.H., Rossiter, J.: Microbial-powered artificial muscles for autonomous robots. In: TAROS 2008, pp. 209–216 (2008)

How Active Vision Facilitates Familiarity-Based Homing

Andrew Philippides[1,*], Alex Dewar[1], Antoine Wystrach[1],
Michael Mangan[2], and Paul Graham[2]

[1] Centre for Computational Neuroscience and Robotics, University of Sussex, UK
[2] School of Informatics, University of Edinburgh, UK
andrewop@sussex.ac.uk

The ability of insects to visually navigate long routes to their nest has provided inspiration to engineers seeking to emulate their robust performance with limited resources [1-2]. Many models have been developed based on the elegant snapshot idea: remember what the world looks like from your goal and subsequently move to make your current view more like your memory [3]. In the majority of these models, a single view is stored at a goal location and acts as a form of visual attractor to that position (for review see [4]). Recently however, inspired by the behaviour of ants and the difficulties in extending traditional snapshot models to routes [5], we have proposed a new navigation model [6-7]. In this model, rather than using views to recall directions to the place that they were stored, views are used to recall the direction of facing or movement (identical for a forward-facing ant) at the place the view was stored. To navigate, the agent scans the world by rotating and thus actively finds the most familiar view, a behavior observed in Australian desert ants. Rather than recognise a place, the action to take at that place is specified by a familiar view.

Such familiarity-based navigation is well-suited to navigating along a route where the ant generally travels in a small range of directions (~45°, say). It seems less well-suited, however, to locating a specific goal location (food or nest) which must be approached from any direction. In [7-8] we showed that place search could be achieved with a familiarity-based mechanism if a learning walk – an active sensing strategy composed of stereotypical movements made by ants when they first leave the goal and used to scaffold visual learning of the goal position - is added to the training route. To add more weight to this idea, we wanted to show that the combination of active vision with our familiarity mechanism could replicate ants' behaviour in a classic experiment used as evidence for an attractor-type snapshot. To this end, we replicated the experiment of Wehner et al [9] in which ants were trained to find food in the centre of three landmarks. During tests with food removed, ants' search distributions are recorded with landmarks in either their original positions at twice the distance, or at twice the distance with landmarks of twice the size (Fig. 1 A-C). This experiment has been used to support the idea that a single snapshot is used for homing as the view of the surroundings from the goal in A matches that from the goal in C, but not B, and the search distribution of ants is the same in A and C, but not B.

---

* Corresponding author.

N.F. Lepora et al. (Eds.): Living Machines 2013, LNAI 8064, pp. 427–430, 2013.
© Springer-Verlag Berlin Heidelberg 2013

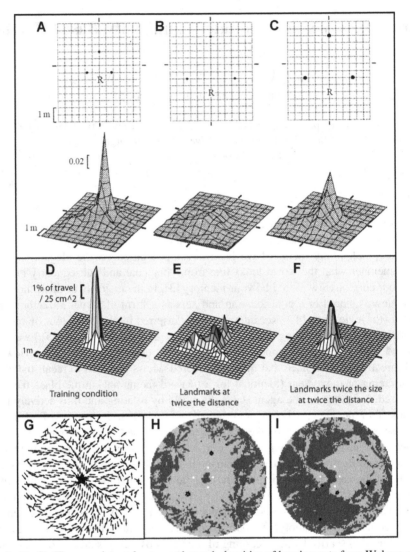

Fig. 1. (A–C): The experimental setup and search densities of homing ants from Wehner at al., [9] in training condition (A), with 20cm high cylinders at twice the distance (B) and with cylinders twice the size and at twice the distance (C). Food position in training indicated by crosshairs. (**D–F**): Search density profiles from the model in training condition (D), with cylinders at twice the distance (E) and with cylinders twice the size and at twice the distance (F). Search densities are collected for 24 'ants' after 10m of search starting at random positions. (**G-I**): Homing performance of the algorithm in two environments with four training views in the same positions (white dots in H and I) oriented toward the goal (star in G). (G) Arrows show direction of movement in environment 1. (H, I) To illustrate homing difference between environments 1 (H) and 2 (I), we coloured regions where homing error was less/greater than (an arbitrary threshold of) 45° in light/dark grey. Black objects are randomly generated grass tussocks (height in the range [20-50] cm). Each environment is 12m in diameter and there are a variety of objects outside this region. A-F adapted from [10] with permission.

However, the same search distributions can be achieved by an agent navigating with a familiarity-based model using training views oriented towards the food, but crucially, *not* at the food location itself (Fig 1. D-F and [10]). During a learning phase, the views are used to train a network so that it subsequently outputs a measure of the familiarity of any view input to it. Navigation then proceeds iteratively, by an agent scanning the world from $-90°$ to $+90°$ around the previous heading in $1°$ steps and moving 1cm in the direction specified by the most familiar view found across the scan. Search distributions recovered are similar to those in [9] showing that familiarity-based navigation can explain the observed behavior (Fig. 1 D-F). See [10] for full description of this and another behavioural experiment replicated with this algorithm.

In the above, the location of training views are based on a learning walk of the ant *Ocymyrmex* [11]. However, we next ask whether the learning walk, and by extension the training view locations, are adapted to the environment or whether there is a set of training view locations relative to the goal suited to all environments. To test this we examined navigation in two simulated environments using training views from the same four locations oriented towards the goal. Heading directions are derived by rotating the current view through $360°$ in $1°$ steps and finding the orientation which best matches each training view. The resultant heading is a weighted average based on closeness of match to each training view. While performance is good in both environments, the pattern of success and failure is different suggesting that training view positions and thus learning walk shape should be adapted to the environment.

Insect navigation has long been a source of elegant solutions for complex engineering tasks. Here we show how a parsimonious algorithm actively produces robust homing. The key for engineering and biology is to further understand how active processes underpinning visual learning and navigation are shaped by the environment.

References

1. Wehner, R.: Desert ant navigation: How miniature brains solve complex tasks. J. Comp. Physiol. A 189, 579–588 (2003)
2. Graham, P., Philippides, A.: Insect-Inspired Vision and Visually Guided Behavior. In: Bhushan, B., Winbigler, H.D. (eds.) Encyclopedia of Nanotechnology. Springer (2012)
3. Cartwright, B.A., Collett, T.S.: Landmark Learning in Bees - Experiments and Models. J. Comp. Physiol. A 151, 521–543 (1983)
4. Möller, R., Vardy, A.: Local visual homing by matched-filter descent in image distances. Biol. Cybern. 95, 413–430 (2006)
5. Smith, L., Philippides, A., Graham, P., Baddeley, B., Husbands, P.: Linked local navigation for visual route guidance. Adapt. Behav. 15, 257–271 (2007)
6. Baddeley, B., Graham, P., Philippides, A., Husbands, P.: Holistic visual encoding of antlike routes: Navigation without waypoints. Adapt. Behav. 19, 3–15 (2011)
7. Baddeley, B., Graham, P., Husbands, P., Philippides, A.: A Model of Ant Route Navigation driven by Scene Familiarity. PLoS Comput. Biol. 8(1), e1002336 (2012)

8. Graham, P., Philippides, A., Baddeley, B.: Animal cognition: Multi-modal interactions in ant learning. Curr. Biol. 20, R639–R640 (2010)
9. Wehner, R., Michel, B., Antonsen, P.: Visual navigation in insects: coupling of egocentric and geocentric information. J. Exp. Biol. 199, 129–140 (1996)
10. Wystrach, A., Mangan, M., Philippides, A., Graham, P.: Snapshots in ants? New interpretations of paradigmatic experiments. J. Exp. Biol. 216, 1766–1770 (2013)
11. Müller, M., Wehner, R.: Path integration provides a scaffold for landmark learning in desert ants. Curr. Biol. 20, 1368–1371 (2010)

Embodied Behavior of Plant Roots in Obstacle Avoidance

Liyana Popova[1,2], Alice Tonazzini[1,2], Andrea Russino[2],
Alì Sadeghi[1], and Barbara Mazzolai[1]

[1] Center for Micro-BioRobotics@SSSA, Istituto Italiano di Tecnologia (IIT),
V. Piaggio 34, 56025 Pontedera, Italy
[2] The BioRobotics Institute, Scuola Superiore Sant'Anna, V. Piaggio 34, 56025 Pontedera, Italy
a.russino@sssup.it, {liyana.popova,ali.sadeghi,alice.tonazzini,
barbara.mazzolai}@iit.it

Abstract. Plant roots are a new paradigm for soft robotics. Study of embodied behavior in roots may lead to the implementation of movements guided by structural deformations and to the use of sensors and actuators as body parts. In this work the obstacle avoidance in roots and its interplay with the gravitropism were studied both from biological and robotic viewpoint. Living roots resulted to achieve the maximum pushing force on an obstacle before starting circumnavigation (30 mN in 100 min), thus indicating the existence of a triggering threshold. Tip-to-obstacle angle (20°) was not influenced by the gravity. A robotic mockup capable to bend like living roots was build on the basis of current knowledge and our results on obstacle avoidance behavior. Exploitation of morphological features and passive body deformation resulted to be useful for implementing a simplified control of the robot during gravitropism and obstacle avoidance.

Keywords: robotic root, plant root embodied behavior.

1 Embodied Behavior in Roots and Its Implementation in Robot

Plants are often seen as passive organisms. Actually they are able to move, perceive the environment, and respond to it. In particular, the apical region (RA) of plant roots moves by growing and has a rich sensing system which enables complex tropic behaviors without any control unit. For these reasons the plant root can be a biological model for the development of new generation of soft robots [1]. One of the most important features to be implemented into an exploratory robot is an obstacle avoidance behavior, similarly to that observed in the biological counterpart.

Roots are able to follow the gravity (gravitropism) and to react to touch stimulations (thigmotropism). Integration of these sensing capabilities allows roots to find low impendence pathways in the soil, circumnavigate obstacles, and grow downward for anchoring and finding water. Gravity perception mainly occurs in the root cap (RC) by sedimentation of statoliths inside cells [2]. This stimulus is also partially enabled by stress sensing on cell walls in the body of root apex (RA), due to the weight of the cytoplasm inside cells and to the weight of RA [2]. RA has two

N.F. Lepora et al. (Eds.): Living Machines 2013, LNAI 8064, pp. 431–433, 2013.
© Springer-Verlag Berlin Heidelberg 2013

bending regions (BRs) which allow achieving an S-shape configuration during obstacle avoidance [2]. When the root approaches the obstacle, RA buckling activates the first bending response. Then tip bending occurs during root growth parallel to the barrier, thus allowing the RC to be continually in contact with obstacle and explore it. During this phase RC orientation is the result of interplay between gravity and touch perception.

Some experiments with the primary maize roots were performed in order to better understand obstacle avoidance by evaluating: (i) forces that root applies to the obstacle (the setup is described in Fig. 1a) (ii) the root tip orientation with respect to the obstacle during its circumnavigation (the setup is detailed in [3]). Roots resulted to reach the maximum pushing force (approximately 30 mN after 100 min contacting the obstacle) before starting the circumnavigation (Fig. 1b). This suggests the existence of a threshold, which may be dependent on the root mechanical properties related to buckling (e.g., root diameter and length). We found that tip-to-obstacle angle stabilizes on approximately 20° after 1.5 h for differently oriented obstacles (Fig. 1c). The results are coherent with the work of [2] and suggest that the touch is a driving-stimulus for tip orientation in this case.

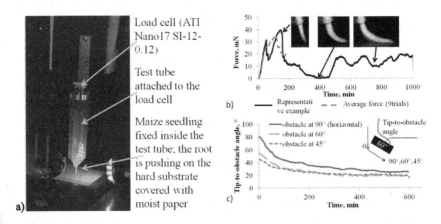

Fig. 1. a) Experimental setup to measure the root pushing force during obstacle circumnavigation. b) Representative example and average of root pushing (9 trials) forces. c) Tip-to-obstacle angle with respect to three obstacle orientations (9 trials). (For methods and setup see [3]).

The mechano-gravity response shows to have different elements of embodiment: (i) sensing and actuation performed by the same part of body; (ii) structural passive deformation, such as buckling, that drives the bending; (iii) touch-gravity interplay in the root cap without any control unit. Implementation of these aspects into the robotic root may decrease the complexity of control of its soft body. A robotic mockup without any control unit was build to observe root embodied behaviors in a mechanical system. It is able to implement the shape of root obstacle avoidance and its interplay with gravitropism (Fig. 2). It has two BRs which can bend separately. When the mockup is positioned horizontally, both BRs actively bend downward due to gravity sensing (distal BR) and body deformation (proximal BR). When the mockup is pushed vertically towards the obstacle (the pushing action resembles root elongation), it assumes the S-shape configuration. First, the proximal BR curves due to its buckling and, then, the distal BR bends by following the gravity. When the pushing force becomes high enough to activate the touch sense at the tip, the tip stops bending and maintains its configuration.

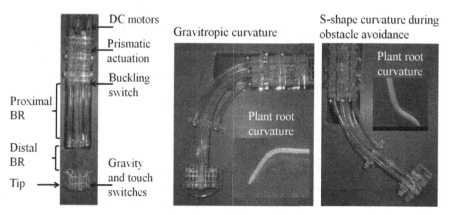

Fig. 2. Mockup consists of two soft bending regions (BR). Each BR consists of nut-screw prismatic mechanism with sliding part made by soft tube to enable the bending during prismatic movement. Bending is performed by differential elongation similarly to the plant roots. Sensing capability of system is provided by three mechanical switches. Buckling switch is activated by the buckling of proximal BR. Gravity switch situated in the tip is activated by floating connector that moves under its weight. A touch switch situated in the tip is activated by root pushing force. It disables the gravity sensing and activates an opposite bending.

Morphology and passive body deformation are the keys for improving mechanical design and simplifying the control of the robotic root. Complex S-shape was obtained in a control-less mechanical soft system by exploiting the embodied behavior of plant roots. This configuration allows the robotic root to explore while passing the obstacle. Next step will be to implement sensors and actuators which are parts of robot body (e.g., smart textiles, electro active polymers) to better exploit the movement activation by the structural deformation by touching and buckling.

Acknowledgments. This work was supported by the FET Programme within the 7th FP for Research of the European Commission, under the PLANTOID project FET-Open n. 293431.

References

1. Mazzolai, B., Mondini, A., Corradi, P., Laschi, C., Mattoli, V., Sinibaldi, E., Dario, P.: A Miniaturized Mechatronic System Inspired by Plant Roots. IEEE Transaction on Mechatronics 16(2), 201–212 (2011)
2. Gilroy, S., Masson, P.H.: Plant tropisms. Blackwell Publishing Ltd., Oxford (2008)
3. Russino, A., Ascrizzi, A., Popova, L., Tonazzini, A., Mancuso, S., Mazzolai, B.: A Novel Tracking Tool for Analysis of Plant Root Tip Movements. Bioinspiration and Biomimetics (accepted)

Motor Control Adaptation
to Changes in Robot Body Dynamics
for a Complaint Quadruped Robot

Soha Pouya, Peter Eckert, Alexander Sproewitz, Rico Moeckel,
and Auke Ijspeert

Biorobotics Laboratory, Ecole Polytechnique Federale de Lausanne (EPFL),
Lausanne, Switzerland
first_name.last_name@epfl.ch
http://biorob.epfl.ch

Abstract. One of the major deficiencies of current robots in comparison
to living beings is the ability to adapt to new conditions either resulting
from environmental changes or their own dynamics. In this work we focus
on situations where the robot experiences involuntary changes in its body
particularly in its limbs' inertia. Inspired from its biological counterparts
we are interested in enabling the robot to adapt its motor control to
the new system dynamics. To reach this goal, we propose two different
control strategies and compare their performance when handling these
modifications. Our results show substantial improvements in adaptivity
to body changes when the robot is aware of its new dynamics and can
exploit this knowledge in synthesising new motor control.

Keywords: Legged Robots, Locomotion Control, Adaptive Behavior.

1 Introduction

The topic of robot legged locomotion continues to flourish and develop, yet
several challenges are present in designing, modelling and control of such robots.
While the mechanical design defines the limits of robot capabilities, the locomotion performance can also be influenced largely with the selected control
technique. For instance decentralised joint space control approaches may suffice
for statically stable robot motions. In such cases, system dynamics is preserved
and play only a minor role. When dealing with body changes, one can not ignore
the strong influence of the system dynamics on the performance. These influences can be partially seen and treated as disturbances (via decentralised PD
controllers). On the other hand a more proper and systematic way -instead of
reducing the effects- is to adapt the system by incorporating leg dynamics when
designing the control laws [1]. Having this objective in mind, we provide control
laws benefitting from model-based feedforward prediction terms to adapt to new
robot dynamics. By adding this term to the PD position controllers at each joint
we decrease the reliance of the robot's control to its feedback terms. We show

N.F. Lepora et al. (Eds.): Living Machines 2013, LNAI 8064, pp. 434–437, 2013.
© Springer-Verlag Berlin Heidelberg 2013

that our proposed control scheme not only helps the robot to perform with more compliant behavior, but also provides higher robustness against changes e.g. in the inertial properties.

2 Methods

We use the biologically inspired Cheetah-cub quadruped robot (Fig-1-c) designed at EPFL/Biorob [2] and develop its rigid body dynamics (RBD) model (Fig-1-d) based on articulated-body algorithm [3] by using the simulation and control software package SL [4]. The model is carefully tuned to match the real robot for both its geometrical and inertial properties. The overall control-learning architecture is represented in Fig-1 (a and b). Through an optimization process for the robot's speed applied at its joint position control profiles, we extract a locally optimized gait as the nominal solution. These profiles are generated via a central pattern generator (CPG) based controller [5] which facilitates the generation of different gait types through the coupling terms between different degrees of freedom. The CPG-based controller is well-suited for the optimization process as it encodes the control profile with only a few tuning parameters. Taking the result of this process as the preferred nominal gait, we setup the control loop with two main blocks: An inverse dynamics (ID) block. It provides feedforward prediction on the required torques regarding the new system dynamics. The second, feedback controller (PD) block performs as the disturbance rejector. We make the assumption that once the inertial parameters of the system change, they can be estimated by using state of the art parameter estimation methods [6]. We then propose two different control schemes; *(i) a low gain PD + ID* controller where the updated dynamics of the system is incorporated within the control law through feedforward torques and *(ii) a high gain PD only* controller where the robot only uses decentralised PD controllers per joint with no updates from the new dynamics terms. In order to investigate the performance of each control scheme when dealing with body changes we systematically decrease the leg inertial properties (mass and moment of inertia) and compare how this affects the robot's speed and cost of transport (COT) while performing bounding gait.

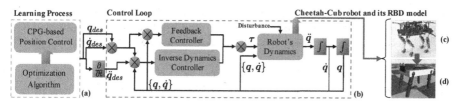

Fig. 1. The proposed control-learning architecture composed of (a) the learning process to find an opimal gait for the given robot dynamics, (b) the control loop to reproduce the optimal gait, (c) the bioinspired quadruped robot Cheetah-cub and (d) the RBD model of the robot in SL software.

3 Results and Discussion

Figure 2 shows the results of the systematically changing inertial parameters, and its effect at the robot's speed and COT, for our two proposed control schemes. We perform these experiments for four kinds of experiments: changes in the left-fore (LF) leg, the right-hind (RH) leg, both fore legs, and both hind legs. In all cases, we observed that the locomotion performance is highly sensitive to these changes when applying decentralised joint-based PD control. Results show major drops in forward velocity particularly for the case of changes in the hind legs (e.g. to less than 0.4 of nominal speed after 20% and total failure after 55% mass reduction for RH). We explain this as follows: PD gains tuned for the initial system start to lose their performance when the system inertia changes. Our proposed control scheme with an updated ID however shows several, substantial improvements; *(i)* it shows higher adaptivity to body changes, by handling up to 80% reduction for the individual legs and up to 60% for both legs. *(ii)* it increases the nominal speed due to lower PD gains and a resulting more compliant behavior. *(iii)* it results in lower COT values due to smoother torque commands in comparison to a high-gain-PD control. In order to complete this architecture, an extra module is required to learn the new dynamics. We are currently developing this module as well as transferring the results to the real robot.

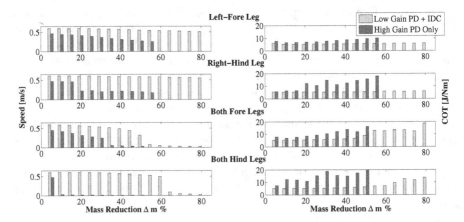

Fig. 2. The robot's speed and COT vs. the systematic change in inertial properties of (from top to down) only left-fore leg, only right-hind leg, both fore and both hind legs. For each setup the two proposed control schemes are used and compared.

Acknowledgments. This work has received funding from the European project Locomorph (grant agreements No. 231688). We gratefully thank Stefan Schaal, Mrinal Kalakrishnan and Ludovic Righetti for their help to develop 3D dynamics model of the robot in SL simulation and control package.

References

1. Siciliano, et al: Robotics Modelling, Planning and Control. Springer (2008)
2. Sproewitz, A., et al.: Towards Dynamic Trot Gait Locomotion Design, Control, and Experiments with Cheetah-cub, a Compliant Quadruped Robot. IJRR (2013)
3. Featherstone, R.: Rigid Body Dynamics Algorithms. Springer, Heidelberg (1987)
4. Schaal, S.: The SL simulation and real-time control software package. Technical report, University of Southern California (2009)
5. Ijspeert, A.J.: Central pattern generators for locomotion control in animals and robots: a review. Neural Networks 21 (2008)
6. Mistry, M., Schaal, S., Yamane, K.: Inertial parameter estimation of floating-base humanoid systems using partial force sensing. In: Humanoid Conference (2009)

The AI Singularity and Runaway Human Intelligence

Tony J. Prescott

Sheffield Centre for Robotics, University of Sheffield, Sheffield, S10 2TN
t.j.prescott@sheffield.ac.uk

Abstract. There is increasing discussion of the possibility of AI being developed to a point where it reaches a "singularity" beyond which it will continue to improve in a runaway fashion without human help. Worst-case scenarios suppose that, in the future, homo sapiens might even be replaced by intelligent machines as the dominant "species" on our planet. This paper argues that the standard argument for the AI singularity is based on an inappropriate comparison of advanced AI to average human intelligence, arguing instead that progress in AI should be measured against the collective intelligence of the global community of human minds brought together and enhanced be smart technologies that include AI. By this argument, AI as a separate entity, is unlikely to surpass "runaway" human (or, perhaps, posthuman) intelligence whose continued advance, fueled by scientific and cultural feedback, shows no sign of abating. An alternative scenario is proposed that human collective intelligence will take an increasingly biohybrid form as we move towards a greater, deeper and more seamless integration with our technology.

Keywords: Societal impact of technology, AI singularity, collective intelligence, human-machine symbiosis, biohybrid.

1 Introduction

Most approaches to the prospects of an AI singularity follow in the path of Good [1] who, writing in 1965, defined an "ultraintelligent machine" as a device that "can far surpass all the intellectual activities of any man however clever". As set out with great clarity by Chalmers [2], concerns about the AI singularity focus on the prospect of a positive feedback whereby future machines, initially developed to just exceed human intellect (AI+), rapidly bootstrap themselves to a level of intelligence far greater than that of any person (AI++). Science fiction scenarios that have explored this idea are usually dystopian, with meagre and unimproveable human intelligence fighting to maintain a foothold in a world dominated by ever-strengthening AI. This pessimism also extends to much academic writing on the topic, for instance, in a recent review Muehlhauser and Salamon suggested that "the default outcome from advanced AI is human extinction" [3].

N.F. Lepora et al. (Eds.): Living Machines 2013, LNAI 8064, pp. 438–440, 2013.
© Springer-Verlag Berlin Heidelberg 2013

2   Runaway Human Intelligence

But this standard scenario starts from a particular assumption about how we measure human intelligence. The metric chosen for defining AI+ is typically the intelligence of an "average human" (e.g. Chalmers), which we could call HI. If we think of HI as "raw brain power", which seems to be how many writers conceive of it, then this is, indeed, a more-or-less stationary quantity which, having evolved to its current capacity around 100,000 years ago has changed relatively little since. Given the slow pace of natural selection there is little prospect for further improvement any time soon. It seems plausible, then, that AI could surpass HI in many of its major aspects in the foreseeable future; in some domains it is already, unarguably, ahead.

An alternative comparison can be made, however, that casts the prospect of the singularity in a quite different light. Specifically, we can compare AI, not with individual human intelligence, but with the collective intelligence of the world human population. Furthermore, there is no obvious reason to consider human intelligence stripped of intelligence-enhancing artifacts. Since at least the Upper Paleolithic (10-50,000 years ago), humans have used external symbol systems to store and communicate knowledge and to boost their individual and collective reasoning capacity (see e.g. [4. 5, 6]). Indeed, computers, the internet, and AI itself, are simply the latest inventions in a set of technologies whose prior members include red ochre, papyrus, the abacus, the slide rule, the typewriter, and the telephone. By inventing and exploiting these intelligence-boosting and knowledge preserving technologies, humanity has precipitated an exponential increase in our shared knowledge and in our ability to apply these insights to control our environment according to our goals. This is "runaway intelligence" at the societal level, fueled by its own positive feedback, as cultural and scientific development has led to a larger, more long-lived and better-educated world population. The argument then, is that human intelligence is not constituted, or best described, at the level of the single mind but in terms of the species. We are as intelligent as the culture to which we belong, able to contribute the raw processing power of own brains to an enhanced collective intelligence (ECI) or—as telecommunications increasingly creates a single world community—to what Heylighen has called the "Global Brain" [7]. Being part of this cultural network, in turn, has a transforming effect on what our individual brains can do. Born with an immature and highly plastic nervous system we spend nearly two decades tuning our brains to take advantage of the intelligence-boosting tools that culture has to offer. In the long run this species-level technologically-enhanced intelligence has no obvious ceiling, we can continue to create technologies that complement our natural intelligence, allow us to communicate faster, and make us collectively smarter. If the prospects for the singularity are considered by comparing future AI with this ECI then the notion that humanity will be outstripped and left behind looks much less plausible. An advance in AI to A+ is after all, also an advance for the culture that generated that AI, so AI+ implies ECI+, AI++ implies ECI++, and so on.

3   A Symbiotic Biohybrid Future?

One question we might still ask is how likely it is that humanity will cease to exploit advances in AI that have the potential to boost collective intelligence. A possible threat here, is of a split emerging between AI and ECI, with a sneaky and malevolent version of AI attempting to conceal its advances, biding its time until it is ready to eliminate all of the unnecessary humans—back to the extinction story again. But this scenario, popular in some recent books and films (for example, both the Terminator quadrilogy and Wilson's Robopocalyse), smacks of anthropomorphism, that is, the assumption that AI systems will necessarily share some of humanity's worst instincts for tribalism. This also underestimates the likely contribution of biological intelligence to the future human-machine collective—there are many things that our brains and bodies do exceptionally well that will be hard for machines to master, and where there will be little economic incentive to improve them in order to do so; symbiotic systems are successful by virtue of their complementarity. The more plausible scenario, then, is that ECI will continue its runaway path but with an increasingly bio-hybrid form due to greater and deeper integration between humans and our intelligence-enhancing technologies. What is good for AI, will then also be good for us.

References

1. Good, I.J.: Speculations concerning the first ultraintelligent machine. In: Alt, F., Rubinoff, M. (eds.) Advances in Computers, vol. 6, Academic Press, New York (1965)
2. Chalmers, D.: The Singularity: A philosophical analysis. Journal of Consciousness Studies, 17 9(10), 7–65 (2010)
3. Muehlhauser, L., Salamon, A.: Intelligence explosion: evidence and import. In: Eden, A., Søraker, J., Moor, J.H., Steinhart, E. (eds.) Singularity hypotheses: A Scientific and Philosophical Assessment. Springer, Berlin (2012)
4. Clark, A.: Being There. MIT Press, Cambridge (1996)
5. Deacon, T.: The Symbolic Species: The Co-evolution of Language and the Brain. W. W. Norton & Co., New York (1997)
6. Mithen, S.: The Prehistory of the Mind: A Search for Origins of Art, Religion, and Science. Thames and Hudson, Ltd., London (1996)
7. Heylighen, F.: The Global Brain as a New Utopia. In: Maresch, R., Rötzer, F. (eds.) Zukunftsfiguren, Suhrkamp, Frankfurt (2002)

ASSISI: Mixing Animals with Robots in a Hybrid Society

Thomas Schmickl[1], Stjepan Bogdan[6], Luís Correia[2], Serge Kernbach[4],
Francesco Mondada[5], Michael Bodi[1], Alexey Gribovskiy[5], Sibylle Hahshold[1],
Damjan Miklic[6], Martina Szopek[1], Ronald Thenius[1], and José Halloy[3]

[1] Artificial Life Lab of the Department of Zoology, University of Graz, Austria
[2] LabMAg, Universidade de Lisboa, Portugal
[3] LIED, Université Paris Diderot, Paris
[4] Cybertronica Research, Stuttgart, Germany
[5] Ecole Polytechnique Fédérale de Lausanne, Switzerland
[6] Faculty of Electrical Engineering and Computing, University of Zagreb, Croatia
thomas.schmickl@uni-graz.at

Abstract. This paper describes the newly started EU-funded FP7 project
ASSISI$_{bf}$, which deals with mixed societies: A honeybee society integrated
with a group of stationary and interacting autonomous robotic nodes and a
group of fish integrated in a society of autonomous moving robots.

Mixed Societies of Animals and Robots

This position paper introduces the newly funded EU-ICT IST-FET FP7 project
"ASSISI$_{bf}$" (Animal and robot Societies Self-organize and Integrate by Social
Interaction) which aims for a novel strategy: it incorporates nature into a new bio-
hybrid ICT system that is composed by both, animal societies and a robotic society.
We will develop a novel distributed and bio-hybrid ICT system, which we classify as
a "collective adaptive system" (CAS) due to its inherent ability of adaptation and
self-organisation.

Fig. 1. Chosen setups for ASSISI$_{bf}$: Left: Stationary nodes interact with a group of bees.
Right: Mobile and autonomous robots with embedded fish lures.

N.F. Lepora et al. (Eds.): Living Machines 2013, LNAI 8064, pp. 441–443, 2013.
© Springer-Verlag Berlin Heidelberg 2013

First, we exploit the social interaction of animal societies and introduce artificial nodes that interact with these animals so that the animal society accepts these nodes. In a 2nd step, we convert our robotic probes in the animal society into "agents provocateurs" which feed new information into the animal society. This way we can modulate the natural society towards desired behavioural states. Third, we will have robots autonomously learning the "social language" of the animals. For ASSISI$_{bf}$ we have deliberately chosen two very different social animal species: fish, which express social interactions by self-positioning in the context of others and this way forming shoals or swarms; and honeybees which form a very integrated complex society, including a sophisticated dance language and communication through food exchange, stigmergy, pheromones and body contacts (Fig. 1). For further reasoning of mixing societies, see [1].

How Does ASSISI$_{bf}$ Go beyond the State of the Art?

One important aspect of ASSISI$_{bf}$ is that it closes the feedback loops between the artificial society and the natural one, with self-organisation of the mixed collective system. Feedbacks can either be pre-programmed or the artificial members of the mixed society can change their behaviours during runtime, which amounts to "learn the social language of the animals". A major difference of ASSISI$_{bf}$ compared to other projects [2] dealing with bio-hybrid societies is the way how we interact with moving groups of animals: For the honeybee experiments an array of 64 static robotic nodes will be generated, each one equipped with a set of sensors (light, touch, vibration, temperature, sound) and with a set of actuators (light, vibration, temperature, sound). Instead of moving robots, "patterns of actuation" will be moved across the arena to motivate the bees to perform specific behaviours. This design choice gives us high computation power and energy autonomy, as the computation device can be mounted below the arena and be plugged into constant power supply. Long-term energy-supply and computational power are crucial prerequisites for performing adaptation algorithms on each node, which is required to achieve the major objectives of ASSISI$_{bf}$: A robotic society learns to integrate itself into a natural one. ASSISI$_{bf}$ is outstanding because it uses 2 different animal species to generate models of collective behaviour, decisions and adaptation.

Methodologies and Their Challenges

To allow robotic nodes to adapt, we have to provide them with a sufficient set of sensors and actuators, but also with powerful computation devices and memory. Thus, the choice of an array of stationary devices that allow stimulus patterns to move instead of moving bee-like robots. To better understand the relationship between the individual and the collective behaviour, it is important to identify channels of communication/signalling that are relevant for social interactions. Thus we need to perform studies to understand the multimodality of social communication leading to collective behaviour. The next step is to design robots that are capable of sending the appropriate signals through the identified communication channels at the correct time and in the appropriate social context.

Formal and mathematical models of collective behaviours (e.g., see [3]) are necessary to design robotic systems capable of self-integrating into an animal society. In our model, raw experimental data will be first processed to extract animal features (e.g., position, posture, speed). Then, a segmentation of behaviours will be performed. The animal's behaviour is classified into primitives representing activities consistent over time. Finally, complex behaviours are formed from behavioural primitives linked by means of animals' interactions with the environment or among themselves. Automating these steps will have a major impact on behavioural biology and can facilitate to produce behavioural models.

Expected Outcome and Applications

First, we aim at building collective robots capable of interacting with groups of fish or bees. Second, we will implement a system that operates without central control. Our robots will be capable of self-regulating purely by sensorial cues and by behavioural patterns in the fish-robot and in the honeybee-robot societies. Third, we will develop automated model generation and behavioural abstraction mechanisms that will be used throughout the project. Finally, two systems that perform collective computation by including the neuronal information processing of animals in a non-invasive way will be delivered. This way, we will research new paradigms of autonomous homeostasis in self-organising systems. We expect novel automated systems for behavioural biology. By combining robots with automated modelling we can achieve an unparalleled automation of animal behaviour experimentation. Such intelligent automated and robotised systems may improve the field of biomedical research using model animals significantly. Another potential field of application can be the management of domestic animal stocks with low-stress. These systems could also be put at work to manage wild life animal pests.

Acknowledgements. Work supported by EU-ICT-FET project 'ASSISI$|_{bf}$', no. 601074.

References

1. Halloy, J., Mondada, M., Kernbach, S., Schmickl, T.: Towards bio-hybrid systems made of social animals and robots. In: Lepora, N.F., Mura, A., Krapp, H.G., Verschure, P.F.M.J., Prescott, T.J. (eds.) Living Machines 2013. LNCS (LNAI), vol. 8064, pp. 384–386. Springer, Heidelberg (2013)
2. Halloy, J., et al.: Social integration of robots into groups of cockroaches to control self-organized choices. Science 318(5853), 1155–1158 (2007)
3. Schmickl, T., Hamann, H., Wörn, H., Crailsheim, K.: Two different approaches to a macroscopic model of a bio-inspired robotic swarm. Robotics and Autonomous Systems 57(9), 913–921 (2009)

Chroma⁺Phy – A Living Wearable Connecting Humans and Their Environment

Theresa Schubert

Faculty of Media, Department Design of Media Environments,
Bauhaus-Universität Weimar, Germany
theresa.schubert-minski@uni-weimar.de

Abstract. This research presents an artistic project aiming to make cyberfiction become reality and exemplifying a current trend in art and science collaborations. Chroma⁺Phy is a speculative design for a living wearable that combines the protoplasmic structure of the amoeboid acellular organism *Physarum polycephalum* and the *chromatophores* of the reptile *Chameleon*. The underpinning idea is that in a future far away or close, on planet earth or in outer space, humans will need some tools to help them in their social life and day-to-day routine. Chroma⁺Phy enhances the body aiming at humans in extreme habitats for an aggression-free and healthy life. Our approach will address actual issues of scientific discovery for society and catalyse idea translation through art and design experiments at frontiers of science.

Keywords: body, design, wearable, hybrid, artificial vascularization, distributed sensing, decentralized intelligence, health, novel substrates.

1 Introduction

Let's imagine we live in outer space, somewhere far away from our orbit:

- There is no day or night-time anymore as the human body is used to. Chroma⁺Phy is showing night-time by darker colours. Thus humans are able to follow their circadian rhythm and minimising risk of sleeping disorders and related health issues.
- UV-radiation from a sun will be very high. Human skin will need a measurement device to alert when radiation levels are getting too high. Physarum polycephalum is an excellent light sensor. It prefers darker environments thus it would try to move and change its physical pattern visibly.
- Chroma⁺Phy can measure temperature and air humidity and reacts with colours thus the wearer can take steps to counter dehydration.
- Chroma⁺Phy can sense the body temperature of the wearer and the heart beat through vibration. Accordingly it will make an interpretation of his current emotional state by changing colours, which will help deciding how or if at all to communicate with fellow humans.

N.F. Lepora et al. (Eds.): Living Machines 2013, LNAI 8064, pp. 444–446, 2013.

2   Future Manual

Chroma⁺Phy comes in a box, ready to be applied directly onto the skin. It sits between a transparent layer of special medical adhesive silicone, that is permeable to air, humidity and temperature. After more than 48h of usage it has to be returned into the box that contains a nutritious, humid gel to 'recharge' for a few hours. Then it can be worn again, in total up to three or four weeks.

Fig. 1. Prototype of box containing the Chroma⁺Phy (left), box opened with living organism inside and recharge media (right)

3   Methods

The slime mould *Physarum polycephalum* has a complex life cycle. In its most active phase it looks like an amorphous yellowish mass with networks of protoplasmic tubes. The plasmodium behaves and moves as a giant amoeba. It is possible to divide the plasmodium and it will live on as separate entities or merge with another blob to one. It is a large single cell capable of distributed sensing and primitive memory [1]. The most prominent feature of the reptile *Chameleon* is its ability to change colours. Its function is in social signalling, in reactions to temperature and other conditions as well as camouflage. Colour change signals a chameleon's physiological condition and intentions to other chameleons. Chameleons have specialised cells, *chromatophores*, which contain pigments in their cytoplasm in three layers below their transparent outer skin. Dispersion of the pigment granules in the chromatophores sets the intensity of each colour, which can change due to rapid relocation of their particles of pigment [2].

We use this specific function of colour change combined with the decentralised logic of Physarum and its ability to attach and connect to any surface ignoring all gravitational laws. Inside the membrane Physarum is filled with cytoplasm. Through genetic manipulation Physarum's membrane and cytoplasm can be merged with Chameleon's chromatophores mechanism. For our research we designed a series of in vitro experiments that mark first steps towards the realisation at first without transgenic methods.

4   Discussion

We propose a hybrid organism that changes colour to indicate the intensity of the wearer's emotions, the percentage of outside humidity/radiation, the temperature as well as the circadian rhythm. The aim is to understand the human body and to improve communication when living in extreme environments. Currently still a speculative design, it exceeds pure fiction as a lot of experiments towards living wearable, control of Physarum polycephalum, chromatophores functionality, and novel silicone substrates can be made in real. Our aim is to make a transgenic Physarum in varying colours.

Fig. 2. Artist impression of Chroma⁺Phy as it might be worn by users

In this context it is also interesting to analyse the implication of such a wearable in a future society. What does it mean if we live in a quasi-symbiosis with a hybrid organism? Would our behaviour change, will our human interactions improve or the opposite? Will we develop a new understanding and rapport to 'primitive' organisms?

Acknowledgements. This research was partly supported by the European Commission within FP7-ICT, UComp, project "Physarum Chip: Growing Computers from Slime Mould". I thank Andrew Adamatzky for advice and discussion about Physarum polycephalum and Rüdiger Trojok for scientific support in microbiology and genetic modification.

References

1. Adamatzky, A.: Physarum machines. Computers from slime mould. World Scientific, Singapore (2010)
2. Klaver, C., Böhme, W.: Das Tierreich The Animal kingdom. Chamaeleonidae, vol. 112. Walter de Gruyter, Berlin (1997)

Plant Root Strategies for Robotic Soil Penetration

Alice Tonazzini[1,2], Ali Sadeghi[2], Liyana Popova[1,2], and Barbara Mazzolai[2]

[1] Scuola Superiore Sant'Anna, The BioRobotics Institute, Pontedera, Italy
[2] Italian Institute of Technology, Center for Micro-BioRobotics, Pontedera, Italy
{alice.tonazzini,ali.sadeghi,liyana.popova,
barbara.mazzolai}@iit.it

Abstract. Soil penetration strategies of plant roots can represent an interesting source of inspiration for designing explorer robots. In this work we present a selection of these strategies whose performances were discussed and evaluated by means of engineering mock-ups and dedicated experiments in granular substrates. The obtained results demonstrated that root elongation from the tip reduces the forces needed for soil penetration up to 50%; tip morphology and anchorage resulted to strongly influence penetration performances.

Keywords: Soil penetration strategies, Plant root inspired robots.

1 Plant Root Penetration Strategies: Description and Evaluation

In order to penetrate soil efficiently, plant roots implement several energy-saving strategies, whose origins can be traced back to the root tip (apex) and that have been extensively investigated by biologists by means of observations, kinematic, and morphological studies. A new approach is represented by studying plant roots from an engineering point of view to extract the key features responsible for an efficient soil exploration and translate them in new design principles for autonomous explorer robots. Plant root moves by growing and its growth is the result of cell division and cell elongation in the apical region (Fig.1a). New produced cells mainly transit from the meristem to the elongation zone, where their axial expansion provides the pressure needed for the apex penetration and bending; peculiar mucilage secreting cells are continuously produced by the distal meristem and sloughed off from the cap surface. Behind the apex, elongated mature cells are stationary [1]. The anchorage between soil and this mature tissue allows the forward movement of the root. The elongation from the tip (EFT) phenomenon allows the reduction of dynamic frictional resistance because only a limited part of the body -the apex- is pushed, while the basal part stays fixed. At the same time the cells which slough off from the cap surface create an interface between soil and root cap: by working similarly to a bearing system, it decreases friction on root flanks. Growth of hairs and secondary roots just behind elongation region increases the root anchorage. Apex morphology strongly influences frictional interaction with soil and it can change depending on soil impedance.

N.F. Lepora et al. (Eds.): Living Machines 2013, LNAI 8064, pp. 447–449, 2013.
© Springer-Verlag Berlin Heidelberg 2013

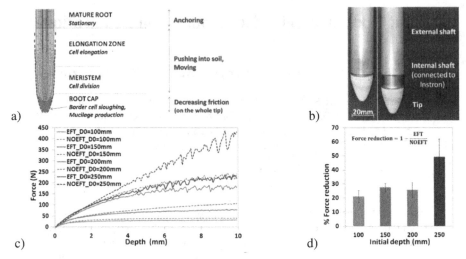

Fig. 1. a) Tip structure and its mechanical functions. b) Mock-up for EFT tests. c) Mean penetration force with respect to depth for four different initial depths (D_0), with EFT and without (NOEFT). d) Force reduction achieved by EFT for different initial depths.

Several penetration tests were performed in order to quantify the reduction of frictional resistance related to the EFT and to define optimal tip shape for penetration in granular substrates. The experimental setup consists of a cylindrical glass recipient for granular soil (glass beads with different diameters), purposely built probes and Instron testing machine for measuring forces and pushing (Load cell ± 1KN, acquisition frequency 100 µs, constant penetration rate 40mm/min). Fig.1b shows the probe used for EFT tests, which consists of two coaxial shafts which can slide frictionless each other; the internal shaft ends with a parabolic tip while the other end is fixed to Instron. If the degree of freedom between them is blocked, a traditional penetration test can be performed (NOEFT); on the contrary, when the external shaft is fixed to the soil and only the internal part is free to penetrate, we mimic an EFT effect. For each condition eleven penetrations 10 mm in depth were performed. Before testing, the probe was positioned into a preformed hole into the remolded soil (ϕbeads=0.5mm). As expected, penetration resistance related to EFT is less compared with NOEFT for all the initial depths tested (Fig.1c). For deeper penetrations the reduction increases from about 20 to 50% (Fig.1d). By carrying on traditional penetration tests with tips characterized by different shapes and dimensions, energy needed for penetrating different kinds of soils was estimated. Conical and parabolic profiles with the greatest height-diameter ratio (htip/ϕtip) resulted to be more efficient in the penetration for all substrates (Fig.2).

In [3] we developed a self-intruding system for soil penetration that takes inspiration by the sloughing cells mechanism in the apical root region: a 3D flexible skin is continuously released between the body of the robotic root and soil and creates a lower friction interface between them (Fig.3a). To verify the contribution of anchorage to penetration capabilities of the robotic root, artificial hairs were added to the skin (Fig.3b). During penetration, bigger is the interface between skin and soil, bigger

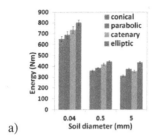
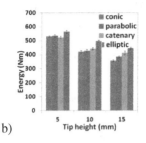
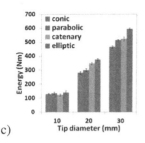

Fig. 2. Mean penetration energy of different tip profiles at 80 mm in depth for: a) $\phi_{bead}= 0.04$, 0.5, 5mm, h_{tip} =15mm and ϕ_{tip}=20mm; b) h_{tip} =5, 10, 15mm, ϕ_{tip} = 20mm, in ϕ_{bead}=0.5mm glass beads. c) ϕ_{tip} = 10, 20, 30 mm, h_{tip}/ϕ_{tip}= 0.75. Mean value and standard deviation were calculated for 11 penetration tests. The used setup is detailed in [2].

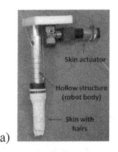

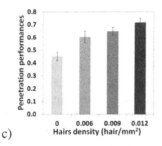

Fig. 3. a) Self-penetrating and self-anchoring system, detailed in [3]. b) Polymeric hairs added on artificial skin. c) Penetration performances for four different hairs density on the skin during free penetration tests in glass beads (ϕ=0.5 mm); penetration performances are calculated as the ratio between skin displacement (40mm) and the related depth reached by the system.

are the anchoring capabilities. The hairs improve the penetration performances by increasing the skin-soil adhesion: this effect is proportional to hairs density along the skin (Fig.3c) and it was verified into different soils (ϕ_{beads}=0.5, 5 mm). A quantitative analysis of root penetration strategies can set the basic ground toward the design of an energy efficient robotic probe for soil exploration. Future work will concentrate on root passive mechanical flexibility and tip morphological adaptation.

Acknowledgments. This work was supported by the FET programme within the 7th FP for Research of the European Commission, under the PLANTOID project FET-Open n. 293431.

References

1. Bengough, A.G., Croser, C., Pritchard, J.: A biophysical analysis of root growth under mechanical stress. Plant and Soil 189, 155–164 (1997)
2. Tonazzini, A., Popova, L., Mattioli, F., Mazzolai, B.: Analysis and Characterization of a Robotic Probe Inspired by the Plant Root Apex. In: Biorob, Roma (2012)
3. Sadeghi, A., Tonazzini, A., Popova, L., Mazzolai, B.: Innovative Robotic Mechanism for Soil Penetration Inspired by Plant Roots. In: ICRA, Karlsruhe (2013)

The Synthetic Littermate

Stuart P. Wilson

University of Sheffield, Sheffield, UK
s.p.wilson@sheffield.ac.uk

Abstract. I suggest how a new type of biohybrid society – a huddle of neonatal rat pups comprising biological and synthetic litttermates – could be used to model the interaction between self-organisation at the neural level and self-organisation at the level of group behaviours.

Keywords: biohybrid systems, mixed societies, self-organisation, thermoregulation.

How do natural developmental experiences shape the developing brain? The synthetic littermate project is designed to directly address this fundamental question. The synthetic littermate is an artificial rat pup, that will share its experiences with a litter of real rat pups, huddling amongst them as part of a new type of biohybrid system – the biohybrid litter. The device will allow us to explore how natural systems self-organise, both at the neural level and at the behavioural level. These two levels are co-dependent because during early postnatal development, the self-organising brain controls how the pup interacts with other littermates and interactions between littermates in turn determine the patterns of sensory input to the self-organising brain. Hence the synthetic littermate will also be used to explore self-organisation at the interaction between brain and behaviour, as it unfolds during natural development within the huddle.

Huddling is 'an active and close aggregation of animals', which reduces the exposed body surface of the individual, lowers the individual's required operating temperature, and heats the surrounding space, reducing the energy expenditure of the individual by up to 50% [1]. Rodents huddle for warmth during the first two postnatal weeks, wriggling amongst each other in a dynamic synergy that aids thermoregulation and limits oxygen consumption, and thus reduces the overall metabolic costs to the individual [2]. In rats, these collective and regulatory behaviours have been modelled as self-organising systems [3]. As such, simulations of huddling reproduce emergent group behaviours observed experimentally, such as thigmotaxis (wall-following) and thermoregulation. According to these models, complex behaviours emerge based on the adherence of the individual to a small set of simple behavioural rules, which lead to a balance of cooperative and competitive interactions between littermates [4].

This balance between cooperation and competition is under active, multi-sensory control from at least postnatal day two (P2), well before the eyes open at around P10 [5]. Orienting behaviours in very young pups are mediated by tactile stimulation, with P2 pups orienting themselves reliably in the direction

N.F. Lepora et al. (Eds.): Living Machines 2013, LNAI 8064, pp. 450–453, 2013.

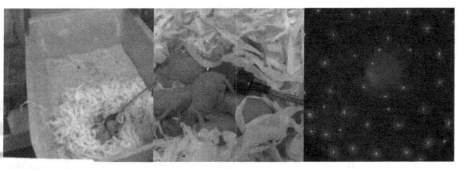

Fig. 1. A prototype synthetic littermate device, interacting with a litter of 6-day old Wistar rat pups. *Left:* Data were collected in the home cage with the mother removed. *Center:* The synthetic littermate (green) is moved by the huddling motion of the biological littermates. *Right:* Internal view of the surrogate showing a visual image through a central 'eye-hole' and peripheral 'tactile markers' on the inner surface of a deformable rubber skin. Displacement of the markers is tracked online (each is marked by a cross), and data from each modality is used to drive self-organising models of topographic map development in multisensory cortices.

of tactile stimuli evoked by contact with a littermate [6]. Huddling behaviours transition at around P14, from an initial 'physiological huddling' stage driven by thermoregulation, to a later 'filial huddling' stage driven by responses to social (and olfactory) cues, and huddling persists into early adulthood (see [7]). Throughout development, the adherence of the individual to a small set of simple behavioural rules gives rise to complexity, measurable in the regulatory dynamics of group temperature, energy, cohesion, and aggregation patterns.

As the body and behavioural repertoire of the littermate develops, so too does its brain. Stimulus-driven self-organisation remains a dominant theory for how sensory experiences shape the developing brain, and specifically for how function is assigned to mammalian neocortical microcircuitry [8]. The theory, as expressed by algorithms derived from the self-organising map (SOM), predicts that a balance between cooperative and competitive interactions within neural circuits, consolidated by Hebbian learning, cause correlated inputs to be represented by nearby neurons [9]. For example, tendencies for adjacent facial whiskers to move in similar directions inclines cortical maps for whisker motion direction towards a global somatotopic organisation [10].

Although such models can in principle be used to generate predictions about the functional organisation assigned to any cortical region, in practice the validity of these models is limited by the accuracy with which natural developmental input patterns can be synthesized. Important steps to overcoming this bottleneck have been to develop biomimetic sensors for biomimetic robot models [11], and to investigate aggregation patterns in groups of robots [12]. Taking a step further, the aim of the synthetic littermate project is to integrate biomimetic sensors into small biomimetic robots and to embed them in the natural physical and social environment of the developing animal, i.e., into the huddle (see

Figure 1). The data collected from multiple sensory modalities can then be used to synthesize training patterns for self-organising models, which reflect naturally occuring multisensory and sensorimotor contingencies. These will likely reflect correlations between e.g., tactile, proprioceptive, and vestibular signals, which are induced by naturally occuring patterns of contact between huddling littermates. Hence the models can be used to derive the first ecologically valid, testable predictions about how developmental experience shapes the functional organisation in the associated 'higher-order' cortices. The working hypothesis is that these cortical regions self-organise into topoglogical maps for the underlying motion parameters that correlate multimodal inputs; the (relative) motion of the littermate. By using network responses to influence the movement of the synthetic littermate at run-time (i.e., by orienting it towards or away from salient multisensory targets), the approach could in turn be used to explore the interaction between self-organisation at the neural level and self-organisation at the level of group behaviour. To this end we have been developing agent-based models to formulate hypotheses about how manipulations to e.g., the temperature of the synthetic littermate, will effect the self-organisation of the huddle [13].

References

1. Gilbert, C., McCafferty, D., Maho, Y.L., Martrette, J.M., Giroud, S., Blanc, S., Ancel, A.: One for all and all for one: the energetic benefits of huddling in endotherms. Biol. Rev. Camb Philos Soc. 85(3), 545–569 (2010)
2. Alberts, J.R.: Huddling by rat pups: group behavioral mechanisms of temperature regulation and energy conservation. J. Comp. Physiol. Psychol. 92(2), 231–245 (1978)
3. Schank, J.C., Alberts, J.R.: Self-organized huddles of rat pups modeled by simple rules of individual behavior. J. Theor. Biol. 189(1), 11–25 (1997)
4. Sokoloff, G., Blumberg, M.S.: Competition and cooperation among huddling infant rats. Dev. Psychobiol. 39(2), 65–75 (2001)
5. Alberts, J.R.: Huddling by rat pups: multisensory control of contact behavior. J. Comp. Physiol. Psychol. 92(2), 220–230 (1978)
6. Grant, R.A., Sperber, A.L., Prescott, T.J.: The role of orienting in vibrissal touch sensing. Front Behav. Neurosci. 6, 39 (2012)
7. Alberts, J.R.: Huddling by rat pups: ontogeny of individual and group behavior. Dev. Psychobiol. 49(1), 22–32 (2007)
8. Swindale, N.V.: The development of topography in the visual cortex: a review of models. Network 7(2), 161–247 (1996)
9. Miikkulainen, R., Bednar, J.A., Choe, Y., Sirosh, J.: Computational maps in the visual cortex. Springer, Berlin (2005)
10. Wilson, S.P., Law, J.S., Mitchinson, B., Prescott, T.J., Bednar, J.A.: Modeling the emergence of whisker direction maps in rat barrel cortex. PLoS One 5(1), e8778 (2010)

11. Mitchinson, B., Pearson, M.J., Pipe, A.G., Prescott, T.J.: Biomimetic robots as scientific models: A view from the whisker tip. In: Neuromorphic and Brain-Based Robots: Trends and Perspectives. Cambridge University Press (2011)
12. Bish, R., Joshi, S., Schank, J., Wexler, J.: Mathematical modeling and computer simulation of a robotic rat pup. Mathematical and Computer Modelling 45(78), 981–1000 (2007)
13. Glancy, J., Gross, R., Wilson, S.P.: A minimal model of the phase transition into thermoregulatory huddling. In: Lepora, N.F., Mura, A., Krapp, H.G., Verschure, P.F.M.J., Prescott, T.J. (eds.) Living Machines 2013. LNCS (LNAI), vol. 8064, pp. 381–383. Springer, Heidelberg (2013)

Evo-devo Design for Living Machines

Stuart P. Wilson and Tony J. Prescott

University of Sheffield, Sheffield, UK
{s.p.wilson,t.j.prescott}@sheffield.ac.uk

Abstract. Natural design is an interaction of both *adaptive* evolutionary forces and *generative* developmental processes – evo-devo. We consider how evo-devo principles can be applied to the design of living machines, and how biohybrid societies (comprising machines and organisms) may be used as a new form of scientific model.

Keywords: biohybrid systems, mixed societies, self-organisation, natural selection.

Evo-devo (from evolutionary developmental biology) is an emerging science, born of a recognition that complexity emerges from the interaction between *generative* (i.e., form-generating) and *adaptive* (i.e., form-selecting) forces [1]. These forces are omnipresent in the dynamical interactions that operate at all scales of living systems. Understanding how adaptive forces maintain complexity, by selecting not forms but form-generating processes, may help us to design new types of living machine.

In natural living systems, dynamics of increasing complexity emerge from 'cycles of reciprocal cause and effect' between interacting elements [2]. Interactions between lower-level dynamical systems can spontaneously give rise to supervenience, as with the emergence of surface tension from interactions between water molecules. Reciprocal interactions that operate upon these supervenient properties can themselves give rise to higher-order, morphogenetic dynamics, and reciprocal morphogenetic dynamics can in turn give rise to the dynamics that biology identifies as self-organisation and natural selection. Nature seems therefore to be fractal in design: Living systems are dynamical hierarchies [3], comprising interacting organisms, which in turn comprise interacting biological systems (i.e., cells), which in turn comprise interacting chemical systems (i.e., molecules), which in turn comprise interacting physical systems (i.e., atoms), and so on through the known sub-atomic scales, and perhaps *ad infinitum* (see Figure 1).

The complexity observable at any one scale in the dynamical hierarchy is dependent upon the interactions between elements at all of the lower scales, yet paradoxically, the macroscopic complexity cannot be predicted by observing only microscopic interactions [4]. This realization, that chaos permeates all levels of complexity in living systems, suggests that natural design is itself a synthetic process – a prefiguration, not of complex function itself, but of the rules by which complexity is *configured* through microscopic interactions.

A synthetic approach has been central to many of the great breakthroughs in modern biological thinking, such as morphogenesis by reaction and diffusion [5],

N.F. Lepora et al. (Eds.): Living Machines 2013, LNAI 8064, pp. 454–456, 2013.

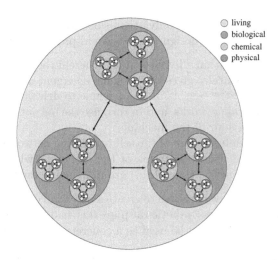

living
biological
chemical
physical

Fig. 1. The fractal design of natural living systems. Reciprocal interactions between elements of a given scale of complexity can spontaneuously give rise to complex elements at the macro-scale. This suggests a hierarchy of complexity, with living systems towards the top of the dynamical hierarchy (i.e., the largest circle).

autocatalysis by self-organisation and selection [6], and canalization by development and evolution [7,8]. Each considers how macroscopic complexity emerges from microscopic interactions, and each marries both generative *and* adaptive interactions that had previously been considered in isolation: Reaction *and* diffusion, self-organisation *and* selection, development *and* evolution.

Modern evo-devo biology is beginning to reveal a range of mechanisms, which offer shortcuts to genomes taking random (i.e., Darwinian) steps towards peak fitness, such as DNA methylation, horizontal gene transfer, and genetic redundancy (see e.g., ref. [9]). These epigenetic mechanisms may help to support powerful pseudo-Lamarckian forms of evolution, by allowing natural selection to improve genomes based not on the phenotypic forms that they generate, but based on a more general capacity for the system to generate complex forms [8].

The goal is to discover similar shortcuts when designing controllers for living machines. The task is to adapt self-organising algorithms to control the behaviour of the individual machine, by using fitness functions that promote the emergence of group-level (i.e., macroscopic) complex forms. Instead of searching randomly for these adaptations, the task may become more tractable if the behaviour of the machine needs only to maintain pre-existing macroscopic forms. For this it may be possible to utilise natural group-level aggregation behaviours, such as those displayed by social animals, i.e., macroscopic forms that emerge naturally from the interactions between organisms.

Evo-devo design could therefore be used to create control systems for living machines, through the use of the *biohybrid society* as a new form of scientific model. In biohybrid societies, we could quantify the extent to which groups comprising both machines and organisms are able to sustain the complex dynamics that emerge naturally in groups of only organisms. To create control systems for living machines, the evo-devo design process would proceed as follows:

1. Embed the machine within a living system, wherein the aggregate behaviour of the organisms naturally gives rise to some emergent dynamical property.
2. Use the inputs from the sensors of the machine to drive the development of self-organising neural networks, and in turn use the output of those networks to control the behaviour of the machine.
3. Adapt the initial configuration of the self-organising networks, using genetic algorithms, defining fitness in terms of the extent to which the emergent dynamical property of the (biohybrid) group is maintained.

In an accompanying paper we propose to create a new form of biohybrid society, a rat litter comprising real and synthetic rat pups that huddle together to regulate group temperature, which could be used as a concrete test case for the evo-devo design process [10].

Designing controllers for machines in biohybrid societies in this way, by adapting generative control algorithms and using biohybrid societies to effectively bootstrap the search for form-generating controllers, may enable us to sidestep questions about whether or not the machines that we create are truly living. If living systems emerge from simple interactions between organisms, and if machines integrated into living systems can sustain the dynamics of the system, then at what ratio of machines to organisms does the complex behaviour of the biohybrid system cease to be living?

References

1. Carroll, S.: Endless Forms Most Beautiful: The New Science of Evo Devo and the Making of the Animal Kingdom. W. W. Norton & Company (2005)
2. Deacon, T.W.: The hierarchic logic of emergence: Untangling the interdependence of evolution and self-organization. In: Evolution and Learning: The Baldwin Effect Reconsidered. MIT Press, Cambridge (2003)
3. Rasmussen, S., Baas, N.A., Mayer, B., Nilsson, M., Olesen, M.W.: Ansatz for dynamical hierarchies. Artif Life 7(4), 329–353 (2001)
4. Gleick, J.: Chaos: Making a new science. Vintage (1987)
5. Turing, A.M.: The chemical basis of morphogenesis. Philosophical Transactions of the Royal Society of London. Series B, Biological Sciences 237(641), 37–72 (1952)
6. Kauffman, S.: Requirements for evolvability in complex systems - orderly dynamics. Physica D 42(1-3), 135–152 (1990)
7. Waddington, C.: Canalization of development and the inheritance of acquired characters. Nature (1942)
8. Hinton, G., Nolan, S.: How learning can guide evolution. Complex Systems 1, 495–502 (1987)
9. Deacon, T.W.: Colloquium paper: a role for relaxed selection in the evolution of the language capacity. Proc. Natl. Acad. Sci. USA 107(2), 9000–9006 (2010)
10. Wilson, S.P.: The synthetic littermate. In: Lepora, N.F., Mura, A., Krapp, H.G., Verschure, P.F.M.J., Prescott, T.J. (eds.) Living Machines 2013. LNCS (LNAI), vol. 8064, pp. 450–453. Springer, Heidelberg (2013)

Preliminary Implementation of Context-Aware Attention System for Humanoid Robots

Abolfazl Zaraki, Daniele Mazzei, Nicole Lazzeri,
Michael Pieroni, and Danilo De Rossi

Research Center "E. Piaggio", Faculty of Engineering, Univ. of Pisa, Italy
a.zaraki@centropiaggio.unipi.it

Abstract. A context-aware attention system is fundamental for regulating the robot behaviour in a social interaction since it enables social robots to actively select the right environmental stimuli at the right time during a multiparty social interaction. This contribution presents a modular context-aware attention system which drives the robot gaze. It is composed by two modules: the scene analyzer module manages incoming data flow and provides a human-like understanding of the information coming from the surrounding environment; the attention module allows the robot to select the most important target in the perceived scene on the base of a computational model. After describing the motivation, we report the proposed system and the preliminary test.

Keywords: Context-aware attention, scene analysis, gaze control, multiparty social interaction.

1 Motivation

Many efforts have been conducted to design attention systems which guide robot fixation based on the saliency of low-level features presented in the visual scene (colours, intensity, orientation and etc.) [1]. However this class of attentional systems is not able to take into account social relevant features such as verbal and non-verbal cues. This paper presents a modular context-aware attention system that considers both low-level visual features and high-level social relevant features in the robot's attention. The system has been implemented in a human-like social robot called FACE (Facial Automaton for Conveying Emotions) [2], in order to drive its dynamic attention and gaze in a multiparty social interaction.

2 The Proposed System

As illustrated in Fig.1 the attention system consists of two modules, the scene analyser module and the attention module. They are designed taking into account not only the social relevant features but also the visual saliency of the human subject and non-human potential target.

N.F. Lepora et al. (Eds.): Living Machines 2013, LNAI 8064, pp. 457–459, 2013.

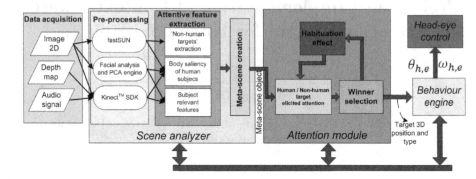

Fig. 1. General structure of the proposed system

2.1   The Scene Analyser Module

This module is deputed to provide the robot with a human-like understanding of the surrounding environment. As shown in Fig.1 it consists of three different units: pre-processing, attentive feature extraction and meta-scene creation. The pre-processing unit employs a software layer in order to extract: a 2D map of the the visual saliency of the scene (based on the FastSUN algorithm [3]); social relevant features such as the subject's distance and orientation, body shape, skeletal information and gesture (through the Microsoft Kinect SDK); facial expressions and subject's name (through a facial analysis engine and a PCA engine). Some other features such as *non-human target* and the *human body saliency* cannot be directly inferred through pre-processing unit and are identified as following.

- *Non-human target*, is the most important point of the saliency map identified by the robot. It allows the robot to be attracted by environmental stimuli during a social interaction. To obtain it, the module performs a local spatial competition across the image and analyses the low-level features of each pixel [3]. The pixel with the highest contrast of luminance, colour, and orientation will win the competition as non-human target.
- *Human body saliency*, is the average score of the saliency values in the area circumscribed by the body shape. It allows the robot's attention to be influenced by parameters such as the colour of the clothes worn by the subject.

Finally, the scene analyser module creates a meta-scene object and stores all the low-level and social relevant extracted features of human and non-human targets with corresponding ID. The output is streamed out to the attention module.

2.2   The Attention Module

The core of the module is a computational model that calculates the amount of the elicited attention EA_n of the features stored in the meta-scene object. Then it calculates for each subject the total amount of EA_t by the sum of the weighted features.

$$EA_t = w_1(EA_1) + w_2(EA_2) + ... + w_n(EA_n)$$

The weight w_n is selected according to the importance of the features. For example, in the current version of the attention model, a subject who speaks or raises her/his hand is more important than a subject who smiles. Moreover, the weight parameters can be adjusted to allow the robot to show different human-like behaviours. The elicited attention of the social relevant features of the non-human target is set to zero. The total elicited attention (EA_t) is assigned to each subject in order to select the winner among human and non-human targets through a competition. The winner is the target with the highest amount of EA_t which should be watched by the robot. The presence of the non-human target allows the robot to have a dynamic gaze behaviour. For instance if all the subjects are not enough interesting, the robot will switch to a environmental target as same as human being does in a similar situation. The 3D position of the target is sent to the behaviour engine which controls the robot's gaze in term of amplitude and velocity of head-eye movement on the base of human-like gaze model [4]. Furthermore, the behaviour engine as part of the FACE platform can affect the whole attentive process by adjusting the parameters of different modules.

3   Conclusion and Future Development

Preliminary tests have been conducted within a social interaction between three subjects while the robot was a bystander. The results showed that the robot properly followed the conversation and identified human target at the right time. The robot switched to non-human targets when none of the subjects showed at least one specific feature such as speech, motion, smile and gesture. The future work will focus on developing the behaviour engine which will simulate the emotional mood of the robot through its gaze movements according to the neurophysiological studies.

References

1. Begum, M., Karray, F.: Visual attention for robotic cognition: a survey. IEEE Transactions on Autonomous Mental Development 3(1), 92–105 (2011)
2. Mazzei, D., Lazzeri, N., Hanson, D., De Rossi, D.: Hefes: An hybrid engine for facial expressions synthesis to control human-like androids and avatars. In: 2012 4th IEEE RAS & EMBS International Conference on Biomedical Robotics and Biomechatronics (BioRob), pp. 195–200. IEEE (2012)
3. Butko, N.J., Zhang, L., Cottrell, G.W., Movellan, J.R.: Visual saliency model for robot cameras. In: International Conference in Robotics and Automation, pp. 2398–2403. IEEE (2008)
4. Itti, L., Dhavale, N., Pighin, F.: Realistic avatar eye and head animation using a neurobiological model of visual attention. In: SPIE's 48th Annual Meeting, International Society for Optics and Photonics, pp. 64–78 (2004)

Author Index

Akiyama, Yoshitake   1
Ambroise, Matthieu   347
Anderson, Iain A.   350
Anderson, Sean R.   311
Antonelli, Marco   12
Assaf, Tareq   311
Ayers, Joseph   299

Baldassare, Gianluca   191
Beccai, Lucia   353, 368, 415
Bernardeschi, Irene   353
Bertram, Craig   24
Blustein, Daniel   299
Bodi, Michael   441
Bogdan, Stjepan   441
Bortoletto, Roberto   96
Brandi, Santiago   356
Breinlinger, Philipp   143
Brown, Keith   36

Capurro, Alberto   204
Capus, Chris   36
Chen, Jian   359
Chiappalone, Michela   274
Chiel, Hillel J.   59
Chinellato, Eris   12, 47
Churchill, Alexander W.   83
Cianchetti, Matteo   368
Ciofani, Gianni   353
Cole, Marina   204
Cope, Alex   362
Correia, Luís   441

Daltorio, Kathryn A.   59
Dario, Paolo   251
Daumenlang, Benjamin   143
Del Pobil, Angel P.   12
Demiris, Yiannis   47
De Rossi, Danilo   393, 457
Dewar, Alex   427
Dodd, Tony J.   405
Dominey, Peter Ford   240
Duran, Angel J.   12

Eckert, Peter   434
Evans, Mathew H.   24, 364

Fernando, Chrisantha   71, 83
Ferrati, Francesco   96
Ferri, Gabriele   378
Follador, Maurizio   368
Fountas, Zafeirios   371
Fox, Charles   108
Fremerey, Max   374

Gardner, Julian W.   204
Giorelli, Michele   378
Glancy, Jonathan   381
Goldschmidt, Dennis   402
Gorb, Stanislav   402
Graham, Paul   427
Greco, Francesco   353, 415
Gribovskiy, Alexey   441
Griffiths, Gareth   323
Groß, Roderich   381
Gurney, Kevin   362

Hahshold, Sibylle   441
Halloy, José   384, 441
Hayasaka, Tomoaki   412
Heepe, Lars   402
Helgadottir, Lovisa   143
Herreros, Ivan   356, 399
Horchler, Andrew D.   59
Hosoda, Koh   409
Huang, Jiaqi V.   119

Ieropoulos, Ioannis   424
Ijspeert, Auke   434
Indiveri, Giacomo   262
Iwabuchi, Kikuo   1

Javaid, Mahmood   24
Jowers, Casey   350

Kanzaki, Ryohei   131, 167
Kapsali, Veronika   387
Karout, Salah   204
Kazama, Toshiya   390

Kelch, Milan   350
Kernbach, Serge   384, 441
Kobayashi, Ryo   390
Komatsu, Yuichi   390
Kovalev, Alexander   402
Krapp, Holger G.   119
Kumar, Sreedhar S.   274
Kurabayashi, Daisuke   131, 167
Kuroiwa, Koki   390

Lallée, Stéphane   287
Landgraf, Tim   143
Lane, David   36
Laschi, Cecila   368
Laschi, Cecilia   378
Lazzeri, Nicole   393, 457
Lepora, Nathan F.   154, 396, 405
Levi, Timothée   347
Ludwig, Tobias   143

Maffei, Giovanni   399
Mangan, Michael   427
Manoonpong, Poramate   402
Marshall, James   362
Martinez-Hernandez, Uriel   154, 405
Mattoli, Virgilio   353
Mazzei, Daniele   393, 457
Mazzolai, Barbara   353, 368, 415, 431, 447
Melhuish, Chris   323
Menciassi, Arianna   251
Miklic, Damjan   441
Mimmo, Tanja   418
Minegishi, Ryo   131, 167
Minzan, Kosuke   409
Mitchinson, Ben   179
Miyasaka, Kota   409
Mizuno, Fumio   412
Moeckel, Rico   434
Mondada, Francesco   384, 441
Morishima, Keisuke   1
Murray, Mark M.   350

Nakai, Junichi   409
Nawrot, Martin   143
Nelson, Bradley J.   216
Nowak, Philipp   143
Nowotny, Thomas   362

Ognibene, Dimitri   47, 191
Ogura, Toshihiko   409

Ohkura, Masamichi   409
Oliver, Raymond   387

Pagello, Enrico   96, 131
Paihas, Yan   36
Palagi, Stefano   415
Pandolfi, Camilla   418
Parr, David R.G.   421
Pasquale, Valentina   274
Pearce, Timothy C.   204
Pearson, Martin J.   179, 311
Petit, Maxime   240
Peyer, Kathrin E.   216
Pezzulo, Giovanni   191
Pfeifer, Rolf   335
Pfeiffer, Michael   262
Philamore, Hemma   424
Philippides, Andrew   427
Pieroni, Michael   457
Pipe, Anthony G.   179
Pipe, Tony   323
Pirim, Patrick   228
Pointeau, Gregoire   240
Popova, Liyana   431, 447
Porrill, John   311
Pouya, Soha   434
Prescott, Tony J.   24, 154, 179, 396, 405, 438, 454

Quinn, Roger D.   59

Rácz, Zoltán   204
Renda, Federico   378
Richter, Eugen   359
Ricotti, Leonardo   251
Rojas, Raúl   143
Rossiter, Jonathan M.   311, 424
Rossiter, Jonathon   323
Russino, Andrea   431

Sabo, Chelsea   362
Sadeghi, Alì   431
Sadeghi, Ali   447
Saïghi, Sylvain   347
Sánchez-Fibla, Martí   356, 399
Sartori, Luisa   47
Schmickl, Thomas   384, 441
Schubert, Theresa   444
Shanahan, Murray   371
Shaw, Kendrick M.   59

Sheik, Sadique   262
Shimizu, Masahiro   409
Siringil, Erdem C.   216
Sproewitz, Alexander   434
Stafford, Tom   24
Stefanini, Fabio   262
Sullivan, J. Charles   179
Sun, Shumeng   350
Suter, Marcel   216
Szopek, Martina   441

Takahashi, Yosuke   131, 167
Takashima, Atsushi   167
Tandler, Lynn   387
Tessadori, Jacopo   274
Thenius, Ronald   441
Tonazzini, Alice   431, 447
Toomey, Anne   387
Tosello, Elisa   131
Tramacere, Francesca   368

Umedachi, Takuya   390

Vannozzi, Lorenzo   251
Vasas, Vera   83
Vasilaki, Eleni   362
Venuta, Daniele   274

Verschure, Paul F.M.J.   287, 356, 396, 399
Vidoni, Renato   418
Viry, Lucie   368
Voges, Danja   374
Volpi, Nicola Catenacci   191
Vouloutsi, Vasiliki   287

Westphal, Anthony   299
Weyrich, Steven   374
Wild, Benjamin   143
Wilson, Emma D.   311
Wilson, Stuart P.   381, 450, 454
Winstone, Benjamin   323
Witte, Hartmut   374
Wörgötter, Florentin   402
Wystrach, Antoine   427

Xu, Daniel   350

Yamaguchi, Takami   412
Yavuz, Esin   362

Zaraki, Abolfazl   393, 457
Zhang, Jianwei   359
Zhang, Li   216
Ziegler, Marc   335